Analytical Methods for
Energy Diversity and Security

Analytical Methods for
Energy Diversity and Security

Portfolio Optimization in the Energy Sector: A Tribute to the work of Dr Shimon Awerbuch

Morgan Bazilian and Fabien Roques

Editors

ELSEVIER

Amsterdam • Boston • Heidelberg • London • New York • Oxford
Paris • San Diego • San Francisco • Singapore • Sydney • Tokyo

Elsevier
The Boulevard, Langford Lane, Kidlington, Oxford OX5 1GB, UK
Radarweg 29, PO Box 211, 1000 AE Amsterdam, The Netherlands

First edition 2008

British Library Cataloguing in Publication Data
Analytical methods for energy diversity and security : mean-variance optimization for
  electric utilities planning : a tribute to the work of Dr. Shimon Awerbuch. – (Elsevier
  global energy policy and economics series; 12) 1.   Electric power systems –
  Management 2.   Electric power production 3.   Electric utilities – Risk management
  I.   Awerbuch, Shimon II.   Bazilian, Morgan III.   Roques, Fabien A.
333.7'932

Library of Congress Catalog Number: 2008933441

ISBN: 978-0-08-056887-4
ISSN: 1571-4985

> For information on all Elsevier publications
> visit our website at www.elsevierdirect.com

Typeset by Charon Tec Ltd., A Macmillan Company.
(www.macmillansolutions.com)

Printed and bound in Hungary

08 09 10 11 12 12 11 10 9 8 7 6 5 4 3 2 1

Cover design: Alexandra Jefferson, [email: alex@propellermedia.com]

CONTENTS

Part III Frontier Applications of the Mean-Variance Optimization Model for Electric Utilities Planning   191

This book is a tribute to our friend and colleague Shimon Awerbuch. It was conceived of by Shimon in the months before he died. We have attempted to implement his vision as a suitable memorial to him with the support of the publisher, the contributors, and a myriad of his colleagues. What has been most rewarding about working on such a project is witnessing the enthusiasm of those participating, and the respect they had for Shimon as a person and as a scholar.

Shimon's work remains vital and resonant in the academic and policy communities. He could not have fully anticipated how quickly the energy landscape has altered since his death. The nexus of energy security and climate change has become the focus of world leaders and public discourse during that time. Of course, implementing change in these areas is a complex matter. Shimon's applied work has laid critical foundations for the numerous political and scientific calls to refocus priorities and policies in the energy sector. His theoretical work continues to inspire many researchers, analysts and decision makers. These innovative contributions represent the type of effort necessary to confront positively the enormity of the task before us.

While Shimon's research spanned a number of related issues, we have focused this book on his work in analytical methods of portfolio optimization in the electricity sector. Fuel mix diversity has long been perceived as contributing to energy security, but researchers and policy makers were lacking a robust methodology to quantify the optimal degree of diversity of an energy system. By applying Mean-Variance Portfolio Theory to weigh the cost and risk implications of greater diversity, Shimon's methodology provided both a novel means to conceptualize the links between energy diversity and security, and some new insights for policy makers. In particular, his work provided a new way to consider how renewable energy and low-carbon technologies can play an integral role in modern energy systems. By helping to alter the vocabulary of policy, his work has already proven to be important in ensuring fair analytical treatment of these critical technologies.

The book is separated into three parts that explore: various approaches to the topic of portfolio optimization, its implementation at national and regional levels, and other applications and expansions of the methodology. The contributions to this tribute book show the depth of analysis possible given a robust financial and economic framework. They represent the cutting-edge of research in the rapidly expanding area. This book and its individual contributions show how innovative academic research can provide new insights on some critical aspects of energy and climate change policy. We trust you will find it a useful tool and reference.

Morgan Bazilian   Fabien Roques
Dublin, Ireland   Paris, France

I am delighted and honored to write this foreword as a tribute to Shimon Awerbuch. Shimon was a remarkable professional whose work covered a range of subjects from utility regulation to energy economics, dealing particularly with innovations and the development of new technologies. These areas are critical if we are to address the challege of climate change. He was a remarkable teacher who commanded great respect from his students by imparting knowledge of very high caliber. As a consultant his advice and analytical insights were of great value not only to the corporate and public sectors.

As a successful leader in the profession with his experience of 30 years covering finance, regulatory and energy economics as well as various aspects of government policy, he has left a major mark on the subjects that he touched and people with whom he came into contact. This volume is a particularly apt tribute to him in its coverage and contents, which provide an extremely valuable source of knowledge for researchers, industry leaders and those involved in government policy. The caliber of authors who have contributed to this volume does justice to the scholarship and the brilliance of Shimon Awerbuch. I am sure this compilation of scholarly writings will be an extremely important contribution to the literature and policy formation in the field that is receiving increasing importance around the world today.

R. K. Pachauri
Director General, The Energy and Resources Institute (TERI) and
Chairman, Intergovernmental Panel on Climate Change (IPCC)
Co-recipient of the 2007 Nobel Peace Prize

Shimon Awerbuch wrote coherently, intelligently and passionately about the impact of renewable energy on different aspects of the economy. He was a believer in opening electricity markets so that the customer could benefit from the innovation that competition brings. In both areas, he contributed to the economic and financial literature.

Shimon perceived of renewables in a unique light, that of facilitating the emergence of a new generation of providers who, by virtue of their technology, were innovative. He was among the first to point out how renewable generation helped reduce the overall price of electricity through consumers. He did this by applying Markowitz's portfolio theory to the costs and risks of electricity systems, and showed how renewables, because of their essentially fixed costs, increase the robustness of power systems in the face of price fluctuations of fossil fuel.

This insight, expanded by many of the researchers in this volume, has been absolutely critical to those of us in the industry. His ability to communicate effectively between academia, policy makers and industry was invaluable.

To everybody who knew Shimon he was much more than a influential theoretical financial economist, he was a most entertaining public speaker and lecturer and, from my own viewpoint, he was a great friend who is missed dearly.

Eddie O'Connor
Founder, Airtricity, CEO Mainstream Renewable Power

The Editors

Energy security and diversity

Energy plays a vital role in society, underpinning all areas of economic activity. The economic impact of supply or price fluctuations can therefore be significant and wide-ranging. This creates an incentive for governments to ensure that secure, reliable and competitively priced energy sources are readily available. Thus, security of supply is a key tenet of energy policy, and is highly coupled with the other two pillars, namely environmental considerations and competitiveness. It is, however, an extremely varied and complex notion, and has no agreed definition (Bazilian et al., 2006). Because of the large amount of related issues encompassed by the term, and the lack of a single analytical framework, security of supply has tended to be an overused and misunderstood term.

A country's energy security policy generally comprises measures taken to reduce the risks of supply disruptions below a certain tolerable level. Such measures need to be balanced to ensure that a supply of affordable energy is available to meet demand. Security of energy supply thus encompasses both issues of *quantity* and *price*. However, *time* is also a key parameter, as a sudden price hike will have very different effects on both society and the economy compared to those of a long-term price increase. Insecurity in energy supply originates in the risks related to the scarcity and uneven geographical distribution of primary fuels and to the operational reliability of energy systems that ensure services are delivered to end users (Blyth and Leferre, 2004).

Too broad a definition of security of supply can make it difficult to develop analytical approaches to quantify and guide policy making. While there is much more to energy security than just diversity (the focus of this book), it is often considered as a key policy to improve energy security.

Finance theory and the energy sector

Finance theory shows that expected generating cost streams are meaningfully valued only in terms of their market risk. Fossil fuel prices have fluctuated considerably over the last several years. When this volatility is reflected through the Capital Asset Pricing Model (CAPM), fossil fuel generation appears significantly more costly than standard engineering estimates that ignore the impact of risk on generation costs. But even CAPM risk-adjusted generation cost estimates are only as good as the underlying price forecasts. Meaningful electricity cost estimates would require *unbiased* price forecasts, but history provides little comfort that today's price projections will prove more reliable than those of the past.

Electricity market liberalisation has modified the allocation of investment risks among the different stakeholders, magnifying the importance of developing a robust analytical framework to conceive of risk diversification for public and private energy planners. This book illustrates how energy planners and policymakers can make use of the practices of financial investors and other risk professionals. No one can predict the stock market's performance, just as no one can predict the price of natural gas. Financial investors deal with market risk by holding efficient, diversified portfolios, which offer the best hedge against an uncertain future. Prudent investors balance their portfolios with a mixture of potentially high yielding securities and low-yielding government obligations and similar "safe" investments. Energy planners can benefit from these principles, and re-evaluate standard practices that identify the *least-cost entrant* – the technology with the lowest *stand-alone cost* – and focus instead on developing optimal generating *portfolios* and strategies.

While many of the risks facing power producers in liberalized electricity markets existed in the regulated industry, the ability to pass these risks on to consumers or taxpayers is no longer automatic. This chapter argues that the change of institutional environment within which utilities operate makes it all the more important to develop portfolio investment valuation methodologies which take into account risks appropriately. In particular, the different contributions illustrate how some investment valuation techniques borrowed from the finance literature, particularly mean-variance portfolio theory, can usefully complement the traditional approach computing the levelized generation costs over the life of the plant. Some general lessons emerge from the different contributions. When added to a risky, fossil-fuel-dominated generating portfolio, renewables and nuclear power can reduce generating cost and risk, as long as the mix can be reshuffled over time. This portfolio effect holds even if renewables or nuclear are believed to cost more on a *stand-alone* basis. Because renewables' generating costs are uncorrelated to fossil prices, their addition diversifies the mix and reduces expected cost and risk, in much the same way as diversification improves the expected performance of financial portfolios.

We hope that the variety and richness of approaches applying MVP theory presented in this book pay an appropriate tribute to one of the pioneer researchers in this field, Dr Shimon Awerbuch. We also hope that this volume will trigger interest in research which will develop and refine Shimon's work further.

Book outline

The *Introductory paper* (Editors) analyzes the concept of diversity in the energy sector and argues that while diversity can be seen as a desirable feature of an electricity system, it is not clear *what* should be diversified, and *how much* diversity is optimal. The introductory paper provides the theoretical background to place the different contributions of the book in perspective. It reviews the different strands of the literature focusing on the quantification (diversity indices) and valuation of diversity (mean-variance portfolio theory and real options theory).

Part I: Analytical approaches to diversity

The first part of the book provides a detailed overview of the different analytical approaches that have been applied to provide a normative framework to the issues of energy diversity. *Chapter 1* (Stirling) provides an in-depth study of the characteristics of a diverse energy system, and identifies three necessary but individually insufficient properties of diversity (variety, balance and disparity). Based on such decomposition of the concept of diversity, Stirling develops a general framework to quantify the diversity of an energy system in a wide range of different energy policy contexts. It can be focused equally at the level of primary energy mixes, electricity supply portfolios or energy service systems. Stirling's approach complements the other analytical approaches of this book which are based on a quantification of risks and benefits associated with diversity based on historical data, in that it also attempts to capture the benefits of diversity in the face of 'unknown' future events.

Chapter 2 (Bolinger and Wiser) concentrates on the main source of risks for investors in the power sector, fuel price risks and their relationship with power prices. Bolinger and Wiser argue that the volatility of fossil fuel prices, particularly gas prices, should be factored in to investment decisions and technology choices. They argue that increasing the market penetration of renewable generation and other technologies that are not exposed to gas price risk may provide economic benefits to ratepayers by displacing gas-fired generation through different channels: first, the displacement of natural gas-fired generation by increased renewable generation reduces ratepayer exposure to natural gas price risk; second, this displacement reduces demand for natural gas among gas-fired generators, which, all else being equal, will put downward pressure on natural gas prices. This chapter then explores analytically how each of these two potential 'hedging' strategies benefits renewable electricity.

Chapter 3 (Editors) argues that diversity is conceived in terms of risk and details the different types of risks that affect utilities investment decisions. Based on the pioneering work of Shimon Awerbuch, it illustrates the potential of an analytical tool borrowed from the financial literature – mean-variance portfolio theory – to balance the risk reduction benefits of various technology portfolios. Mean-variance portfolio theory analysis for the power generation sector provides a consistent framework to gain better insight into the portfolio (cost) risk associated with alternative technology deployment portfolios.

Chapter 4 (Blyth) explores the potential of another analytical framework – real options – borrowed from the financial literature to complement MVP in the analysis of the different risks facing a power investor. Real options theory captures the impact of investment timing and managerial flexibility on investment valuation. This gives a rather different set of policy insights. The issue here is concerned with not so much a 'top–down' definition of an optimal generation mix, but a 'bottom–up' understanding of the incentives facing an individual investment decision in the face of uncertainty and flexibility. The chapter illustrates these concepts through an application of real options theory to the case of investment in power generation plant when faced with uncertain climate change regulation.

Part II: Portfolio optimization using MVP theory

The second part of the book explores the potential of MVP theory as a tool for decision-makers to quantify and value the costs and benefits of a more diverse power generation mix. This part illustrates, through five case studies, how MVP can be applied to identify socially optimal fuel mix for a national/regional electricity system [European Union (EU), the USA, the Netherlands, Scotland and Ireland]. The focus is on electricity consumers' welfare, hence focusing on fuel price risk. The five case studies provide a large overview of the specific issues and constraints related to smaller (The Netherlands, Scotland and Ireland) as well as larger and more diversified electricity systems (EU, USA).

Some papers consider other types of risks/uncertainties. *Chapter 5* (Awerbuch and Yang) takes into account fuel price risk, but also capital cost risk, operations and maintenance (O&M) risk, as well as carbon dioxide (CO_2) allowance price risk in the European Trading Scheme context. Chapter 5 produces an expository evaluation of the 2020 projected EU-BAU (business-as-usual) electricity generating mix, and explores generating portfolios that could reduce cost and market risk as well as CO_2 emissions relative to the BAU mix. Optimal generating portfolio mixes generally include greater shares of wind, nuclear and other non-fossil technologies that often cost more on a stand-alone engineering basis, but overall costs and risks are reduced because of the effect of portfolio diversification. They also enhance energy security.

Chapter 6 (Jansen and Beurskens) presents results of an application of MVP to the future portfolio of electricity generating technologies in the Netherlands in the year 2030, based on base-case scenarios designed by the Dutch Central Planning Office. This chapter focuses in on the electricity cost–risk dimension of the Dutch portfolio of generating technologies and the potential for additional deployment of renewable generating technologies to enhance the efficiency of base-case generating mixes in 2030. The results show that promotion of renewable energy can greatly decrease the portfolio risk with relatively small impact on portfolio costs.

Chapter 7 (Awerbuch et al.) explores further the potential contribution of renewables to the security of the power generation mix by applying MVP to Scotland's electricity generating mix. The chapter illustrates how wind and other renewables can benefit the Scottish generating mix. The optimal results indicate that compared with National Grid projected mixes, there exist generating mixes with larger wind shares at equal or lower expected cost and risk.

Chapter 8 (Doherty et al.) constitutes an interesting integrated case study of a portfolio in an isolated electricity system (Ireland). The optimal generation portfolios need to take into account some specific constraints and issues, such as the maximum share of intermittent generation technologies to guarantee the security of supply of the system.

Chapters 6, 7 and 8 clearly demonstrate the value of renewable energy other low-carbon technologies. The quantification of the risk hedging qualities of these technologies should provide policy-makers with ample grounds to encourage greater diversification of power systems. *Chapter 9* (Hogendorn and Kleindorfer) explores the different approaches for policy makers to improve investment incentives in renewable technologies, including Renewable Portfolio Standards,

Renewable Resource Credits, and various hybrid approaches. The chapter compares the efficiency and institutional issues associated with the implementation of the different support schemes aiming at increasing the share of renewables in power generation portfolios.

Part III: Frontier applications of portfolio optimization

The third part of the book presents some recent 'frontier' applications of MVP theory to the power sector. The seminal literature using MVP techniques in the power sector (some of which are included in this text) focused on minimizing generation costs. Such an approach can be interpreted as the maximization of social welfare in a central planning paradigm under the former regulated industry structure. As the literature applying MVP to optimizing power generation portfolios is growing, several avenues to refine and develop further this approach have emerged. This includes consideration of liberalised and competitive electricity markets with dispersed decision-making paradigms.

The chapters in the third part create a varied menu of approaches, some focusing on energy security, others on maximizing corporate profitability. The focus ranges from improving societal welfare and sustainable energy to enhancing profitability and power network congestion. The chapters are also varied in their methodological approaches, some focusing in detail on generating costs for individual technologies, other focused more directly on improvements to the procedure to derive optimal portfolios.

Chapter 10 (Krey and Zweifel) concentrates on ways to improve the stability over time of the correlation estimates between different fuel prices, and thereby to improve the robustness of the MVP approach. A specific econometric approach, the Seemingly Unrelated Regression Estimation (SURE), is applied to filter out the systematic components of the covariance matrix. The approach is illustrated by computing optimal generation portfolios for Switzerland and the USA.

Chapter 11 (Roques et al.) introduces MVP portfolio analysis for investment planning in liberalized markets from the perspective of (large) electricity generators. The objective function is different from other chapters, as optimal portfolios for private electricity generators maximize profit (while optimal portfolios from a social planners' perspective minimize system generation costs). Chapter 11 demonstrates that high degrees of correlation between gas and electricity prices – as observed in most European markets – reduce gas plant risks and make portfolios dominated by gas plant more attractive. Long-term power purchase contracts and/or a lower cost of capital can rebalance optimal portfolios toward more diversified portfolios with larger shares of nuclear and coal plants.

Chapter 12 (Liu and Wu) applies portfolio optimization from the point of view of a power generator to optimize trading decisions. Contrary to previous chapters which focused on long-term physical diversification through the plant generation mix, application of MVP theory to identify optimal trading strategies in the short term also requires taking into account financial diversification possibilities through, for example, contracts. Chapter 12 illustrates how a power generator can use MVP theory to optimize its trading schedule/trading portfolio under a specific strategy.

Chapter 13 (Kotsan and Douglas) applies MVP theory to analyze the locational value of generating assets in a power grid setting. Assuming that risk-averse investors can freely create a portfolio of shares in generation located on buses of the electrical network, it determines portfolios that minimize the variance of the weighted average locational marginal price (LMP) and maximize its expected value. The chapter illustrates the potential of such an approach by conducting simulations based on the New York–New England system and calculates LMPs in accordance with the PJM methodology for a fully optimal AC power flow solution. Results indicate that the network topology is a crucial determinant of the investment decision as line congestion makes it difficult to deliver power to certain nodes at system peak load. Determining those nodes is an important task for an investor in generation as well as the transmission system operator.

Chapter 14 (van Zon and Fuss) determines optimal portfolios for the UK, with volatile fuel prices and uncertainty concerning technological progress in a context of embodied technical change and irreversible investment. The chapter presents a model which combines a clay–clay vintage framework with elements from financial portfolio theory and thus captures both dynamic investment aspects and the irreversibilities associated with large sunk costs. This model is illustrated in the context of current UK policy, and demonstrates that the reduction of risk in generation portfolios is accompanied by an increase in total costs. Moreover, the embodiment of technical change – in combination with the expectation of a future switch toward another technology – can reduce current investment in that technology (while temporarily increasing current investment in competing technologies). This enables rational but risk-averse investors to maximize productivity gains by waiting for ongoing technical change to materialize until they plan to switch and subsequently investing more heavily in the most recent vintages.

There remain important research avenues to enhance the reliability and widen the scope of applications for the MVP approach in the domain of electricity and energy mix portfolios related to energy security. For instance, work is ongoing to use some kind of market segmentation model or dispatch model to distinguish generation technologies with different capacity factors. The contribution to short-term ancillary power provision services is also important, and the question of the intermittency of renewable resources would warrant more sophisticated electricity market modelling. Finally, the chapters in this volume use a one-period analysis, and a useful extension for further research would be to develop a framework for a multiperiod analysis, permitting not only the identification of efficient portfolios in a certain target year, but also the determination of optimal trajectories for rebalancing portfolios from the base year to the target year.

References

Bazilian, M. and O'Leary, F. (2006). Security of Supply Metrics. Energy Policy and Statistical Support Unit, Ireland.
Blyth and Leferre (2004). Energy Security and Climate Change Interactions. IEA.

Analytical Approaches to Quantify and Value Fuel Mix Diversity

The Editors*

Acknowledgements

The authors would like to thank Shimon Awerbuch, David Newbery, William Nuttall, Paul Twomey and anonymous referees for their helpful comments.

1 Introduction

This chapter aims to explore the concept of diversity as applied to electricity systems. It argues that greater diversity enhances the robustness of an electricity system to fossil fuel supply shocks, and hence yields economic and security of supply benefits. However, a diverse electricity system is not a necessary nor sufficient condition to guarantee security of supply. This chapter argues that the concept of diversity as applied to electricity systems remains ill-defined. It is not clear *what* should be diversified, nor is it straightforward to quantify the *costs* and *benefits* of increased diversity.

The chapter reviews different analytical approaches to *quantify* and *value* the diversity of an electricity system. It argues that while there have been many attempts to design diversity indicators that serve as useful proxies to quantify diversity, such indicators suffer from not appropriately accounting for the costs of increased diversity. It identifies that more research is required to identify the economic costs associated with greater diversity, as well as to weigh the costs and benefits of increased diversity. The chapter discusses how analytical tools borrowed from the financial literature can be used to consider the costs and

---

* International Energy Agency, Economic Analysis Division, Paris, France.

Analytical Methods for Energy Diversity and Security © 2008 Elsevier Ltd.
978-0-08-056887-4 All rights reserved.

benefits of energy sector diversity. These include both *static* valuation methods such as Mean-Variance Portfolio Theory, and *dynamic* valuation methods such as Real Options which take into account the option value of diversity as a hedge against potential fossil fuel supply or price shocks. The different contributions in this volume present innovative applications of such financial tools to quantify and value optimal fuel mix diversity for policy makers and utilities investors.

2 Defining the diversity of the electricity system

The concept of diversity as applied to electricity generation is intuitively appealing at times when the resurgence of political tensions raises questions about the reliability of fossil fuel imports. However, the diversity of an electricity system remains ill-defined, both qualitatively and quantitatively (Roques, 2003). The basic principle of diversity is straightforward – not putting all one's eggs in one basket. But this can apply to a wide range of characteristic features of the electricity system, including the mix of fuels used to generate electricity, plant technology and manufacturers, or plant operators. This chapter concentrates on fuel mix diversity, which appears as the most important source of diversification in the electricity generation.

Greater fuel import dependency has different potential economic and security of supply consequences in the short- and long-term. As an example, a partial or complete sudden gas supply disruption would affect differently a gas importing country economy depending on the length of the interruption. Besides, the potential benefits of fuel mix diversity hinge on the practical feasibility of fuel mix diversification. Most electricity infrastructure is long lived, such that in the short-term, a utility is limited to selecting power sources from its existing portfolio of generating facilities and third-party power purchases. In the long-term, the utility would contemplate what fuels it would burn in new power plants or what fuels are contained in future power purchases. The next sections examine accordingly the feasibility and potential benefits of greater fuel mix diversity under two different time scales: in the short term, through improved system resilience to sudden supply disruptions, and in the long term, through a lower macroeconomic impact of high or volatile fossil fuel prices.

2.1 Diversity and resilience to supply shocks

In the short term, a more diverse electricity generation system is likely to be less affected by fuel supply disruptions, because of its greater ability to switch fuels. Fuel mix diversity is believed to provide a hedge against any shock that could render some fuel suddenly unavailable or extremely expensive. In particular, relying on imports for gas exposes countries to any disruptive event either in the exporting countries, or on the transit routes of the fuel. The diversity of the fuel mix is a multifaceted issue: not only does the primary choice of fuels matter, but also the geographical source of the fuel imports, as well as the transit routes of such fuel imports. Such considerations have a critical impact on the relationship between fuel mix diversity and security of supply. While coal can be bought on a global market, gas production and reserves are concentrated in a few regions, mainly Russia and North African countries for gas imported in the EU (IEA, 2006). Besides, whereas coal can be shipped easily, gas is mainly imported by pipelines through a few critical transit routes which are vulnerable to political instability or terrorist actions in the transit countries. In this perspective, the expected development of a global liquefied natural gas (LNG) market could greatly contribute to diversifying the transit supply risks associated with gas.

Energy price elasticities are generally much higher in the long term than the short term, and vary largely by fuel and region. Price elasticities are particularly low for transport fuels, as few practical substitutes are yet available for oil-based fuels for cars and trucks. In a recent study, the International Energy Agency (IEA) estimates that the weighted average crude oil price elasticity of total oil demand across all regions is -0.03 in the short term and -0.15 in the long term (IEA, 2006). Similarly, demand for electricity is highly price inelastic, with estimates ranging from -0.01 to -0.14 in the long term and even lower in the short term (IEA, 2006). Different fuels – gas, coal and oil products – can provide non-electricity stationary services (such as fuel for heating boilers), so demand for these fuels in these sectors is generally more sensitive to changes in price, especially where multifiring equipment is widespread. Power generators may also be able to switch more quickly to cheaper fuels if they have dual-firing capability or reserve capacity.

Much debate remains, however, with regard to the link between energy dependency and security of supply. The threats to energy security are more subtle and varied than portrayed in the crude expression of concerns about import dependence (Grubb et al., 2006). Counter-arguments include the co-dependence of importers and exporters, and the nature of international markets as reasons not to fear over dependency-related threats. Bohi and Toman (1993) provide a detailed discussion of the conceptual arguments and empirical evidence related to the potential sources of market failure for energy security. There are many possible sources of interruption to supply: from unreliable political sources, from disruptions to transit routes and facilities, and even from the possibility of stalled European energy market liberalization. Grubb et al. (2006) emphasize, for instance, that the major interruptions of the UK energy system in the past three decades have arisen from miners' strikes, domestic fuel blockades and occasional power cuts – not from foreign supply disruptions.

In short, diversity helps to manage the risks that are associated with individual energy technologies or sources, but diversity is not a necessary characteristic of a secure system. For instance, the French electricity supply system, based on nuclear energy, is little diversified: there is a strong focus on one fuel, one technology, and a small number of related designs. In some respects it is very secure, being robust to external political events and economic changes. In other respects it could be argued to be insecure to generic technical faults, terrorist threats or a serious nuclear accident. At the other extreme, the old UK coal-based system was also apparently secure, based on indigenous coal and a limited number of technologies. Because of its exposure to the action of trade unions, it was a non-diversified system, with a single vulnerability that turned out to be critical (Costello, 2005).

2.2 Diversity reduces the macroeconomic sensitivity to oil and gas prices

The growing share of gas-fired generation after liberalization in many electricity markets has raised concerns over the adverse macroeconomic effects of the decrease of fuel mix diversity and greater gas imports for gas-importing countries. An important question is whether increasing reliance on gas-fired generation – hence greater gas import dependency for gas-importing countries – will increase their economy sensitivity to the level and volatility of oil and gas prices.

For oil-importing countries, the immediate magnitude of the direct effect of a given oil price increase on national income can be conceptualized as depending on the ratio of oil imports to gross domestic product (GDP) (IEA, 2006). This, in turn, is a function of the amount of oil consumed for a given level of national income (oil intensity) and the degree of dependence on imported oil (import dependency).[1] It also depends on the extent to which gas prices rise in response to an oil price increase, the gas intensity and gas import dependency of the economy and the impact of higher prices on other forms of energy that compete with or, in the case of electricity, are generated from oil and gas. The impact of a given change in oil and gas prices on the economy is proportionally linked to the size of the shift in the terms of trade. That shift, in turn, depends on energy-import intensity. The impact of a given change in energy prices on the economy is linked to the size of the shift in the terms of trade. Levels of and historical trends in intensity vary among countries and regions. Some regions have seen a substantial decline in oil-import intensity since the 1980s, notably Europe and the Pacific region, while import intensity has risen in some developing countries, including China and India (IEA, 2006).

As Awerbuch and Sauter (2006) note, a large body of academic literature surveyed suggests that oil price increases and volatility dampen macroeconomic growth by raising inflation and unemployment and depressing the value of financial and other assets in oil-consuming nations. The so-called 'oil–GDP relationship' has been statistically studied since the late 1940s (Awerbuch and Sauter, 2006; Greene and Tishchishyna, 2000).[2] The impact of oil price movements on economic growth depends largely upon the country considered. The quantitative relationship between oil price changes and economic activity and inflation can be decomposed as follows (IEA, 2004):

- *Terms of trade effects*: the first, and principal, impact of oil price shifts on activity arises from changes in purchasing power between oil-importing and oil-exporting nations.

---

[1] Oil import intensity (Net oil imports/GDP) = Import dependency (Net oil imports/Total oil use) × Oil intensity [(Total oil use/Total energy use) × (Total energy use/GDP)].
[2] IEA (2004) notices, however, that the negative correlation between oil prices and macroeconomic indicators seems to have substantially weakened over time. It gives three main reasons: first, the weight of oil and oil products in domestic production has dropped, so that terms of trade shifts are less important. Second, the wage formation process has become less responsive to fluctuations in oil prices. Third, heightened competition has helped to reduce the secondary impact on core inflation from changes in oil prices.

- *Effect on domestic prices and inflation*: whether the increase in the price level translates into a shift in core inflation depends on the 'second round' effects – i.e. whether workers and/or enterprises are able to compensate for the income loss through higher wages and prices – which, in turn, depend on the monetary policy regime in place.
- *Domestic demand effects*: since oil is an input into many goods both consumers and producers would bear losses.
- *Supply-side implications: impact on output and employment*: the impact on output and employment is determined by the relative supply responses of labor and capital.
- *Longer-term outcomes*: the negative impact of an oil price rise on domestic demand and income will diminish over time as consumers and producers modify their behavior. However, research indicates that there is an asymmetric effect, insofar as oil demand does not revert to its initial level as oil prices fall. In that case, the income losses experienced by energy importers may eventually be partly reversed. Where fluctuations in oil prices create uncertainty, there may be a reduction in trend investment activity, but it is less clear that the effects on profitability or capacity utilization are asymmetric (Awerbuch and Sauter, 2006; Gately and Huntingdon, 2002).

While the mechanism by which oil prices affect economic performance is generally well understood, the precise dynamics and magnitude of these effects – especially the adjustments to the shift in the terms of trade – are very uncertain (IEA, 2006). Quantitative estimates of the overall macroeconomic damage caused to the economies of oil-importing countries by the oil price shocks of 1973–1974, 1979–1980 and 1990–1991, as well as the gains from the 1986 price collapse, vary substantially. This is partly due to differences in the models used to examine the issue, reflecting the difficulty of capturing all the interacting effects. For the same reason, the results of models used to predict the impact of an increase of oil prices on the GDP vary greatly.[3] IEA (2006) estimates, as a rule of thumb, that the impact of a sustained $10 per barrel oil price increase would now cut average real GDP by around 0.3% in the OECD (Organization for Economic Co-operation and Development) and by about 0.5% in non-OECD countries. Awerbuch and Sauter (2006) point out that the oil–GDP effect has significant ramifications for policies reducing fuel import dependency, such as increasing fuel mix diversity (e.g. through greater use of renewable or nuclear energies) and demand-side energy efficiency and flexibility (e.g. through greater fuel-switching possibilities). These policies mitigate exposure to fossil fuel risk and therefore help nations to avoid costly economic losses; these arguments will be discussed further in Section 3 of this chapter.

Turning now to the impact of gas prices level and volatility on the economy, it is important to note that the price of gas tends to be highly correlated with international oil prices. This is because of both explicit price indexation and interfuel competition at the burner tip. The EC Sector Inquiry looked at the indexation

---

[3] See, for example, Hunt et al. (2002), Barrell and Pomerantz (2004), Hamilton (2005) and IMF (2005).

according to the region of the purchasing company (EC, 2005).[4] Figure 1 shows
the indexation of long-term gas supply contracts depending on whether the
buyer was from the UK, Western Europe or Eastern Europe.[5] Interestingly,
the indexation present in long-term contracts for gas supply to continental
Europe is very different to that found in the UK, where over 40% of the price
volatility of gas under long-term contracts is determined by changes to the actual
hub price of gas (usually the National Balancing Point (NBP) or International
Petroleum Exchange (IPE) prices). For Western Europe, changes in hub gas prices
account for only around 5% of indexation. Conversely, the importance of heavy
fuel oil and light fuel oil to determine the price level paid under long-term contracts is much higher in Western Europe (over 80% of indexation) and Eastern
Europe (around 95% of indexation) than in the UK (around 30% of indexation).

Even in North America and Britain, where most contracts no longer include
any formal links to oil prices, gas prices tend to move in line with oil prices
because of fuel switching by industrial end-users and power plants (IEA, 2006).
Opportunities for arbitrage with continental Europe, by LNG and, in the case of
Britain, via the Bacton–Zeebrugge Interconnector, also tend to make oil and gas

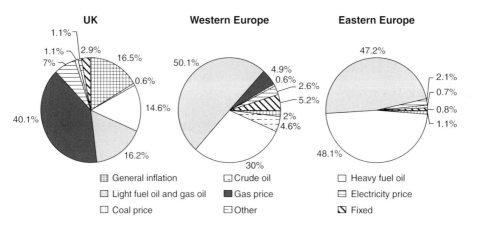

FIGURE 1 Indexation of long-term gas supply contracts by origin of the purchasing company.
Source: EC, 2005.

[4] The results of the EU Inquiry are based on analysis of long-term purchase agreements
(i.e. over 12 months) of 30 major producers and wholesalers of gas. The analysis is based
on data for the calendar year 2004 and indicates the average volume-weighted indexation
found in the sample of over 500 long-term contracts, representing around 400 billion cubic
metres of contracted gas. These contracts include those between companies exporting gas
to Europe and major EU gas wholesalers, as well as contracts between different EU gas
wholesalers.
[5] The Western Europe sample consists of long-term gas supply contracts to companies
in Austria, Belgium, Denmark, France, Germany, Italy and the Netherlands. The Eastern
Europe sample consists of long-term gas supply contracts to companies in the Czech
Republic, Hungary, Poland, Slovakia and Slovenia.

prices converge.[6] This explicit price indexation and interfuel competition results in wholesale gas prices reflecting the developments of the oil market, and in particular the market for oil derivatives such as heavy or light fuel oil. The EC Inquiry estimates that these account for around three-quarters of gas price volatility (EC, 2005). As a consequence, greater dependence on gas-fired power generation can be expected to amplify the sensitivity of the European countries' economies to oil and gas price fluctuations and shocks.

3 Quantifying and valuing the benefits of diversity

While it seems relatively difficult to compute the macroeconomic value of diversifying the electricity generation mix, it is important to advance research in this area to provide a normative approach for policy makers. As pointed out by Costello (2005), care must indeed be taken that arguments in favor of diversity are not used opportunistically by those seeking (via political mechanisms) to protect particular firms and industries. This underlines the need to develop analytical tools to quantify the costs and benefits of increased fuel mix diversity. This section introduces various indices that can be used to quantify fuel mix diversity, and then discusses how new analytical tools borrowed from the financial literature (such that mean-variance portfolio theory and real options theory) can be used to value the costs and benefits of generation mix diversity.

3.1 Quantifying fuel mix diversity

Stirling (1994, 1998, 2001) pioneered research in the application of diversity concepts to the energy sector. Stirling argues that *uncertainty* and *ignorance*, rather than *risk*, dominate real electricity investment decisions, and conceptualizes *diversification* as a response to these more intractable knowledge deficiencies. In addition to difficulties in definitely characterizing or partitioning the possibilities, there is a prospect of unexpected outcomes, arising entirely outside the domain of prior possibilities.[7] Stirling (1998) shows that diversity can be considered from different angles, notably *variety* (the number of available options, categories, species), *balance*

---

[6] Term contracts – often covering very long terms of 20 or more years – account for well over 95% of bulk gas trade in continental Europe (almost 100% outside Belgium and the Netherlands). Almost all these contracts include oil price indexation. In Britain, term contracts – which are generally much shorter in duration than in the rest of Europe – account for 90% of all bulk trade. In contrast to the rest of Europe, they almost always price the gas on the basis of spot or futures gas prices, usually at the National Balancing Point. Nonetheless, a small number of contracts may have some limited degree of oil price indexation (IEA, 2006).

[7] Stirling distinguishes three basic states of incertitude: risk: 'a probability density function may meaningfully be defined for a range of possible outcomes'; uncertainty: 'there exists no basis for the assignment of probabilities'; and ignorance: 'there exists no basis for the assignment of probabilities to outcomes, nor knowledge about many of the possible outcomes themselves …'.

(the spread among options) and *disparity* (the nature and degree to which options are different from each other). Variety, balance and disparity constitute 'three necessary but individually insufficient conditions for diversity' (Stirling, 1998). He, however, points out that inclusion of disparity remains cumbersome, as the concept of disparity differs from variety and balance in that it is inherently qualitative.

In a seminal contribution to mathematical ecology, Hill (1973) directly addresses the fundamental issue of the tradeoff between variety and balance in the measurement of diversity. Based on the characterization of diversity in terms of 'proportional abundance', Hill (1973) identifies and orders an entire family of possible quantitative measures of diversity. Each is subject to the same general form:

$$\Delta_a = \left(\sum_{i=1}^{I} \left(p_i^a \right) \right)^{1/(1-a)}, a \neq 1$$

$$= \sum_{i=1}^{I} -p_i \ln(p_i), a = 1$$

where Δ_a specifies a particular index of diversity, p_i represents (in economic terms) the proportional representation of option i in the portfolio under scrutiny, and a is a parameter which effectively governs the relative weighting placed on variety and balance. The greater the value of the parameter a, the smaller the relative sensitivity of the resulting index to the presence of lower contributing options.

For $a = 2$, the reciprocal of the function is referred to in ecology as the *Simpson diversity index* and in economics as the *Herfindahl–Hirschman concentration index*. Assuming that p_i is the market share of the ith firm or the proportion of generation met by one particular fuel source, then the Herfindahl–Hirschman concentration index is calculated according to $\Delta_2 = \Sigma_{i=1}^{I} p_i^2$. The Herfindahl–Hirschman index takes into account both the relative size and distribution of each source, increasing as the number of firms falls and the disparity in the size of those firms increases.[8]

For $a = 1$, the result is the *Shannon–Wiener diversity index* (Stirling, 1998). The Shannon–Wiener diversity index is the most attractive simple index reflecting both variety and balance in an even way (Stirling, 1998). The reasons are that this index is insensitive to final ordering (changes of the base of logarithms do not change the rank orderings of different system) and is additive in the case of a refining of the taxonomy (the index value for a system of options, which has been disaggregated according to a combined taxonomy, should be equal to the sum of the index values obtained for the same system classified under each taxonomy individually). The higher the value taken by the index, the more diverse the system.

An intuitive rationale for the use of the Shannon–Wiener function as an index of electricity supply system security is to think of it as a measure of the probability that a hypothetical unit of electricity sampled from the system at random

---

[8] The maximum value of the index is 10000 in the case of a monopoly, falling toward zero as the market moves toward a situation of perfect competition.

has been generated by any particular option. The more diverse the system, the greater the uncertainty over which option will have generated the next sampled unit of electricity. Jansen et al. (2004) elaborate on the Shannon–Wiener diversity index to design a macroindicator for long-run energy supply security. Four long-term energy security indices are presented, allowing for an increasing number of long-term supply security aspects, and then applied to reference year 2030 of four long-term sustainability outlook scenarios. Aspects introduced in their indicators on a stepwise additional basis are successively:

- Diversification of energy sources in energy supply: this corresponds to the basic Shannon–Wiener diversity index.
- Diversification of imports with respect to imported energy source: this second indicator provides for an adjustment of the basic indicator for the net import dependency.
- Long-term political stability in import regions: the third additional adjustment to the indicator accounts for the level of long-term political stability in regions of origin, using the United Nations Development Programme (UNDP) Human Development Indicator as an index for long-term socioeconomic stability.
- Allowance for resource depletion: the fourth indicator allows for the level of resource depletion on an additional basis.

3.2 From quantification to valuation of fuel mix diversity

While Stirling's (1994, 1998, 2001) pioneering work on diversity indices greatly contributed to defining the diversity of an electricity system, it does not inform the question as to *how much* diversity is needed.[9] The extent to which diversity is to be pursued depends on the balance between the extra costs and the degree of risk reduction achieved. Fuel diversity should not be perceived as an end, but only as a means that has the capability to generate benefits less costly than other alternatives in achieving the same objectives. For example, financial instruments may have lower costs than fuel diversity, which can be viewed as a physical hedge in reducing price risk to a tolerable level. Fuel diversity may also create costs from the loss of scale economies associated with traditional generation technologies, and from owning and operating a portfolio of power sources that include several fuels and technologies, some of which may not have the lowest expected costs.

The diversity indices presented before do not exploit statistical information. Thinking about fuel mix diversity in terms of risk, e.g. price risk for fossil fuel supplies, one can make use of other analytical approaches using statistical data to identify the optimal degree of diversity of an electricity system, by trading off the degree of risk reduction achieved by diversifying away from gas-fired generation

---

[9] See also Lucas et al. (1995) for a critique of Stirling's diversity index.

against the extra cost of doing so. As argued by Awerbuch and Berger (2003), such approaches rely on the assumption that while these precise outcomes may never be perfectly repeated in the future, they at least provide a guide to the future.[10] The strength of such approaches rests on the presumption that the past is a reliable guide to the future. This is not to say that unexpected events will not happen, only that the effects of these events are already known from past experience (Awerbuch and Berger, 2003).

Fuel mix diversity provides a hedge against potential price shocks affecting one type of fuel, e.g. imported gas, or supply shocks due to physical disruption in the supply chain. Investing in generation technologies which help a country (or a utility) to mitigate its exposure to fossil fuel supply disruptions or price risks can be thought as an *insurance*. Calculating the value associated with such insurance requires a different approach from the traditional static valuations of the 'least cost option' on a stand-alone basis (e.g. see Roques, 2008, for a critique of the traditional levelized cost methodology). Power generation investment valuations need to capture both the *portfolio effects* – the complementarity of one additional unit with the existing portfolio of plants of a country or utility – and the *option value effects* arising out of uncertainties in fossil fuel prices and volatility – e.g. the option value of operating renewables and nuclear plants in case gas prices increase. In other words, identifying the optimal degree of fuel mix diversity for a country or utility requires valuation approaches of power generation investments which trade off the risks and returns of increased portfolio diversification, in both a *static* and a *dynamic* perspective. The following subsections introduce successively static (value-at-risk and portfolio theory) and dynamic approaches (real options) to value fuel mix diversity.

3.2.1 Mean-variance portfolio theory

The *value-at-risk* (VaR) approach gained increasing popularity in banking and assets and liabilities management applications by the end of the 1990s.[11] The value at risk calculates the maximum loss expected (or worst case scenario) on an investment, over a given period and given a specified degree of confidence (Brealey and Myers, 2000). The VaR approach can be applied to any portfolio of assets and liabilities, whose market values are available on a periodic basis. Typically, normal distributions are assumed with values for price volatility, based on past statistics. Using calculated parameter values for the whole portfolio, the maximum portfolio loss can be projected provided a specific unlikely event does not occur, for example a

---

[10] While no particular random event may ever be precisely duplicated, nonetheless, historic variability is widely considered to be a useful indicator of future volatility (e.g. in the case of equity stocks).

[11] VaR is based on the common-sense fact that for investors, risk is about the odds of losing money. By assuming investors care about the odds of a big loss, VaR addresses one of the main issues with the traditional measure of risk, volatility. The main problem with volatility, indeed, is that it does not address the direction of an investment's movement: a stock can be volatile because it suddenly jumps higher. But investors are not distressed by gains.

5% chance of an adverse price movement within the next holding period. However, to implement it, the probability distribution of price changes for each portfolio instrument should be known, and the VaR approach depends critically on reasonable estimates of price volatility and correlations among financial assets, as well as the assumed distribution of price changes. Kleindorfer and Li (2005) provide a recent review of progress in the VaR theoretical literature, and apply it to characterize multiperiod VaR-constrained portfolios of real and contractual assets in the power sector.

Another probabilistic approach to value and optimize fuel mix diversity is Markowitz's *mean-variance portfolio theory* (Markowitz, 1952).[12] Mean-variance portfolio theory (hereafter MVP) defines portfolio risk as *total risk* – the sum of random and systematic fluctuations – measured as the standard deviation of periodic historic returns.[13] An efficient portfolio is one which has the smallest attainable portfolio risk for a given level of expected return (or the largest expected return for a given level of risk). The process for establishing an optimal (or efficient) portfolio generally uses historical measures for returns, risk (standard deviation) and the correlation coefficients between the different assets to be used in the portfolio.

By computer processing the returns, risk (standard deviation of returns) and correlation coefficients data, it is possible to establish a number of portfolios for varying levels of return, each having the least amount of risk achievable from the asset classes included. These are known as optimal portfolios, which lie on the *efficient frontier*. Figure 2 shows the efficient frontier for a portfolio of two risky assets. The graph visualizes the set of optimal portfolios. Optimality refers to Pareto optimality in the tradeoff between portfolio risk and portfolio return. For each portfolio on the efficient frontier:

- the expected portfolio holding period return (HPR) cannot be improved without increasing expected portfolio HPR risk;
- the expected portfolio HPR risk cannot be reduced without reducing expected portfolio HPR.

The investor then simply has to choose which level of risk is appropriate for their particular circumstances (or preference) and allocate their portfolio accordingly. In other words, MVP theory prescribes not a single optimal portfolio combination, but a range of efficient choices. Investors will choose a risk–return combination based on their own preferences and risk aversion.

Mean-variance portfolio (MVP) theory, initially developed for financial securities, can be applied to generation assets to determine the optimal portfolio for a country or generation company. This theory makes assumptions on the assets

---

[12] See e.g. Fabozzi et al. (2002) for a recent review of the developments of portfolio theory.

[13] Modern portfolio theory makes some assumptions about investors. It assumes they dislike risk and like returns, will act rationally in making decisions and make decisions based on maximizing their return for the level of risk that is acceptable for them. When making asset allocation decisions based on asset classes it is assumed that each asset class is diversified sufficiently to eliminate specific or non-market risk.

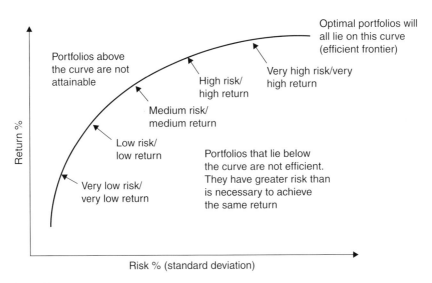

FIGURE 2   Efficient frontier for a portfolio of two risky assets.

considered and investors' behavior (such as risk aversion), which are discussed in detail in Awerbuch and Berger (2003) and Roques et al. (2008) in the context of investment in electricity markets. As Awerbuch and Berger (2003, p. 5) observe, 'the important implication of portfolio-based analysis is that the relative value of generating assets must be determined not by evaluating alternative assets, but by evaluating alternative asset portfolios. Energy planning therefore needs to focus less on finding the single lowest cost alternative and more on developing efficient (i.e. optimal) generating portfolios.'

Bar-Lev and Katz (1976) pioneered the application of MVP theory to fossil fuel procurement in the US electricity industry. By applying an MVP approach on a regional basis, they determined the theoretical efficient frontier of fossil fuel mix for various regulated utilities and compared it with the actual experience of the electric utilities. Bar-Lev and Katz (1976) shows that generally the electric utilities efficiently diversified, but that their portfolios were generally characterized by a relatively high rate of return and risk, which they interpreted as being a consequence of the 'cost-plus' regulatory regime encouraging utilities to behave in a risky way. Humphreys and McClain (1998) use MVP theory to demonstrate how the energy mix in the USA could be chosen given a national goal to reduce the risks to the domestic macroeconomy of unanticipated energy price shocks. They note that the electric utility industry has moved toward more efficient points of production since the 1980s, and that the switch toward natural gas in the 1990s might be driven by the desire for higher returns to energy investment in the industry.

Awerbuch (2000) evaluates the US gas–coal generation mix and shows that adding wind, photovoltaics and other fixed-cost renewables to a portfolio of conventional generating assets serves to reduce overall portfolio cost and risk, even through their stand-alone generating costs may be higher. Awerbuch and Berger (2003) use MVP to identify the optimal European technology mix, considering not only fuel price risk but also O&M, as well as construction period risks. They find that compared with the EU-2000 generation mix, the projected EU-2010 mix exhibits a higher risk coupled with higher return. Jansen et al. (2006) use portfolio theory to explore the risk and returns of various portfolio mixes corresponding to different scenarios of the electricity system development in the Netherlands. The general conclusion of these early portfolio theory applications which were based on production costs and concentrated on fuel price uncertainty, i.e. taking a national or societal perspective, is that more diverse generation portfolios are in general associated with lower risks for the same returns. In particular, optimal portfolios contain a substantial share of fixed-cost (when considering only fuel price uncertainty) renewables and nuclear, whose costs have a low covariance with the production costs of fossil fuel technologies.

Some recent studies have taken a different approach by taking a private investor perspective. These studies therefore also take into account electricity price risk (and in Europe CO_2 price risk), and the covariance of electricity, fuel and CO_2 prices. Roques et al. (2008) conclude that in the absence of long-term power purchase agreements, optimal portfolios for a private investor in the UK differ substantially from socially optimal portfolios, as there is little diversification value for a private investor in a portfolio of mixed technologies, because of the high empirical correlation between electricity, gas and carbon prices. Moreover, their results suggest that the current UK industry framework is unlikely to reward fuel mix diversification sufficiently so as to lead private investors' technology choices to be aligned with the socially optimal fuel mix, unless investors can find counter-parties with complementary risk profiles to sign long-term power purchase agreements. These findings raise questions as to whether and how policy makers or regulators should modify the market framework, given the macroeconomic and security of supply benefits of a diverse fuel mix.

3.2.2 Dynamic valuation approaches: the option value of diversity

Another concept borrowed from the finance literature, called *real options*, can be applied to supplement the information provided by static discounted cash flow analysis. In its simplest terms, real options theory says that when the future is uncertain, it pays to have a broad range of options available and to maintain the flexibility to exercise those options. Real options theory has pointed to the shortcomings of the static valuation approaches for inputting a value on the ability of a utility to react dynamically to changing market and other conditions. Specifically, static approaches can understate, if not ignore, *managerial flexibility*. Real options valuation allows for adjustment of the *timing* of the investment decision. It is therefore particularly well suited to evaluate investments with uncertain payoffs and costs, as it can capture the option value contained in managerial flexibility in the face of future uncertain developments: the greater the uncertainty that can be resolved, the more advantageous it is to wait and thus the higher the option value (Dixit and Pindyck, 1994; Trigeorgis, 1996).

Blyth (Chapter 4 in this volume) details how real options can be used to inform investment timing and technology choices under uncertainty about future climate policy and hence CO2 price. Real options theory can be applied to analyze the economics of renewable energy or nuclear power versus fossil fuel generation technologies when CO2 prices, fuel prices and/or electricity prices are uncertain.[14] There are potentially two attributes of non fossil-fuel technologies such as renewables and nuclear power generation that could improve their value to society or investors in a dynamic perspective. First, production costs of such technologies are insensitive to both gas and carbon prices.[15] Therefore, rising gas prices and carbon trading or carbon taxes will make nuclear and renewables more competitive against combined cycle gas turbines (CCGTs) and coal-fired plants. Second, investing in non fossil-fuel technologies can be thought of as a hedge against the volatility and risk of gas, coal and carbon prices for a country or a (large) generating company. The uncertainty over the evolution of gas and carbon prices implies that there is an option value associated with being able to choose between non fossil-fuel generation technologies and fossil fuel technologies in the future.

Real options theory can therefore rationalize embarking upon a power-plant project that is not expected to be economical for a period of years but offers the possibility of benefits in the longer term. For instance, Murto and Nese (2002) compare natural gas-fired plant economics with biomass plants and show that natural gas price uncertainty considerably improves the competitiveness of the biomass plant, when taking into account the option value associated with input cost uncertainty. Roques et al. (2006) compute the option value to a company of the ability to choose between a nuclear and a gas-fired plant investment at successive moments in the future, when the company faces stochastic gas, carbon and electricity prices. They show that this option value depends sensitively on the degree of correlation between electricity, gas and carbon prices, and conclude that there is little private company value in retaining the option to choose between nuclear and CCGT technologies in the future in liberalized European electricity markets, which exhibit a strong correlation between electricity, gas and carbon prices.

Real options analysis can also be applied to other benefits associated with a more diverse electricity generation system, as the value of real options is closely linked to the benefits of having more flexibility. The concept can, for instance, apply to whether a utility should buy a new power plant or purchase power. An

[14]Quantifying the option value requires, however, restrictive assumptions on the stochastic behavior of the electricity and natural gas market prices, and reliance on data from relatively illiquid forward markets. See Frayer and Uludere (2001) for a description of the limits of applying real options analysis to power investments.

[15]Nuclear fuel price have relatively little effect on electricity generation costs: a doubling of the uranium oxide price would increase the fuel cost for a light water reactor by 30% and the electricity cost by only about 7%, whereas doubling the gas price would add 70% to the price of electricity.

illustration of a failure to retain an option would be where a utility signs a long-term purchased power contract with rigid take and price provisions (Costello, 2005). If subsequent to the signing of the contract the market price of electricity plummeted or expected load growth failed to materialize, or both, the utility could suffer large contractual liability. Real options theory could also justify staggering the timing of capital expenditures for new generation facilities under uncertainty, committing to new construction in stages.[16] By waiting for new information, and in the meantime initiating development of promising technologies (e.g. on a pilot or demonstration basis), the utility would have more flexibility in adapting to the new conditions as they unfold. An example of a project involving an option is shown in Figure 3.

Another application of interest in Real Option modelling concerns the valuation of research, development, demonstration and deployment (RDDD) programs of new power generation technologies. Cost–benefit analysis of such publicly funded programs typically employs a deterministic forecast of the cost and performance of renewable and non-renewable fuels which ignores uncertainty in the cost of non-renewable energy, the possibility of adjustment to the RDDD effort commensurate

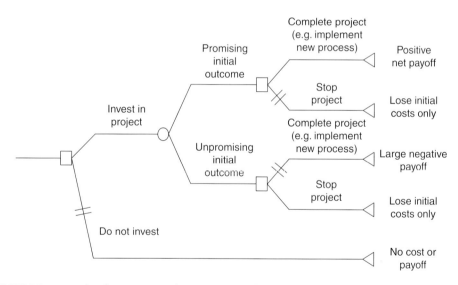

FIGURE 3   Example of a project involving an option: the initial decision to do R&D provides the option to invest in completing development, for instance by building a plant.

[16] Gollier et al. (2005) compare the benefit of a large nuclear power plant project coming from increasing returns to scale, to the benefit of a modular sequence of smaller, modular, nuclear power plants on the same site. They show that under price uncertainty only, the benefit of modularity is equivalent in terms of profitability to a reduction in the cost of electricity by one-thousandth of a euro per kWh.

with the evolving state of the world, and the underlying technical risk associated with RDDD. Siddiqui et al. (2007) find that the total option value of renewable energies is dominated by the value of existing renewable technologies, while the value of enhancements to renewable technologies from future RDDD is a modest 10% of the total, and the value of the abandonment option is insignificant. Davis and Owens (2003) use a similar real options approach to estimate the value of the US renewable electric technologies R&D program in the face of uncertain fossil fuel prices. They estimate the current value of expected future supply from renewable electric technologies, net of federal R&D expenditures, at $30.6 billion (in 2000 dollars).[17] While these two models' estimates of renewable technologies option value are sensitive to the selected parameters values, which are subject to debate, the results of Davis and Owens (2003) and Siddiqui et al. (2007) demonstrate that renewable technologies hold a significant amount of value that cannot be detected by using traditional static valuation techniques.

4 Conclusions

This chapter discussed potential adverse consequences of a power generation mix dominated by gas-fired power plants in terms of security of supply, and macroeconomic resilience to oil and gas price movements. Importantly, it remains unclear as to exactly what should be diversified, and how much diversity is optimal. Because the generation mix diversity is a multi-faceted issue, it is difficult to quantify the costs and benefits associated with greater fuel mix diversity.

Analytical tools borrowed from the financial literature are powerful in valuing the costs and benefits of reducing some risks, and can be applied to optimise power generation portfolios. The literature surveyed in the introductory chapter and the different contributions in this volume provide some cutting-edge insights about the value of diversifying the electricity generation mix. In particular, recent studies using such analytical approaches to value diversity show that non-fossil fuel technologies have a significant 'hedging value' from a societal perspective vis-à-vis fuel and CO2 price risks. Valuation approaches can therefore rationalise embarking upon a power-plant project that is not expected to be economical for a period of years (using standard financial metrics), but offers the possibility of benefits in the longer term. Most importantly, contrasting the societal value of diversity with the results from studies quantifying the value of fuel mix diversity to private investors casts doubt as to whether the current liberalised market framework provides adequate diversification incentives.

---

[17] The model assumes a current ratio of renewables to non-renewables electricity generating costs of 1.29, and a 1–4% annual rate of decline of renewable technologies generating costs, depending on the level of R&D funding. The cash flows of renewable and non-renewable technologies are discounted using the risk-free interest rate.

References

Awerbuch, S. (2000). Investing in photovoltaics: risk, accounting and the value of new technology. *Energy Policy*, 28, 1023–1035.

Awerbuch, S. and Berger, M. (2003). *Energy Security and Diversity in the EU: A Mean-Variance Portfolio Approach*. IEA Research Paper. Paris: IEA (February). www.iea.org/techno/renew/port.pdf

Awerbuch, S. and Sauter, R. (2006). Exploiting the oil–GDP effect to support renewables deployment. *Energy Policy*, 34, 2008–2819.

Bar-Lev, D. and Katz, S. (1976). A portfolio approach to fossil fuel procurement in the electric utility industry. *Journal of Finance*, 31(3), 933–947.

Barrel, R. and Pomerantz, O. (2004). *Oil Prices and the World Economy*. London: NIESR.

Bazilian, M., O'Leary, F., O'Gallachoir, B., Howley, M. (2007). Security of Supply Metrics for Ireland, . EPSSU, Dublin.

Bohi, D. and Toman, M. (1993). Energy security: externalities and policies. *Energy Policy, (November)*, 1093–1109.

Brealey, R. and Myers, S. (2000). *Principles of Corporate Finance*, 6th edn. Irwin McGraw-Hill.

Costello, K. (2005). A perspective on fuel diversity. *The Electricity Journal*, 18(4), 28–47.

Davis, G. and Owens, B. (2003). Optimizing the level of renewable electric R&D expenditures using real options analysis. *Energy Policy*, 31, 1589–1608.

Dixit, A. and Pindyck, R. (1994). *Investment Under Uncertainty*. Princeton, NJ: Princeton University Press.

European Commission (2005). *Competition Directorate, Energy Sector Inquiry*. Draft Preliminary Report. Available at www.europa.eu.int/comm/competition/antitrust/others/sector_inquiries/energy

Fabozzi, F., Gupta, F. and Markowitz, H. (2002). The legacy of modern portfolio theory. *Journal of Investing, Institutional Investor* (Fall), 7–22.

Frayer, J. and Uludere, N. (2001). What is it worth? Application of real options theory to the valuation of generation assets. *The Electricity Journal*, 14(October), 40–51.

Gately, D. and Huntingdon, H. (2002). The asymmetric effects of changes in price and income on energy and oil demand. *The Energy Journal*, 23(1).

Gollier, C., Proult, D., Thais, F. and Walgenwitz, G. (2005). Choice of nuclear power investments under price uncertainty: valuing modularity. *Energy Economics*, 27(4), 667–685.

Greene, D. and Tishchishyna, N. (2000). *Costs of Oil Dependence: A 2000 Update*. Prepared for Department of Energy, ORNL TM-2000/152 (May).

Grubb, M., Butler, L. and Twomey, P. (2006). Diversity and security in UK electricity generation: the influence of low-carbon objectives. *Energy Policy*, 34(18), 4050–4062.

Hamilton, J. (2005). *The Macroeconomic Consequences of Higher Crude Oil Prices*. Final Report (EMF SR 9) to US Department of Energy, Energy Modeling Forum, Stanford.

Hill, M. (1973). Diversity and evenness: a unifying notation and its consequences. *Ecology*, 54(2), 427–432.

Humphreys, H. and McClain, K. (1998). Reducing the impacts of energy price volatility through dynamic portfolio selection. *The Energy Journal*, 19(3), 107–131.

Hunt, B., Isard, P. and Laxton, D. (2002). The macroeconomic effects of higher oil prices. *National Institute Economic Review*, 179. London: NIESR (January).

International Energy Agency (2004). *The Impact of High Oil Prices on the Global Economy*. Economic Analysis Division Working Paper. Paris: OECD/IEA (May). Available at http://www.iea.org/textbase/papers/2004/high_oil_prices.pdf

International Energy Agency (2006). *World Energy Outlook*. Paris: OECD/IEA.

International Monetary Fund (IMF) (2005). *Oil Market Developments and Issues*. Washington DC: Policy Development and Review Department.

Jansen, J., van Arkel, W. and Boots, M. (2004). *Designing Indicators of Long-Term Energy Supply Security*. Report C-04-007. Energy Research Center at the Netherlands (ECN) (January).

Jansen, J., Beurskens, L. and van Tilburg, X. (2006). *Application of Portfolio Analysis to the Dutch Generating Mix*. Report C-05-100. Energy Research Center at the Netherlands (ECN) (February).

Kleindorfer, P. and Li, L. (2005). Multi-period VaR-constrained portfolio optimisation with applications to the electric power sector. *The Energy Journal*, 26(1).

Lucas, N., Price, T. and Tompkins, R. (1995). Diversity and ignorance in electricity supply investment – a reply to Andrew Stirling. *Energy Policy*, 23(1).

Markowitz, H. (1952). Portfolio selection. *Journal of Finance*, 7(1), 77–91.

Murto, P. and Nese, G. (2002). *Input Price Risk and Optimal Timing of Energy Investment: Choice Between Fossil- and Biofuels*. Working Paper No. 25/02. Bergen: Institute for Research in Economics and Business Administration (May).

Roques, F. (2003). *Security of Electricity Supplies*. UK Parliamentary Office of Science and Technology Report and Postnote 203 (September). Downloadable at www.parliament.uk/post/pn203.pdf

Roques, F. (2008). Technology choices for new entrants in liberalized markets: The value of operating flexibility and contractual arrangements, Utilities Policy (2008), doi:10.1016/j.jup.2008.04.004.

Roques, F., Newbery, D., and Nuttall, W. (2008). Fuel mix diversification incentives in liberalized electricity markets: a mean-variance portfolio theory approach. *Energy Economics* 30(4), 1831–1849.

Roques, F., Newbery, D., Nuttall, W., de Neufville, R. and Connors, S. (2006). Nuclear power: a hedge against uncertain gas and carbon prices? *The Energy Journal*, 27(4), 1–24.

Siddiqui, A. S., Marnay, C. and Wiser, R. H. (2007). Real options valuation of US federal renewable energy research, development, demonstration, and deployment. *Energy Policy*, 35(1), 262–279.

Stirling, A. (1994). Diversity and ignorance in electricity supply investment. Addressing the solution rather than the problem. *Energy Policy*, 22 (March), 195–216.

Stirling, A. (1998). *On the Economics and Analysis of Diversity*. SPRU Electronic Working Paper 28 (October). http://www.sussex.ac.uk/spru/publications/imprint/sewps/sewp28/sewp28.html

Stirling, A. (2001). Science and precaution in the appraisal of electricity supply options. *Journal of Hazardous Materials*, 86, 55–75.

Trigeorgis, L. (1996). *Real Options: Managerial Flexibility and Strategy in Resource Allocation*. Cambridge, MA: MIT Press.

PART I

Assessing Risks, Costs and Fuel Mix Diversity for Electric Utilities

CHAPTER 1

Diversity and Sustainable Energy Transitions

Multicriteria Diversity Analysis of Electricity Portfolios

Andy Stirling*

Acknowledgements

First and foremost, for reasons discussed in the chapter, I would like to acknowledge Shimon Awerbuch for the many stimulating conversations on issues addressed here. I am also very grateful to comments from colleagues in the Sussex Energy Group, especially Raphael Sauter. The work reported is funded by the ESRC.

1.1 Diversity, security, sustainability and wider energy policy

As I know from happy memories of invigorating discussions, two great passions in Shimon Awerbuch's work lay in pushing for greater attention to be paid to diversity and sustainability in energy policy (Awerbuch, 1976, 2000a; Awerbuch and Yang, 2007). In a way that I hope Shimon would have appreciated, this chapter seeks to address both themes. Here, my approach, which often amused us in our dealings together, contrasts quite markedly with Shimon's own favored methods. Yet, in attempting an interdisciplinary effort of this kind, much would have been gained from Shimon's distinguishing qualities. His lively questioning of received assumptions, his openness to new ideas and his enthusiasm for positive change would all have made for a better chapter – and are sorely missed.

In this sadly felt absence, the discussion will be more prosaic. The present chapter will begin by reviewing some underexamined dimensions of the links

*SPRU – Science and Technology Policy Research Freeman Centre, University of Sussex, UK

between diversity and sustainability in energy policy. This will involve a brief survey of the remarkable variety of broad strategic reasons for an interest in energy diversity. Against this background, discussion will then turn to the definition of energy diversity itself, concentrating on three quite concrete properties and their treatment in various analytical approaches. It is on this basis that the chapter will then propose a new conceptual framework for addressing these attributes, in a way that remains flexible to different interpretations and priorities, but which allows for more systematic and transparent analysis of alternative strategies for securing energy diversity. Finally, a novel quantitative heuristic tool will be presented, with which to explore the tradeoffs between portfolio diversity and other aspects of economic and wider strategic performance, including system-level properties. It will be concluded that this new approach may offer a potentially fruitful framework for analysis under conditions (which fascinated Shimon), where conventional portfolio analysis is not applicable.

The starting point for this discussion, then, lies in the fundamental nature of the links between diversity and sustainability. Despite the high profiles of energy policy debates over diversification and transitions to sustainability, it is curious that, as general concepts, these remain ostensibly quite separate (DTI, 2006; CEC, 2007). In particular, there is a dearth of attention to their inherent (rather than circumstantial) interrelationships (Helm, 2007; IEA, 2007). Historically, discussion of energy diversity has been preoccupied with supply security, driven by successive 'oil shocks' and geopolitical concerns over the distribution and transit of fossil fuels (IEA, 1985; CEC, 1990; Verrastro and Ladislaw, 2007). Debates on sustainable energy, in contrast, have tended to focus on environmental imperatives and, especially latterly, abatement of carbon emissions (DTI, 2006). With important exceptions in the more detailed literature (Awerbuch and Yang, 2007; Helm, 2007; IEA, 2007), these two themes for the most part coexist in parallel as distinct 'pillars' in mainstream energy policy discourses (CEC, 2007).

Of course, it is important to acknowledge at the outset that there exist many more security-of-supply strategies than diversification alone. Although diversity is sometimes treated as the dominant means to energy security (PIU, 2001; DTI, 2006), a wide range of different measures is discussed in the literature. For instance, the ubiquitous aspiration to increase reliance on indigenous resources may sometimes be advocated as an energy security strategy, even if this reduces diversity (IEA, 1980). Likewise, states seek to deter threats of disruption by fostering economic interdependence on the part of supplier interests (European Energy Charter Secretariat, 2004). Intergovernmental bodies are established [such as the International Energy Agency (IEA) of the Organization for Economic Co-operation and Development (OECD)] to provide for more effective international planning of responses to disruption (Adelman, 1995). Expensive stockpiles are maintained for key resources, such as European and US strategic reserves (respectively) of refined and crude oil (Greene et al., 1998). Efforts are made to exercise greater control over energy supply chains, as by the UK Conservative government's action against the British miners in the 1980s (Lawson, 1992). Investments are made in redundant infrastructure, such as 'mothballed capacity' and 'spinning reserve' in electricity transmission and network reinforcement in distribution systems (Farrell et al., 2004). A premium is often paid for more flexible options, like dual fuel firing capability in electricity generation (Costello, 2004). Demand-side efficiency programs are supported in part to reduce the economic impacts of supply disruptions (Lovins and Lovins, 1982). Key infrastructures, such as that for long-range transport of natural gas, are increasingly audited for properties of 'resilience' in the face of particular or unspecified threats (JESS, 2004). Offensive military action remains a major energy security strategy option in many influential circles (Plummer 1983), although acknowledged with reticence. Finally, at the most general of levels, the effective functioning of energy markets themselves is often presented as the principal means toward optimizing energy security (Helm, 2007).

In all these ways, security of supply is clearly about more than diversity. Yet what is less well recognized is that diversity is equally more than just a security of supply strategy (Stirling, 1994). It is here that we encounter the sustainability agenda as a major parallel area for fundamental interests in energy diversity (Grubb et al., 2006; Bird, 2007). Before reviewing the linkages, however, it should be noted that some of the most prominent apparent connections actually tend on closer inspection to be quite contingent. For instance, one obvious resonance lies in considering resource depletion. This hinges on the seminal formulation of the Brundtland Commission, in which sustainability is about 'development that meets the needs of the present without compromising the ability of future generations to meet their own needs' (Brundtland, 1987). Although energy security concerns are often more proximate and political, all are conditioned by the fact that fossil fuels are scarce and depleting. Assuming their continued value, persistent unsustainable use of these finite resources can therefore be seen both as a consequence and as a driver of marginalization of interests of future generations. To diversify away from present dependences on scarce, diminishing fossil fuel supplies thus addresses both security and sustainability agendas.

The key point here is that this link between sustainability and diversity, although important, is more an artifact of history (and geology) than a fundamental interrelationship. On close inspection, the object of interest is not the property of diversity itself, but a contingent feature of the particular resources on which we presently happen to be overwhelmingly dependent. Were our incumbent energy technologies to be indefinitely sustainable into the future through harnessing of renewable (or, as sometimes claimed, nuclear) resources, then the depletion argument for diversification would not apply. A similar picture arises where (as is increasingly the case) sustainability concerns are restricted to the relatively circumscribed challenge of 'emissions reduction', and even further to the single indicator of carbon intensity. This reduced scope again prompts attention to diversification only incidentally, as a reflection of the currently marginal status of low-carbon options in the energy mix (DTI, 2006). Under scenarios where such options are dominant (again, as in systems based on renewables or nuclear), then diversity and 'emission reduction' agendas appear quite distinct. The conjunctions of diversity and sustainability in emission reduction agendas thus hinge more on contingent correlations of attributes among specific options rather than on any intrinsic inter-relationships between the qualities of diversity and sustainability themselves.

Without diminishing the enormous political and economic significance of these circumstantial correlations between properties of particular energy options, it is important that they not be allowed to obscure appreciation of the deeper underlying linkages between diversity and sustainability. These more fundamental connections persist irrespective of any specific features of particular options. At root, the real interlinkages stem from the crucial (and often neglected) fact that diversity is an intrinsic and irreducible property of an energy system taken as a whole, rather than an attribute of any individual option in that system (Stirling, 1994). As a consequence, it is meaningless to speak of any single option offering 'diversity', without being explicit about the energy mix and other potential substitute option contributions by reference to which this diversity is appraised. Associating diversity with certain currently marginal options rather than others, begs crucial questions over the relative scale or efficiency of diversity benefits offered by other less privileged options (MacKerron and Scrase 2008). This has long been the norm in UK energy policy, for instance, where nuclear power has tended to be treated as a proxy for diversity, in ways that a former Secretary of State for Energy later acknowledged as an expedient 'code' (Lawson, 1992). The question has always been: how does nuclear compare with other means to secure equivalent diversity benefits (Stirling, 1994)? Whatever the specifics, a shorthand that conflates the general property of diversity with more partisan technology strategies is little more than rhetorical special pleading.

An understanding of diversity as an irreducible 'system property' takes a more comprehensive perspective and so avoids this problem. In so doing, it points to a more fundamental series of convergences between diversity and sustainability agendas. These link to the multitude of complexities conveyed in the Brundtland Commission's broad focus on *'development'*, explicitly encompassing dimensions of economic well-being and social equity, as well as conventional emphases on environmental quality (including carbon emissions) (Brundtland, 1987). This challenging agenda invokes a range of wider intended *outcomes* of more sustainable energy pathways, such as reducing health impacts, alleviating fuel poverty, strengthening local livelihoods and enhancing social cohesion, as well as the crucial goal of combating climate change. But it also focuses on the governance *processes* through which these more sustainable pathways are pursued, involving greater democratic accountability in technology choice and improved community empowerment in the managing of local energy resources and infrastructures. Likewise, in introducing the principle of precaution, this broader sustainability agenda also recognizes the pervasive role of *uncertainty*. Taken together, this wider (well-established and internationally recognized) understanding of sustainability presents a formidable imperative for industrial and social change. Transitions to truly sustainable energy systems will therefore require an unprecedented intensity of *innovation*, in technology, institutions and behavioral practices alike (Stirling, 2007b).

It is in these more ambitious and general challenges of sustainability that we find the truly fundamental interlinkages between energy sustainability and energy diversity as an intrinsic system-level property. Quite apart from the role in supply security, diversification of energy systems presents an intriguing common denominator in many of these further demanding strategies. For instance, 'putting eggs in different energy baskets' does not only help to hedge against uncertainty over energy disruptions. It also offers a *precautionary* strategy for greater resilience, flexibility and adaptiveness in the face of other potential surprises, such as those arising from unexpected developments in environmental or engineering performance (Brooks, 1986; Stirling, 1999). Likewise, pursuit of a diversity of energy resources and technologies offers a way to achieve greater sensitivity to local geographical and cultural *context* (Landau et al., 1996). Set-piece confrontations between advocates of contending energy technologies typically undermine the credibility of social consensus around any single 'best' option. Deliberate diversification thus also offers a means for society at large better to *accommodate* the plural social values and polarized interests, which so often attend sustainability strategies (Stirling, 1997; JESS, 2004).

Finally, there are the formidable imperatives presented by these ambitious sustainability goals for enhanced technological and institutional *innovation*. Here, as in supply security, the agenda is shared between sustainability (Kaijser et al., 1991) and mainstream policy concerns with competitiveness (Aoki, 1996). For it has long been recognized in the economics and business literature that institutional and technological diversity are key factors in achieving more effective and socially robust *innovation* (Rosenberg, 1996; Landau et al., 1996; Grabher and Stark, 1997). It is also through careful strategies of diversification that we can guard against *premature lock-in* to what in the long run might prove to be suboptimal technological trajectories (Arthur, 1989). Such a dynamic has, for instance, frequently been observed by sceptics and proponents alike in the case of early closure around light water reactor designs in nuclear power (Cowan, 1991). High power densities and reliance on active control systems were necessary for the confined early priority of submarine propulsion, but not such a good basis for civilian power production (Stirling, 2008b). These lessons apply with particular intensity where sustainability objectives demand rapid development of new configurations in a wide variety of technological areas from presently small niches to possible future mainstream roles. It is only when this broader picture is appreciated that we can see the true multiple significance of energy diversity.

In short, the real ('fundamental') interlinkages between diversity and sustainability in energy strategies lie in nurturing sensitivity to context, accommodating plural values, hedging against ignorance, mitigating lock-in and fostering innovation (Stirling, 2008a). It is these latter qualities that are essential to sustainability transitions and which are intrinsic to diversity. This is generally so, in a fashion that is *quite independent of the specific attributes of individual energy options.*

So much for the breadth of the potential general benefits of energy diversity. But, of course, it is important not to get carried away. Despite the attractions, diversity is rarely a 'free lunch' (Weitzman, 1992). Indeed, deliberate diversification by definition involves prioritizing options that are otherwise assigned relatively low performance (David and Rothwell, 1996). In addition, there are typically trade-offs between diversity and transaction costs (Williamson, 1993) and with forgone benefits such as coherence (Cohendet et al., 1992), accountability (Grabher and Stark, 1997), standardization (Cowan, 1991) and economies of scale (Matthews and McGowan, 1992). Diversification may retard learning effects for incumbents in favor of more marginal options. The crucial challenge thus lies in striking a balance between the benefits of diversity and these countervailing aspects of portfolio performance (Geroski, 1989). The value of the 'diversity premium' (Ulph, 1988) that is warranted in any particular energy mix will be a function of the performance attributed to individual energy options and the contributions that each makes to system diversity (Stirling, 1994). In the end, all such issues are judgemental, offering ample scope for legitimate disagreement. Recognition of these complexities underscores the importance of a systematic general framework for exploring different perspectives on the implications of energy diversity (Bruno et al., 1991; Mercier and McGowan, 1996). It is to this challenge that discussion will now turn.

1.2 General properties of energy diversity: variety, balance and disparity

At root, diversity is a property of any system whose elements may be apportioned into categories. Energy systems are no exception. Disciplines such as ecology, economics, taxonomy, paleontology, complexity and information theory have all developed sophisticated frameworks for analyzing various aspects in contrasting contexts (Stirling, 1998). The analysis of energy diversity may therefore gain much through building on the approaches developed in other fields. This is all the more the case, because the parameters of interest are, as we have seen, so wide ranging within the energy policy field itself. This breadth in the policy salience of energy diversity presents an inherent advantage for the most generally applicable frameworks.

To take the electricity sector as an example, discussions of diversity span an array of disparate supply- and demand-side technologies and primary resources. The scope of diversity analysis is further extended by a variety of other relevant factors, including: the regional sourcing of fuel and associated supply routes; concentration among trading, supplier or service companies; reliance on generic equipment or component vendors; dependencies on monopoly utilities, shareowners or labour unions; and the configurations and spatial distribution of infrastructures (PIU, 2001; Farrell et al., 2004; Verrastro and Ladislaw, 2007; Helm, 2007; CEC, 2007). These are all prominent features of debates over diversity in energy security. Likewise, each is potentially relevant to diversity as a means to hedge ignorance, foster innovation, mitigate lock-in or accommodate plural values and interests, in the broader senses discussed above in relation to transitions to energy sustainability. It is therefore desirable that any framework for the analysis of energy diversity be equally applicable in principle across all these aspects.

Fortunately, it is precisely when approached in this most general fashion (as a fundamental property of any system apportioned among elements) that experience in these other disciplines holds the clearest lessons for analyzing energy diversity. In short, diversity concepts employed across the full range of sciences mentioned above display some combination of just three basic properties: 'variety', 'balance' and 'disparity' (Stirling, 1994). Each is a necessary but individually insufficient property of diversity (Stirling, 1998). Although addressed in different vocabularies, each is applicable across a range of contexts. Each is aggregated in various permutations and degrees in quantitative indices (Hill, 1973). Despite the multiple disciplines and divergent empirical details, there seems no other obvious candidate for a fourth important general property of diversity beyond these three (Stirling, 1998, p. 47). They are summarized schematically in Figure 1.1.

In terms of electricity supply portfolios, *variety* is the number of diverse categories of 'option' into which a system may be apportioned. It is the answer to the question: 'how many options do we have?' This aspect of diversity is highlighted (for instance) in conventional approaches based on the simple counting of named categories such as 'coal', 'gas', 'nuclear' and 'renewable energy'. *All else being equal, the greater the variety of distinct types of energy option, the greater the system diversity.* For instance, in 1990, standard OECD statistics partitioned national member state electricity supply systems among six options: 'coal', 'oil', 'gas', 'nuclear', 'hydro/geothermal' and 'other' (IEA, 1991). For more specific purposes in 2001, the resolution of reporting increased to 11 options: 'coal', 'oil', 'gas', 'nuclear', 'hydro', 'geothermal', 'solar', 'tide/wave/ocean', 'wind', 'combustion renewables and waste' and 'other (e.g. fuel cells)' (IEA, 2002). Each scheme provides a different basis for counting variety.

Balance is a function of the apportionment of the energy system across the identified options. It is the answer to the question: 'how much do we rely on each option?' The denominator here may (depending on the context) be expressed as

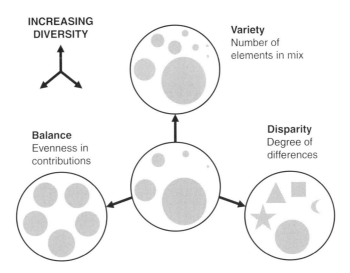

FIGURE 1.1   Schematic summary of three co-constituting properties of energy diversity.

energy inputs or outputs, power capacity or economic value. Either way, balance (like statistical variance; Pielou, 1977) is simply represented by a set of positive fractions, which sum to one (Laxton, 1978). This dimension appears most frequently in energy debates, in discussions around a possible role for designating the contributions of different supply options (Helm, 2007). It is captured in more detail in a variety of 'concentration' or 'entropy' measures that are nowadays quite widely applied in energy policy, such as the Shannon–Wiener (Stirling, 1994; DTI, 1995, 2003; Scheepers et al., 2007) and/or (Grubb et al., 2006) Herfindahl–Hirschman (IEA, 2007; Hubberke, 2007) indices. *All else being equal, the more even the balance across energy options, the greater the system diversity.* An example of the importance of considering balance lies in the contrasting stories of Japanese and French electricity systems following the 1973 'oil shock'. Over the 27-year period up to 2000, both Japan and France moved away from oil-dominated systems with nuclear at the margin (2% in Japan, 8% in France). In the Japanese case, the diversification strategy led to a roughly even balance across nuclear, coal and gas as modal options (Suzuki, 2001). In France, however, the 'diversification' strategy simply substituted an initial 40% dependence on oil for an even greater 77% dependence on nuclear at the end (IEA, 2002), involving a cumulative *decrease* in diversity over this period (Stirling, 1994).

Disparity refers to the manner and degree in which energy options may be distinguished (Runnegar, 1987). It is the answer to the question: 'how different are our options from each other?' This is the most fundamental, and yet most frequently neglected, aspect of energy diversity. After all, it is judgements over disparity that necessarily govern the resolving of the categories of energy option, which underlie characterizations of variety and balance. One valuable attempt to address this property in the energy policy literature lies in the work of Shimon Awerbuch, who (among others, e.g. Ulph, 1988; ERM, 1995; NERA, 1995 and including many in the present volume) took fossil fuel price covariance as a stochastic proxy for wider disparities, such as those mentioned above. This approach will be returned to below. Alternative approaches to this property in other disciplines are usually based on some more general form of scalar distance measure. *Either way, all else being equal, the more disparate the energy options, the greater the system diversity.* In other words, an electricity supply system divided equally among gas, nuclear, wind and biomass power is more disparate than one divided equally among coal, oil, and Norwegian and Russian gas.

The consequence of this threefold understanding of diversity, is a recognition that, although disparity is fundamental, each property helps to constitute the other two (Stirling, 1998). This in turn highlights difficulties with diversity concepts and associated indices that focus exclusively on subsets of these properties (Eldredge, 1992), an illustrative selection of which is displayed in Table 1.1. The resulting ambiguities or hidden assumptions can exacerbate the tendency already noted for insufficiently rigorous treatments of diversity to serve as a vehicle for special pleading.

Variety and balance, for instance, cannot be characterized without first partitioning the system on the basis of disparity (May, 1990). An electricity system may be assigned a nominal variety of four, if it is divided into categories labelled

Table 1.1   Selected indices of contrasting subordinate properties of diversity

Property	Name and/or reference	Form
Variety	Category count (MacArthur, 1965)	N
Balance	Shannon evenness (Pielou, 1969)	$\dfrac{-\sum_i p_i \ln p_i}{\ln N}$
Variety/balance	Shannon–Wiener (Shannon and Weaver, 1962)	$-\sum_i p_i \ln p_i$
	Herfindahl/Simpson (Simpson, 1949)	$\sum_i p_i^2$
	Gini (1912)	$1 - \sum_i p_i^2$
	Hill (1973)	$\sum_i \left(p_i^a\right)^{1/(1-a)}$
Disparity	Weitzman (1992)	$\max_{i \in S}\{\mathbf{D}_W(S\backslash i) + \mathbf{d}_W(i, S\backslash i)\}$
	Solow and Polasky (1994)	$f(\mathbf{d}_{ij})$
Variety/balance/disparity	Junge (1994)	$\left(\dfrac{\sigma}{\mu \cdot \sqrt{n-1}}\right)\left(\dfrac{1}{\sqrt{N}}\right) \cdot$ $\left(\sqrt{N-1} - \sqrt{N\sum_i p_i^2 - 1}\right)$
	Stirling (2007a)	$\sum_{ij} \mathbf{d}_{ij}^{\alpha} \cdot (p_i \cdot p_j)^{\beta}$

Notation	Interpretation in terms of energy portfolios
N	Number of categories of energy options
$\ln$	Logarithm (usually natural)
p_i	Proportion of energy system comprised of option category i
a	A parameter governing relative weighting on variety and balance
n	Number of attributes displayed by options
σ	Standard deviation of attributes within option categories
μ	Mean of attributes within option categories
$f(\mathbf{d}_{ij})$	Function of disparity distance between option categories i and j ($\mathbf{d}_{ij}$)
$\mathbf{D}_W(S)$	Aggregate disparity of energy system S

(Continued)

Table 1.1   (Continued)

Notation	Interpretation in terms of energy portfolios
$\mathbf{d}_W(i, S\backslash i)$	Disparity distance between option i and nearest option in S if i is excluded
α, β	Parameters governing relative weightings on variety, balance and disparity

'coal', 'gas', 'nuclear' and 'renewable energy' (Stirling, 1994). Yet 'renewables' might readily be further divided into numerous other nested categories (e.g. 'hydro', 'wind', 'biomass' and 'tide'). The mutual disparities between many of these ostensibly subordinate taxa might reasonably be thought greater than those between large centralized thermal nuclear and fossil fuel plant.

For similar reasons, considerations of disparity also hold crucial importance for the resolving of balance. These hinge on the simple fact that the structures of proportional contributions are – like counting the categories themselves – determined by the ways and degrees in which options are divided up. Although the more complex quantitative form may confer an apparent authority, an index of balance (e.g. Shannon or Herfindahl) is no less arbitrary than the simple counting of variety. It will yield radically different results depending on the partitioning of options. The implications may be addressed by systematic sensitivity analysis and by adopting explicitly conservative assumptions on disparity with respect to the hypothesis under test (Stirling, 1994). However, the fact remains that taking measures of variety and/or balance as proxies for diversity is highly sensitive to tacitly subjective taxonomies and arbitrary linguistic conventions concerning the implicit bounding of categories.

Conversely, the importance of disparity to energy diversity is itself typically qualified by the pattern of apportionment across options. For instance, an electricity supply portfolio comprising a 95% contribution from one of three highly disparate resources (e.g. Russian gas with the residual 5% made up of nuclear, wind and hydro) might reasonably under some perspectives be judged *less* diverse than a portfolio comprising even contributions from four much less disparate options (e.g. piped Russian, Norwegian and UK gas with internationally traded and transported liquefied natural gas) (PIU, 2001). Likewise, the balance of a portfolio is neglected where attention is restricted to variety alone, as is the case in much of the literature. At what scale of contribution is an option considered to add to system diversity? Does the installation of the first household rooftop photovoltaic panel increase the diversity of a national energy portfolio by the same degree as the construction of the first 1.5 GWe new nuclear power station? If not, at what scale of contribution does any given option begin to be counted? How do we avoid perverse threshold effects? Different indices of balance treat this crucial threshold issue in quite radically different ways (J. Skea, Note on behaviour of diversity indices, personal communication to A. Stirling,

October 2007). Taking disparity (or variety) as proxies for diversity ignores the balance with which a system is apportioned. It therefore seems that the only robust approach to thinking about energy diversity is to think about variety, balance and disparity together.

1.3 Aggregating, accommodating and articulating different aspects of energy diversity

Thus far, we have established a definition of energy diversity in terms of three necessary but individually insufficient properties of disparity, variety and balance. A series of methodological questions follows from this. How can these quite distinct aspects of diversity be *aggregated* into a single coherent framework? How might such an analytical framework be applied such as to *accommodate* the range of relevant perspectives typically engaged in real debates over energy strategy? And how can the results of any diversity analysis on these lines be *articulated* with wider policy considerations, such as the performance of individual generating options under criteria of economic efficiency, environmental quality, social impact and security of supply raised earlier in relation to broad sustainability goals? The present section will consider these challenges of aggregation, accommodation and articulation.

With regard to *aggregation*, most contemporary approaches to analyzing energy diversity focus on variety and/or balance alone (e.g. Stirling, 1994; DTI, 2003; Grubb et al., 2006; ECN, 2007; Hubberke, 2007; IEA, 2007). Even where disparity is thus neglected, however, it is far from straightforward what relative emphasis to place on variety as compared with balance. How much weight should be assigned to small contributions from additional options, compared with enhanced balance among dominant options? It is a little-recognized property of widely used sum-of-the-squares concentration indices (e.g. Herfindahl–Hirschman/Simpson and Gini; Table 1.1) that different exponent powers yield divergent rank orderings for portfolios displaying different patterns of composition across marginal and dominant options (Stirling, 1998, p. 56). That such divergent rankings may not arise in practice for certain particular portfolios (Grubb et al., 2006) does not resolve this concern. Yet there exists no firmly grounded theoretical or empirical reason for taking the commonly used exponent value of two rather than, say, three, four or so on.

Logarithmic entropy functions (e.g. Shannon–Wiener) avoid similar ranking sensitivities across different logarithm bases (Stirling, 1998, p. 56). But there still arises the question as to why the particular implicit weighting embodied in such indices should necessarily reflect the appropriate weighting for real energy systems or stakeholder perspectives. Theoretical work in mathematical ecology derives generalizations of these kinds of index, in which this crucial issue is dealt with by explicit weighting parameters (e.g. Hill, 1973, in Table 1.1; Kempton, 1979). Just because such parameters are not recognized in the conventional indices used in the energy sector does not remove this problem. To ignore this risks straying into mathematical mysticism, where contingently privileged algorithms are ascribed transcendent authority concerning appropriate interests and priorities in the real world.

Beyond this, however, the most serious difficulty with conventional variety-balance indices concerns the neglect of the crucial property of disparity. Where attention is restricted to the enumeration or concentration of options, the assumption is effectively made that all options are equally disparate. This can yield manifestly perverse results, such as that represented in Figure 1.2. Here, any index restricted to variety and/or balance alone will entirely fail to address the fact that the enhanced representation of the more disparate option makes Portfolio A more diverse than Portfolio B. This point is returned to in comparing Figures 1.4 and 1.5 (see below).

Of course, the salience of the example illustrated in Figure 1.2 rests on the understanding (illustrated with the stylized dendrograms) that wind is more disparate from coal and gas than either of these fossil fuels are from the other. This may seem reasonable at an immediate intuitive level. But how can we be sure? And any practically useful diversity analysis must also be clear about the *degree* to which this is the case. In other words, in contemplating an increase in diversity in moving from Portfolio B to Portfolio A where gas is rated as the best performing option, exactly what value of performance on the part of gas might be sacrificed as a premium for the additional diversity conferred by wind? This raises the second challenge identified above: that of *accommodating* the divergent perspectives on disparity (as well as performance) that typically characterize even the most specialist discussions of energy strategy.

This is essentially the problem that Shimon Awerbuch and others have sought to address through the exploratory application of portfolio theory to the analysis of contending possible electricity supply mixes (Ulph, 1988; NERA, 1995; ERM, 1995; Awerbuch and Yang, 2007). In brief, what this does is take the single parameter of covariance in fuel price risk as a stochastic proxy for a range of multidimensional economic, physical, environmental, technical, institutional and geographical disparities between electricity options. Such elegant shorthand may suffice for

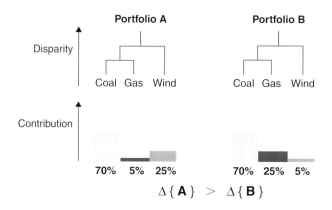

FIGURE 1.2   A case where varying diversity is not discriminated by conventional indices.

short-term decisions by private firms, dominated simply by financial risks due to fuel price shocks (Awerbuch, 2000b). It has the benefit, at least in principle, of requiring only relatively objectively attested data. But, as Shimon Awerbuch acknowledged (Awerbuch et al., 2006), as strategic interests grow wider and time-frames longer, serious questions arise over the sufficiency (and even validity) of this approach. To what extent can the historic behavior of fuel prices be taken as a reliable guide even to the future trajectories of this one parameter, let alone the host of other factors invoked by wider sustainability agendas? How do we address options whose attributes are simply not reflected in fossil fuel markets? Where diversity is undertaken as a response to strict uncertainty and ignorance, the use of probabilistic concepts such as covariance coefficients is by definition invalid (Stirling, 2003). Although the origins in Nobel Prize-winning work in financial economics may lend a certain authority, portfolio theory is a highly circumscribed basis for addressing the full strategic scope of energy diversity.

Although rarely integrated with consideration of variety and balance, the challenge of accommodating divergent possible understandings of disparity is quite well addressed in other disciplines. Approaches in fields such as taxonomy, paleontology, archeology and conservation biology all routinely adopt a framework based on the notion of distance. Often, there exist in such fields some well-established or even objectively determinable criteria of disparity. This is the case, for instance, with genetic distance measures in evolutionary ecology, which can be assumed to display a strict branching form (Weitzman, 1992; Solow and Polasky, 1994). Under such circumstances, the analysis of diversity can be quite strictly codified in terms of 'disparity distances' between elements and relatively unambiguous answers derived.

Elsewhere, however, the picture is much more similar to that in the energy field, where options are not differentiated by strictly branching genetic processes and where there typically exists a variety of different views over the relevant aspects of disparity and their respective degrees of importance (Stirling, 1998). Even here, however, it is possible to use the simple concept of distance. Options are characterized in terms of whatever are held to be the salient attributes of difference, such that each can be represented as a coordinate in a multidimensional 'disparity space' (Stirling, 2007a). With the different disparity attributes weighted to reflect their relative importance, the simple Euclidean distances separating options in this space can be taken as a reflection of their mutual disparity. By appropriate normalization and weighting procedures, a 'disparity space' of this kind can be constructed accurately to reflect any conceivable perspective on the distinguishing features of different energy options (Kruskal, 1964). Indeed, fuel price covariance might be seen just as one such possible dimension of disparity.

As a starting point for implementing this broader disparity-distance approach in the analysis of energy diversity, it is useful to consider the nature of existing datasets concerning the economic and/or wider sustainability performance of different energy options. Encompassing different aspects of financial, operational, environmental, health and broader social impacts, many such datasets have been generated over past years by a range of different disciplines, including cost–benefit, environmental impact and technology assessment and comparative risk, life cycle and multicriteria analysis. To the extent that the performance of each energy option is structured differently under the various performance criteria, each of these data-sets contains potentially useful information over their disparities. If the different criteria are normalized such that all options are reassigned nominally equal performance, then the distances in the resulting multidimensional space will provide a robust indication of broader disparities according to the perspective in question on salient strategic attributes. Figure 1.3 provides a schematic illustration of this approach. Of course, none of this precludes use of non-normalized performance data, as a basis for determining tradeoffs between performance and diversity (Stirling, 2007a).

The kind of disparity structures that pilot work indicates may be yielded by such performance data are illustrated in Figure 1.4. Distances between successively more remote pairs of options in disparity space can be represented using standard cluster analysis techniques and represented as a dendrogram (with disparity distance indicated on the horizontal axis labelled 'd'). The actual underlying distances used in diversity analysis will not be affected by the sometimes slightly contrasting representations yielded by different clustering metrics, algorithms and procedures (Sneath and Sokal, 1973).

More importantly, however, different performance datasets *can* be expected to generate different disparity structures. Similarly, given the broad strategic scope required in considering diversity, it may also be expected that the different dimensions of such analysis will be weighted differently under different perspectives. This is a matter for further investigation, currently underway by the author in collaboration with colleagues Go Yoshizawa and Sigrid Stagl. For the moment, early indications are that direct elicitation from interviewees, by reference to typical detailed sustainability performance datasets, consistently yields intuitively meaningful structures, such as those indicated in Figure 1.4. Existing performance datasets provide a useful starting point to the construction

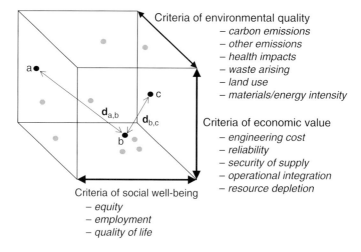

FIGURE 1.3   A disparity space generated by normalized sustainability performance data.

of meaningful patterns of disparity, which may readily be followed up by direct elicitation of further more-specific disparity attributes in an intensive interview or a deliberative group setting.

In considering the many possible queries that might be raised over detailed disparity structures such as that illustrated in Figure 1.4, two things must be remembered. The first is that the point here is not to assert that there exists any single well-defined 'objective' disparity structure that applies irrespective of context or perspective across real-world energy options. Instead, the value of this general approach lies in the possibility of more systematic and transparent ways of accommodating inevitably divergent viewpoints on disparity. It is interesting and potentially significant that existing performance datasets, backed up by direct elicitation and deliberation involving specialists, may quite readily yield intuitively robust disparity structures.

For those to whom this seems like an uncomfortably subjective or open-ended approach, the second point is that these challenges are unavoidably intrinsic to the complexities of energy disparity itself. Simply to ignore this challenge – for instance through the conventional restriction of attention to variety and balance alone or a single circumscribed parameter such as fuel price covariance – does not avoid assumptions over option disparity. It simply conceals them. If indices such as Shannon or Herfindahl are used to analyze diversity among a set of electricity supply options like that resolved in Figure 1.4, for instance, this simply amounts to treating all identified options as equally disparate, as represented in Figure 1.5.

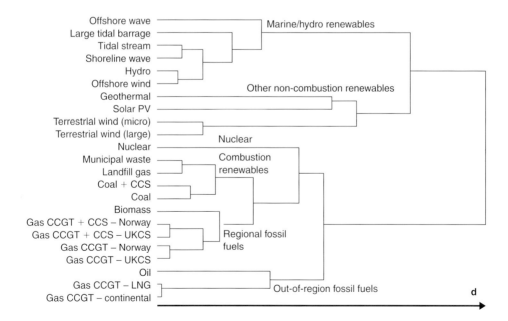

FIGURE 1.4 Indicative disparities in multicriteria performance of electricity options. PV: photovoltaic; CCS: carbon capture and storage; CCGT: combined cycle gas turbine; UKCS: UK Continental Shelf; LNG: liquefied natural gas.

The choice is therefore not *whether* to respond to the challenge of accommodating divergent perspectives on disparity but *how* – and with what degree of rigor and openness. The resulting questions concern not the absolute precision of any given directly elicited disparity structure, but its relative plausibility in relation to a default picture like that in Figure 1.5.

The third and final challenge raised at the beginning of this section concerns the *articulation* of diversity with other properties of strategic interest in the appraisal of energy portfolios. As already noted, the key dimensions of portfolio performance will often be quite distinct from diversity, concerning criteria such as operational efficacy, financial performance, environmental impacts and other aspects of supply security. The challenge is particularly acute as attention extends to the full range of pressing issues invoked by sustainability agendas. Aggregate option performance under such criteria (under any perspective) will typically be a function of the performance of the individual options. This will therefore be subject to important tradeoffs as deliberate diversification draws larger contributions from what appear under the perspective in question to be lower performing options. In order to be useful, therefore, any practical framework for the analysis of energy diversity should not be applied in isolation, but should articulate directly with these broader appraisal criteria, such as to allow systematic exploration of relevant tradeoffs, interactions and operational effects. It is to one possible framework for addressing these challenges of aggregation, accommodation and articulation that attention will now turn.

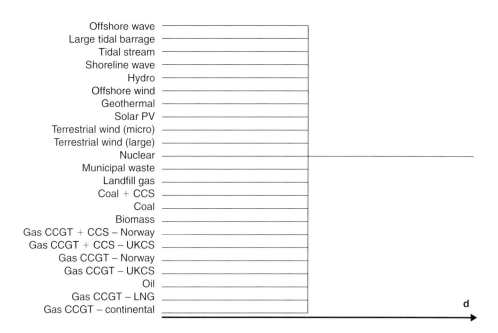

FIGURE 1.5   Indicative disparities implicit in use of Shannon or Herfindahl indices. PV: photovoltaic; CCS: carbon capture and storage; CCGT: combined cycle gas turbine; UKCS: UK Continental Shelf; LNG: liquefied natural gas.

1.4 A novel diversity heuristic for strategic appraisal of energy portfolios

To take seriously these problems of aggregation, accommodation and articulation does not necessarily lead to a counsel of despair over the potential for systematic general characterizations – or even quantifications – of energy diversity. A more positive starting point is the observation that the futility of deriving a single definitive diversity *index* need not preclude the possibility of a flexible general *heuristic*. Like an index, a heuristic may be quantitative. But rather than aiming to measure diversity in some unconditional objective fashion, it offers an explicit, systematic basis for exploring sensitivities to the assumptions conditioning aggregation, accommodation and articulation (Stirling, 2007a).

For any particular perspective on the appropriate weightings for variety and balance and the salient dimensions of disparity, such a heuristic would behave as an index. It would accommodate different views on the relevant attributes of disparity, aggregate these with consideration of variety and balance and allow systematic articulation with important system-level properties other than diversity (like portfolio interactions). For applications involving a range of perspectives, this heuristic would allow clear comparisons to be made between the implications of contending judgements. In other words, a heuristic characterization of energy diversity aims to combine the rigor, transparency and specificity of quantification with the scope, applicability, flexibility and symmetry of qualitative approaches. The real challenge lies in achieving this, while minimizing the introduction of further complexity and contingency.

No existing diversity index addresses all three properties of variety, balance and disparity in an unproblematic way. However, based on well-established criteria applied to the treatment of these individual diversity properties by researchers such as Hill (1973), Pielou (1977), Laxton (1978), Weitzman (1992) and Solow and Polasky (1994), a series of non-trivial requirements is quite readily developed. For instance, some significant desirable features of a general diversity heuristic Δ that help explicitly to address challenges of aggregation, accommodation and articulation as defined here are outlined in Table 1.2.

No established diversity index satisfies all the criteria summarized in Table 1.2. Yet there is one relatively straightforward quantitative heuristic, independently derived in different disciplines (Rao, 1982; Stirling, 1998), which does offer a starting point. This is the sum of pairwise option disparities, weighted in proportion to option contributions (D):

$$D = \Sigma_{ij(i \neq j)} d_{ij} \cdot p_i \cdot p_j \qquad (1.1)$$

where p_i and p_j are proportional representations of options i and j in the energy system (balance), and d_{ij} is the distance separating options i and j in a particular disparity space (Figure 1.3). The summation is across the half-matrix of $((n^2 - n)/2)$ non-identical pairs of n options (i ≠ j). In the special case where all d_{ij} are equal (scalable to unity), D reduces to half the value that would be obtained by the Gini index (Table 1.1).

It is readily demonstrated that this heuristic, D, complies with criteria 1–7. Compliance with criterion 8 remains a matter of judgement, but it is difficult to

Table 1.2   Formal conditions for a robust general heuristic of energy diversity, Δ

Condition	Description (see references in Stirling, 2007a)
1. Scaling of variety	Where option variety is equal to one, Δ takes a value of zero
2. Monotonicity of variety	Where energy options are evenly balanced and equally disparate, Δ increases monotonically with variety
3. Monotonicity of balance	For given option variety and disparity, Δ increases monotonically with balance (i.e. Δ is maximal for equal reliance on all options)
4. Monotonicity of disparity	For given variety and balance, Δ increases monotonically with the aggregate disparity between energy options
5. Scaling of disparity	Where aggregate disparity is zero (i.e. where all energy options are effectively identical), Δ takes a value of zero
6. Open accommodation	Δ accommodates any perspective on salient dimensions of disparity under which energy options can be differentiated
7. Insensitivity to partitioning	For any given perspective on taxonomy, Δ is insensitive to alternative partitionings of options into categories
8. Parsimony of form	Δ is as uncomplicated in structure and parsimonious in form as necessary to fulfil the above conditions
9. Explicit aggregation	Δ permits explicit aggregation of variety, balance and disparity, by reflecting divergent contexts or perspectives using weightings
10. Ready articulation	Δ allows unconstrained articulations of diversity with other salient properties of individual options or the energy system as a whole

imagine a solution to these criteria that is simpler or more parsimonious. As to criterion 9, this raises a final notable feature of D, which can be illustrated by introducing just two further terms (Stirling, 2007a):

$$\Delta = \Sigma_{ij(i \neq j)}(d_{ij})^{\alpha} \cdot (p_i \cdot p_j)^{\beta} \tag{1.2}$$

If exponents α and β are allowed to take all possible permutations of the values 0 and 1, this yields four variants of the heuristic Δ. Each of these usefully captures

Table 1.3 Four variants of Δ and their relationships with diversity properties

Property	α	β	$\Delta =$	Equivalents (see Table 1.1)	Interpretation
Variety	0	0	$\Sigma_{ij}\, d_{ij}^0$	([Category count] $-\,1)^2/2$	Scaled variety
Balance	0	1	$\Sigma_{ij}\, p_i \cdot p_j$	[Gini]/2	Variety-weighted balance
Disparity	1	0	$\Sigma_{ij}\, d_{ij}$	[Solow and Polasky]	Variety-weighted disparity
Diversity	1	1	$\Sigma_{ij}\, d_{ij} \cdot p_i \cdot p_j$	D	Balance-weighted disparity

one of the four properties of interest: variety, balance, disparity and diversity (Table 1.3).

Shifting values of exponents α and β between 0 and 1 yields further variants of Δ, collectively addressing all possible relative weightings on variety, balance and disparity. Of these, the reference case, D ($\alpha = \beta = 1$), does the same job as other widely used non-parametric measures such as those of Gini, Shannon and Simpson, but with the major additional feature that it also captures disparity. Unlike the disparity measures proposed by Weitzman or Solow and Polasky (Table 1.1), Δ also addresses variety and balance. Unlike the 'triple concept' proposed by Junge (1994; see Table 1.1), Δ accommodates radically divergent perspectives on disparity itself, yet is relatively parsimonious in form. An entirely novel feature of Δ is that it systematically addresses alternative possible aggregations of these subordinate properties, according to perspective and context.

1.5 Articulating energy diversity with other aspects of strategic performance

Of the formal criteria identified for a general heuristic of energy diversity in Table 1.2, the discussion in the last section leaves only criterion 10 unaddressed, concerning the articulation of diversity with other relevant system-level properties. As already mentioned, energy diversity is rarely a free lunch. The overall strategic performance of a portfolio as a whole will be a function not only of diversity but of other system properties and the performance of individual options. For instance, there will typically be portfolio effects resulting from interactions between subsets of options, such as potentially negative feedback between high penetrations from intermittent renewables and the inflexibility of large, predominantly base-load plant like nuclear power (Gross et al., 2006) and competition between contending technology strategies or more specific institutional and industrial tensions through which, for instance, nuclear power can 'crowd out' large-scale commitments to renewable energy (Mitchell and Woodman, 2006). Of course, there can also be positive synergies between disparate energy options, such as those often argued to benefit joint pursuit of distributed generation and demand-side energy efficiency measures (Prindle et al., 2007). Diversity itself may provide a strategic response to supply security challenges, as well as hedging more general sources of ignorance, fostering innovation, mitigating lock-in and accommodating pluralism. But it will typically require some compromise on other aspects of performance, such as financial costs, operational efficacy, environmental impacts or wider economic factors.

In addition, many energy options will be constrained in their possible contributions. Several renewable energy sources, for instance, are in this position. Rather than being static in nature, such constraints will take the form of a dynamic resource curve, under which successive increments are available at varying levels of performance. The shape of this curve will reflect the performance attributed to successive incremental 'tranches' for these options. This will be a function of two contending factors. On one hand, there are the combined negative effects of using successively less favorable sites (OXERA, 2004; DTI, 2005). On the other hand, there are learning effects and other increasing returns processes, which will yield countervailing positive improvements as experience accumulates with increasing use (Jacobsson and Johnson, 2000). To take account of these factors, then, the value assigned under any given perspective to any particular energy system under specific conditions ($V\{S\}$) can be expressed as the sum of the value due to the aggregate performance of individual options ($V\{E\}$) and an incremental value attached to irreducible portfolio-level properties including diversity ($V\{P\}$). If the net implications of diversity are adverse, then $V\{P\}$ can be negative.

$$V\{S\} = V\{E\} + V\{P\} \tag{1.3}$$

It has already been mentioned in discussing disparity distances that there exist numerous well-tried methods – and extensive bodies of data – addressing the multicriteria performance of energy options. Long experience in the field of decision analysis (Vincke et al., 1992) shows that – just as divergent notions of difference can be represented as coordinates in an n-dimensional Euclidean disparity space (Figure 1.2) – so can divergent valuations of individual system elements be represented as coordinates in an m-dimensional Euclidean performance space (Stirling, 2006). The dimensions of this space represent any set of m relevant performance criteria, each weighted to reflect their respective importance (Stirling, 1998, p. 81). As with disparity, the selection, characterization and scaling of these criteria will vary across context and perspective (Stirling, 1997). Although it is difficult to justify any single approach to aggregating *across* perspectives, decision analysis has shown that any single perspective can be uniquely captured by means of the following expression for the overall value attached to the performance of individual system elements $V\{E\}$:

$$V\{E\} = \Sigma_i \Sigma_c (w_c \cdot s_{ic}) \cdot p_i \tag{1.4}$$

where s_{ic} is the value attached to the performance of option i under criterion c; w_c is a scalar weighting reflecting the effective relative importance of criterion c

(under the perspective and context in question); and p_i is (as in Equations 1.1 and 1.2) the proportional representation of option i in the energy system in question. It follows from Equation (1.2) that the corresponding value attached to irreducible portfolio-level properties including diversity ($V\{P\}$) can then be expressed follows:

$$V\{P\} = \delta \cdot \Delta'$$
$$= \delta \cdot \Sigma_{ij(i \neq j)}(d_{ij})^\alpha \cdot (p_i \cdot p_j)^\beta \cdot \iota_{ij} \tag{1.5}$$

where Δ' represents an augmented form of the diversity heuristic Δ given in Equation (1.2), which includes an additional term to reflect portfolio interactions (ι_{ij}). ι_{ij} is an array of scalar multipliers exploiting the pairwise structure of Δ' to express the effect on system value of synergies or tensions between options i and j, respectively, as marginal positive or negative departures from a default of unity ($\iota_{ij} = 1 \pm \partial\iota$: for most systems $\partial\iota \ll 1$). This serves as a means to capture a variety of system-level properties that, like diversity, are irreducible to individual options. The coefficient δ scales expressions of portfolio value to render them commensurable with aggregate values of individual options in Equation (1.4). For positive assessments of portfolio value, $0 < \delta < \infty$. From Equations (1.3)–(1.5), we therefore obtain the following heuristic system-level articulation ($V\{S\}$) of the value attached to diversity together with that assigned to other portfolio properties ($V\{P\}$) and the value attached to the performance of individual energy options ($V\{E\}$).

$$V\{S\} = \Sigma_i \Sigma_c (w_c \cdot s_{ic}) \cdot p_i + \delta \cdot \Sigma_{ij(i \neq j)}(d_{ij})^\alpha \cdot (p_i \cdot p_j)^\beta \cdot \iota_{ij} \tag{1.6}$$

It is in $V\{S\}$ that we have a means to address the final criterion (10) developed above for a heuristic of energy diversity, in that it should allow systematic unconstrained articulation of system diversity with alternative characterizations of other salient properties of the energy system as a whole (interactions) or its component options (individual performance). Under such an approach, the 'systems' and 'options' in question may equally be defined to address contexts such as primary energy mixes, electricity supply portfolios, energy service provision and transport fuel systems. The approach can as readily focus on specific economic performance or broader criteria of energy sustainability.

The interest of the heuristic $V\{S\}$ lies not in any attempt to derive some unconditional 'optimal' balance between the performance of individual options, system interactions and portfolio diversity. Instead, $V\{S\}$ can be used systematically to explore different possible perspectives and assumptions concerning the contributions of these components to the overall value of an energy system. For each perspective on the available options, their individual performance, dynamic resource curves, joint interactions, mutual disparities, aggregations of diversity properties and the performance–diversity tradeoff, there will exists a particular apportionment of options that yields some maximum overall value. By varying δ between zero and infinity, $V\{S\}$ yields a set of all possible conditionally optimal energy systems ranging (respectively) from those that maximize value due to aggregate performance of individual options, to those that maximize positive value due to portfolio interactions and system diversity.

Reflecting work currently in progress, Figure 1.6 provides a schematic illustration of the kind of picture that arises from this analysis. For purposes of exposition, it focuses on the mix of generating technologies at the level of the UK electricity supply system taken as a whole. It is constructed on the basis of economic and resource data developed for the UK Government's recent Energy Reviews (PIU, 2001; DTI, 2005). Broader sustainability criteria are also included under one particular perspective on the weighting of different aspects of performance (Stirling, 2007a). Such an approach might as easily be addressed with respect to regional or international systems, to primary energy mixes in a broad sense, or to include information on demand-side options for the provision of energy services. Either way, disparities will be conceived in a fashion similar to that represented in Figure 1.2: concerning a wide range of attributes of the

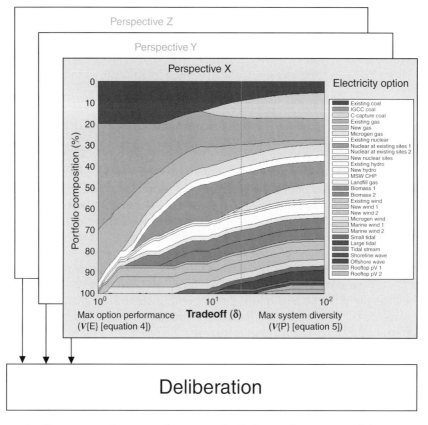

FIGURE 1.6   Illustrative performance–diversity tradeoffs for UK electricity portfolios. IGCC: integrated gasification combined cycle; MSW: municipal solid waste; CHP: combined heat and power; PV: photovoltaic.

resources and technologies involved, together with their geographical, commercial, institutional and sociopolitical contexts (Stirling, 1996). Positive and negative economic, organizational and operational synergies between different technologies inform the modelling of interactions (using term ι_{ij} in Equations 1.5 and 1.6). Certain options are tightly constrained in terms of the available resource, or display reductions (from learning or scale) or increases (from depletion) in costs or impacts as the contributions rise. Illustrative data on all these aspects are provided elsewhere (Annex to Stirling, 2007a). For now, the point is simply that Figure 1.6 shows, for a particular hypothetical perspective, how the resulting conditionally optimal electricity portfolios vary as greater or lesser priority is placed on diversity.

Low values of δ in Figure 1.6 may express high confidence in performance appraisals of individual technologies, with little concern over deep uncertainties and ignorance as to which diversity is a reasonable response (Awerbuch et al., 2006). Likewise, low values of δ may imply that priority is attached to maximizing this performance, rather than the other benefits of diversity (in fostering innovation, mitigating lock-in or accommodating pluralism). High values of δ reflect a dominant interest in these benefits of diversity, with little concern over the resulting compromises on performance. Again, the value of this kind of heuristic framework is as a means more explicitly and systematically to inform analysis under individual perspectives, and to provide a basis for more effective and transparent deliberation between contending positions.

1.6 Conclusion

This chapter has outlined a novel general framework for the analysis of energy diversity in relation to broader sustainability goals. The discussion began by noting many different reasons for an interest in diversity, including well-rehearsed issues in supply security debates. For most current systems based on fossil fuels, diversification into distributed and renewable energy and demand-side energy services offers important ways to address environmental imperatives like the mitigation of climate change. Beyond this, as recognized in Shimon Awerbuch's own work in this area, the benefits of diversity also assist in addressing further fundamental challenges of sustainability. Here, as in other areas of mainstream industrial strategy, appropriate tradeoffs between diversity and wider portfolio performance offer means to help promote innovation, hedge ignorance, mitigate lock-in and accommodate pluralism. Above all, they offer a uniquely important opportunity for achieving qualities of precaution, resilience and robustness that are valued in common across sustainability, supply security and other high-profile areas of energy and industrial policy.

Based on the recognition of three necessary but individually insufficient properties of diversity (variety, balance and disparity), the present general framework is applicable in a wide range of different energy policy contexts. It can be focused equally at the level of primary energy mixes, electricity supply portfolios or energy service systems. Unlike other approaches, it is not constrained to address only certain specific performance criteria (like the fuel prices highlighted in portfolio theory). Whilst it may be applied in purely economic terms if this is thought appropriate, the method can also be readily extended to encompass any range of issues addressed in well-established multicriteria appraisal techniques. Nor is this approach dependent on assumptions that past experience necessarily provides a reliable guide to future performance. The pictures of disparity that underlie this analysis can be informed by extensive bodies of existing multicriteria performance data, building only on the general underlying structures in this performance, rather than the specific values. Finally, it can be seen that the framework allows attention to be given to more detailed properties such as dynamic resource curves and system-level interactions between options.

Perhaps most importantly, the present framework for multicriteria energy diversity analysis can be applied flexibly to an unconstrained array of different specialist, institutional or stakeholder perspectives. This amenability to more open and plural processes of engagement is also central to sustainability agendas (Stirling, 2005). As a framework for quantification, the approach is compliant with a series of rigorous formal quality criteria (Table 1.2). However, as a heuristic approach, it does not purport to provide a basis for deriving a single objectively 'optimal' energy portfolio, irrespective of perspective. In any case, such aims (and claims) might be regarded as misleading and spurious in fields as complex, uncertain, dynamic and contested as energy and sustainability policy.

Instead, the present framework for multicriteria diversity analysis offers a means to be more systematic and transparent in articulating a range of different salient perspectives. It does this by avoiding the imposition of hidden parameters that privilege one perspective over another. In the end, the value of such a framework lies not in prescribing decisions, but simply in informing more robust, rigorous and accountable policy deliberation. As such, it builds on some of the central motivations underlying the work of Shimon Awerbuch, which the present author (along with many others) is proud to continue pursuing.

References

Adelman, M. (1995). *The Genie Out of the Bottle: World Oil Since 1970*. Cambridge, MA: MIT Press.

Aoki, M. (1996). An evolutionary parable of the gains from international organisational diversity. In Landau, R., Taylor, T. and Wright, G. (Eds), *The Mosaic of Economic Growth*. Stanford, CT: Stanford University Press, p. 263.

Arthur, W. B. (1989). Competing technologies, increasing returns, and lock-in by historical events. *Economic Journal*, 99.

Awerbuch, S. (1976). *Policy Evaluation for Community Development: Decision Tools for Local Government (with W. A. Wallace)*. New York: Praeger.

Awerbuch, S. (2000a). Investing in photovoltaics: risk, accounting and the value of new technology. *Energy Policy*, 28, 1023–1035.

Awerbuch, S. (2000b). Investing in distributed generation. In Bietry, E., Donaldson, J., Gururaja, J., Hurt, J. and Mubayi, V. (Eds), *Decentralized Energy Alternatives: Proceedings of the Decentralized Energy Alternative Symposium*. Columbia School of Business, Sustainable Development Initiative.

Awerbuch, S. and Yang, S. (2007). Efficient electricity generating portfolios for Europe: maximising energy security and climate change mitigation. *EIB Papers*, 12(2), 8–37.

Awerbuch, S., Stirling, S. A., Jansen, J. and Beurskens, L. (2006). Full-spectrum portfolio and diversity analysis of energy technologies. In Leggio, K. B., Bodde, D. L. and Taylor, M. L. (Eds), *Managing*

Enterprise Risk: What the Electric Industry Experience Implies for Contemporary Business. Amsterdam: Elsevier, pp. 202–222.

Bird, J. (2007). *Energy Security in the UK.* London: Institute for Public Policy Research.

Brooks, H. (1986). The typology of surprises in technology, institutions and development. In Clark, W. and Munn, R. (Eds), *Sustainable Development of the Biosphere.* Cambridge: Cambridge University Press.

Brundtland, G. (Ed.) (1987). *Our Common Future: Report of the United Nations Commission on Environment and Development.* Oxford: Oxford University Press.

Bruno, S., Cohendet, P., Desmartin, F., Llerena, D., Llerena, P. and Sorge, A. (1991). *Modes of Usage and Diffusion of New Technologies and New Knowledge: A Synthesis Report.* Project Report. FOP 227, Prospective Dossier 1. Brussels: European Commission.

Cohendet, P., Llerena, P. and Sorge, A. (1992). Technological Diversity and Coherence in Europe: an analytical overview. *Revue d'Economie Industrielle,* 59, 1re Trimestre 92. Paris: CNRS.

Commission of the European Communities (1990). *Security of Supply.* SEC(90)548 Final. Brussels: European Commission.

Commission of the European Communities (2007). *An Energy Policy for Europe. Communication from the Commission to the European Council and the European Parliament.* SEC(2007)12. Brussels: European Commission.

Costello, K. (2004). *Increased Dependence on Natural Gas for Electric Generation: Meeting the Challenge.* National Regulatory Research Institute, Ohio State University.

Cowan, R. (1991). Tortoises and hares: choice among technologies of unknown merit. *Economic Journal,* 101, 801–814.

David, P. and Rothwell, G. (1996). Standardisation, diversity and learning: strategies for the coevolution of technology and industrial capacity. *International Journal of Industrial Organization,* 14, 181–201.

Department of Trade and Industry (1995). *The Prospects for Nuclear Power in the UK: Conclusions of the Government's Nuclear Review.* Cmnd 2860. London: HMSO.

Department of Trade and Industry (2003). *Digest of UK Energy Statistics.* London: HMSO.

Department of Trade and Industry (2005). *The Role of Fossil Fuel Carbon Abatement Technologies (CATs) in a Low Carbon Energy System – A Report on the Analysis Undertaken to Advise the DTI's CAT Strategy.* London: HMSO (December).

Department of Trade and Industry (2006). Our Energy Future – Creating a Low Carbon Economy. Energy White Paper. London: HMSO.

ECN (2007). Scheepers, M., Seebregts, A., de Long, J., Maters, H. *EU Standards for Energy Security of Supply: Updates on the Crisis Capability Index and the Supply/Demand Index, Quantification for EU-27.* Netherlands Energy Research Centre, Petten, 2007.

Eldredge, N. (1992). *Systematics, Ecology and the Biodiversity Crisis.* New York: Columbia University Press.

Environmental Resources Management (1995). Diversity in UK Electricity Generation: A Portfolio Analysis. Report commissioned by Scottish Nuclear Ltd, Environmental Resources Management (December).

European Energy Charter Secretariat (2004). *The European Energy Charter Treaty and Related Documents: A Legal Framework for International Co-operation.* Brussels: European Commission. Available (12/7) at http://www.encharter.org/fileadmin/user_upload/document/EN.pdf#page=211

Farrell, A., Zerriffi, H. and Dowlatabadi, H. (2004). Energy infrastructure and security. *Annual Review of Environment and Resources,* 29, 421–469.

Geroski, P. (1989). The choice between diversity and scale. In Davis, E. (Ed.), *1992: Myths and Realities.* London: Centre for Business Strategy, London Business School, pp. 29–45.

Gini, C. (1912). Variabilita e mutabilita. *Studi Economica-Giuridici della R. Universita di Cagliari,* 3, 3–159.

Grabher, G. and Stark, D. (1997). Organizing diversity: evolutionary theory, network analysis and postsocialism. *Regional Studies,* 31(5), 533–544.

Greene, D., Jones, D. and Leiby, P. (1998). The outlook for US oil dependence. *Energy Policy,* 26, 55–69.

Gross, R., Heptonstall, P., Anderson, D., Green, T., Leach, M. and Skea, J. (2006). *The Costs and Impacts of Intermittency: An Assessment of the Evidence on the Costs and Impacts of Intermittent Generation on the British Electricity Network.* Report for the UK Energy Research Centre, London (March).

Grubb, M., Butler, L. and Twomey, P. (2006). Diversity and security in UK electricity generation: the influence of low-carbon objectives. *Energy Policy,* 34(18), 4050–4062.

Helm, D. (2007). European energy policy: meeting the security of supply and climate change challenges. *EIB Papers*, 12(1), 30–49.

Hill, M. (1973). Diversity and evenness: a unifying notation and its consequences. *Ecology*, 54, 2.

Hubberke, D. (2007). *Indicators of Energy Security in Industrialised Countries*. University of Chemnitz (December).

IEA (2002). *World Energy Outlook 2003*. Paris: OECD.

International Energy Agency (1980). *A Group Strategy for Energy Research, Development and Demonstration*. Paris: IEA.

International Energy Agency (1985). *Energy Technology Policy*. Paris: IEA.

International Energy Agency (1991). *Energy Policies of IEA Countries: 1990 Review*. Paris: IEA.

International Energy Agency (2002). *Electricity Information 2002*. Paris: IEA.

International Energy Agency (2007). *Energy Security and Climate Policy – Assessing Interactions*. Paris: IEA.

Jacobsson, S. and Johnson, A. (2000). The diffusion of renewable energy technology: an analytical framework and key issues for research. *Energy Policy*, 28(9), 625–640.

Joint Energy Security of Supply Working Group (2004). *Fourth Report*. London: DTI (May).

Junge, K. (1994). Diversity of ideas about diversity measurement. *Scandinavian Journal of Psychology*, 35, 16–26.

Kaijser, A., Mogren, A. and Steen, P. (1991). *Changing Direction: Energy Policy and New Technology*. Stockholm: Statens Energiverk.

Kempton, R. (1979). The structure of species abundance and measurement of diversity. *Biometrics*, 35, 307–321.

Kruskal, J. (1964). Nonmetric multidimensional scaling of a numerical method. *Psychometrika*, 29, 115–129.

Landau, R., Taylor, T. and Wright, G. (1996). *The Mosaic of Economic Growth*. Stanford, CT: Stanford University Press.

Lawson, N. (1992). *The View from Number 11: Memoirs of a Tory Radical*. London: Bantam.

Laxton, R. (1978). The measure of diversity. *Journal of Theoretical Biology*, 70, 51–67.

Lovins, A. and Lovins, L. (1982). *Brittle Power: Energy Strategy for National Security*. Andover, MA: Brick House.

MacArthur, R. (1965). Patterns of species diversity. *Biological Review*, 40, 510–533.

MacKerron, G. and Scrase, I. (Eds) (2008). *Climate of Urgency: Empowering Energy Policy*. London: Palgrave.

Matthews, M. and McGowan, F. (1992). Reconciling diversity and scale: some questions of method in the assessment of the costs and benefits of European integration. *Revue d Economie Industrielle*, 59, 1–31.

May, R. (1990). Taxonomy as destiny. *Nature*, 347, 129–130.

Mercier, J. and McGowan, R. (1996). The greening of organisations. *Administration and Society*, 27(4), 459–482.

Mitchell, C. and Woodman, B. (2006). *New Nuclear Power: Implications for a Sustainable Energy System*. Coventry: Green Alliance, University of Warwick.

National Economic Research Associates (1995). *Diversity and Security of Supply in the UK Electricity Market*. Report to the Department of Trade and Industry, London.

OXERA (2004). *Results of Renewables Market Modelling*. London: Study conducted for UK Department of Trade and Industry, London.

Performance and Innovation Unit (2001). *Working Paper on Generating Technologies: Potentials and Cost Reductions to 2020*. London: UK Cabinet Office.

Pielou, E. (1969). *An Introduction to Mathematical Ecology*. New York: Wiley.

Pielou, E. (1977). *Mathematical Ecology*. New York: Wiley.

Plummer, J. (Ed.) (1983). *Energy Vulnerability*. Cambridge: Ballinger.

Prindle, B., Eldridge, M., Eckhardt, M. and Frederick, A. (2007). *The Twin Pillars of Sustainable Energy: Synergies between Energy Efficiency and Renewable Energy Technology and Policy*. ACEEE Report No. E074. Washington DC: American Council for an Energy Efficient Economy.

Rao, C. R. (1982). Diversity and dissimilarity coefficients: a unified approach. *Theoretical Population Biology*, 21, 24–43.

Rosenberg, N. (1982). *Inside the Black Box: technology and economics*. Cambridge University Press, Cambridge.

Rosenberg, N. (1996). Uncertainty and technological change. In Landau, R., Taylor, T. and Wright, G. (Eds), *The Mosaic of Economic Growth*. Stanford, CT: Stanford University Press.

Runnegar, B. (1987). Rates and modes of evolution in the Mollusca. In Campbell, M. and May, R. (Eds), *Rates of Evolution*. London: Allen and Unwin.

Scheepers, M., Seebregts, A., de Jong, J. and Maters, H. (2007). *EU Standards for Energy Security of Supply: Updates on the Crisis Capability Index and the Supply/Demand Index, Quantification for EU-27*. Petten: Netherlands Energy Research Centre.

Shannon, C. and Weaver, W. (1962). *The Mathematical Theory of Communication*. Urbana, IL: University of Illinois Press.

Simpson, E. (1949). Measurement of diversity. *Nature*, 163, 41–48.

Sneath, P. and Sokal, R. (1973). *Numerical Taxonomy: The Principles and Practice of Numerical Classification*. San Francisco, CA: Freeman.

Solow, A. and Polasky, S. (1994). Measuring biological diversity. *Environmental and Ecological Statistics*, 1, 95–107.

Stirling, A. (1994). *Diversity and ignorance in electricity supply investment:* addressing the solution rather than the problem. *Energy Policy*, 22(3), 195–216.

Stirling, A. (1996). Optimising UK electricity portfolio diversity. In MacKerron, G. and Pearson, P. (Eds), *The UK Energy Experience: A Model or a Warning?*. London: Imperial College Press (March).

Stirling, A. (1997). Multicriteria mapping: mitigating the problems of environmental valuation?. In Foster, J. (Ed.), *Valuing Nature: Economics, Ethics and Environment*. London: Routledge.

Stirling, A. (1998). *On the Economics and Analysis of Diversity*. SPRU Electronic Working Paper No. 28. University of Sussex. (May 2006) Available at http://www.sussex.ac.uk/Units/spru/publications/imprint/sewps/sewp28/sewp28.pdf

Stirling, A. (1999). *On Science and Precaution in the Management of Technological Risk: Volume I – A Synthesis Report of Case Studies*. EUR 19056 EN. Seville: European Commission Institute for Prospective Technological Studies (May). Available at ftp://ftp.jrc.es/pub/EURdoc/eur19056en.pdf

Stirling, A. (2003). Risk, uncertainty and precaution: some instrumental implications from the social sciences. In Scoones, I., Leach, M. and Berkhout, F. (Eds), *Negotiating Change: Perspectives in Environmental Social Science*. London: Edward Elgar, pp. 33–76.

Stirling, A. (2005). Opening up or closing down: analysis, participation and power in the social appraisal of technology. In Leach, M., Scoones, I. and Wynne, B. (Eds), *Science and Citizens: Globalization and the Challenge of Engagement*. London: Zed, pp. 218–231.

Stirling, A. (2006). Analysis, participation and power: justification and closure in participatory multi-criteria analysis. *Land Use Policy*, 23(1), 95–107.

Stirling, A. (2007a). A general framework for analysing diversity in science, technology and society. *Journal of the Royal Society Interface*, 4(15), 707–719. Published electronically by FirstCite, February 2007. Available (4/7) at http://www.journals.royalsoc.ac.uk/content/a773814672145764/fulltext.pdf

Stirling, A. (2007b). Deliberate futures: precaution and progress in social choice of sustainable technology. *Sustainable Development*, 15, 286–295.

Stirling, A. (2008a). 'Framing', 'lock-in' and diversity in social choice of energy futures. In Dorfman, P. (Ed.), *Nuclear Consultation: Public Trust in Government*. Coventry: Nuclear Consultation Working Group, University of Warwick (January).

Stirling, A. (2008b). Foreword. In MacKerron, G. and Scrase, I. (Eds), *Climate of Urgency: Empowering Energy Policy*. London: Palgrave.

Suzuki, T. (2001). *Energy Security and the Role of Nuclear Power in Japan*. Central Research Institute of the Electric Power Industry.

Ulph, A. (1988). *Quantification of Benefits of Diversity from Reducing Exposure to Volatility of Fossil Fuel Prices*. Evidence to Hinkley Point C Planning Inquiry for Central Electricity Generating Board, 25 October.

Verrastro, F. and Ladislaw, S. (2007). Providing energy security in an interdependent world. *The Washington Quarterly*, 30(4), 95–104.

Vincke, M., Gassner, M. and Roy, B. (1992). *Multicriteria Decision-Aid*. Chichester: John Wiley & Sons.

Weitzman, M. (1992). On diversity. *Quarterly Journal of Economics*, 107, 363–405.

Williamson, O. (1993). Transaction cost economics and organisation theory. *Industrial Economics and Corporate Change*, 2, 107–156.

CHAPTER 2

The Value of Renewable Energy as a Hedge Against Fuel Price Risk

Analytical Contributions from Economic and Finance Theory

Mark Bolinger and **Ryan Wiser**[*]

2.1 Introduction

For better or worse, natural gas has become the fuel of choice for new power plants being built across the USA. According to the Energy Information Administration (EIA), natural-gas-fired units accounted for nearly 90% of the total generating capacity added in the USA between 1999 and 2005 (EIA, 2006b), bringing the nationwide market share of gas-fired generation to 19%. Looking ahead over the next decade, the EIA expects this trend to continue, increasing the market share of gas-fired generation to 22% by 2015 (EIA, 2007a). Although these numbers are specific to the USA, natural-gas-fired generation is making similar advances in many other countries as well.

A large percentage of the total cost of gas-fired generation is attributable to fuel costs, i.e. natural gas prices. For example, at current spot prices of more than $7/MMBtu, fuel costs account for more than 75% of the levelized cost of energy from a new combined cycle gas turbine, and more than 90% of its operating costs (EIA, 2007a). Furthermore, given that gas-fired plants are often the marginal supply units that set the market-clearing price for *all* generators in a competitive wholesale market, there is a direct link between natural gas prices and wholesale electricity prices.

In this light, the dramatic increase in natural gas prices since the 1990s should be a cause for ratepayer concern. Figure 2.1 shows the daily price history of the

[*]Lawrence Berkeley National Laboratory, Berkeley, CA, USA

Analytical Methods for Energy Diversity and Security © 2008 Elsevier Ltd.

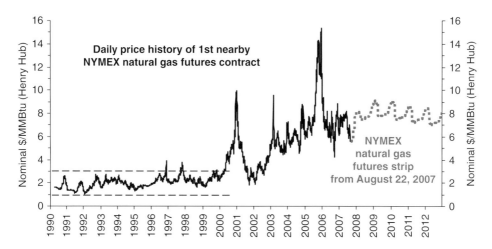

FIGURE 2.1   Historical NYMEX natural gas futures prices (first-nearby contract) and current NYMEX natural gas strip. Source: LBNL.

'first-nearby' (i.e. closest to expiration) NYMEX natural gas futures contract (solid line) at Henry Hub, along with the futures strip (i.e. the full series of futures contracts) from 22 August 2007 (dotted line). At that time first-nearby prices, which closely track spot prices, were trading within a $7–9/MMBtu range in the USA and, as shown by the futures strip, and were expected to remain there through 2012. These price levels are $6/MMBtu higher than the $1–3/MMBtu range seen throughout most of the 1990s, demonstrating significant price escalation for natural gas in the USA over a relatively brief period.

Perhaps of most concern is that this dramatic price increase was largely unforeseen. Figure 2.2 compares the EIA's natural gas wellhead price forecast from each year's *Annual Energy Outlook* (AEO) going back to 1985 against the average US wellhead price that actually transpired. As shown, forecasting abilities have proven rather dismal over time, as overforecasts made in the late 1980s eventually yielded to underforecasts that have persisted to this day. This historical experience demonstrates that little weight should be placed on any one forecast of future natural gas prices, and that a broad range of future price conditions ought to be considered in planning and investment decisions.

Against this backdrop of high, volatile and unpredictable natural gas prices, increasing the market penetration of renewable generation such as wind, solar and geothermal power may provide economic benefits to ratepayers by displacing gas-fired generation. These benefits may manifest themselves in several ways. First, the displacement of natural gas-fired generation by increased renewable generation reduces ratepayer exposure to natural gas price risk, i.e. the risk that future gas prices (and by extension future electricity prices) may end up markedly different than expected. Second, this displacement reduces demand for natural gas among gas-fired generators, which, all else being equal, will put

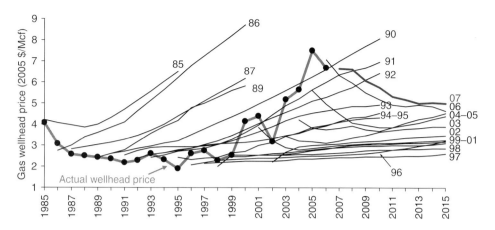

FIGURE 2.2 Historical Annual Energy Outlook (AEO) wellhead gas price forecasts versus actual wellhead price. Source: EIA.

downward pressure on natural gas prices. Lower natural gas prices in turn benefit both electric ratepayers and other end-users of natural gas.

Using analytical approaches that build upon, yet differ from, the past work of others, including Awerbuch (1993, 1994, 2003), Kahn and Stoft (1993) and Humphreys and McClain (1998), this chapter explores each of these two potential 'hedging' benefits of renewable electricity.[1] Although we do not seek to judge whether these two specific benefits outweigh any incremental cost of renewable energy (relative to conventional fuels), we do seek to quantify the magnitude of these two individual benefits. Note that these benefits are not unique to renewable electricity: other generation (or demand-side) resources whose costs are not tied to natural gas would provide similar benefits.

Section 2.2 explores the first potential hedge benefit of renewable energy – the value of the price risk mitigation provided by renewables – by comparing natural gas prices that can be locked in through long-term hedging instruments (e.g. gas futures, swaps and fixed-price physical supply contracts) with those contained in contemporaneous government spot-price forecasts. When used to calculate the levelized cost of gas-fired generation, the former yield results that are most directly comparable to the fixed-price nature of renewable generation, but utility resource planners and policymakers nevertheless often use the latter when making resource decisions. This practice can result in apples-to-oranges comparisons – at least with respect to fuel price risk – if the two gas price streams differ in a significant and systematic fashion. Although the data are limited, it is found that

---

[1] A few other potential benefits are mentioned later, in Section 2.3.1.2. In addition to these, others include enhanced security of energy supply, and a potential reduction in wholesale power prices as low-operating-cost renewable generation displaces high-operating-cost gas-fired generation from the supply curve.

from November 2000 through 2006, forward gas prices for terms of five to ten years have been considerably higher than most natural gas spot price forecasts in the USA. This implies that resource planning and modelling exercises based on these forecasts over this period have yielded results that may be biased (with respect to fuel price risk) in favor of gas-fired generation, assuming that rate stability is valued by ratepayers.

Section 2.3 turns to the second potential hedge benefit of renewable energy – the value of putting downward pressure on natural gas prices – by reviewing economic theory and summarizing the results of past modelling studies to determine the impact that increased renewables penetration might be expected to have on natural gas demand and prices. An attempt is also made to benchmark those modelling results against empirical measurements. Although results vary and uncertainty in the value of this benefit remains, current knowledge suggests that a 1% long-term reduction in demand for natural gas (whether caused by renewables or something else) in the USA may, conservatively, lead to a 0.8–2% long-term reduction in US natural gas prices. This price reduction would be expected to flow through to *all* sectors of the economy (i.e. not just the electricity sector), resulting in significant consumer savings.

Although using different analytical techniques, much of the analysis presented in this chapter was inspired by the work of the late Shimon Awerbuch. For example, although the approach in Section 2.2 does not directly depend on the applicability of modern portfolio theory or the Capital Asset Pricing Model (CAPM) to the electricity sector, the empirical results presented in Section 2.2 are nevertheless consistent with what one would expect from CAPM, and therefore could be used to support Shimon's more theoretical work in this area (indeed, CAPM is discussed as one potential explanation for the observed premiums). Similarly, the analysis in Section 2.3 is related to Shimon's later work examining the interrelationship between natural gas and oil prices. In both cases, the present authors' work reaches the same basic conclusion that Shimon so tirelessly championed throughout his distinguished career: that renewables can reduce certain electric-sector risks, and therefore have an economically justifiable place in generating portfolios.

2.2 Renewable energy reduces exposure to natural gas price risk

Unlike many contracts for natural-gas-fired electricity generation, renewable generation is typically sold under long-term, fixed-price contracts that eliminate price risk for 15–20 years or longer.[2] Contract pricing for natural-gas-fired generation, in contrast, is often indexed to the spot price of natural gas. Assuming that electricity consumers prefer the less risky of two otherwise similar expected

---

[2] Although the analysis and results are presented in the context of comparing renewable to gas-fired generation, they are equally applicable to comparisons of other forms of generation (e.g. coal or nuclear power) or demand reduction (e.g. energy efficiency) to variable-price gas-fired generation.

cash flows (or, said another way, that they prefer the cheaper of two similarly risky cash flows), a retail electricity supplier that is looking to expand its resource portfolio (or a policy maker interested in evaluating different resource options) should arguably compare the cost of fixed-price renewable generation to the *hedged* or *guaranteed* cost of natural-gas-fired generation, rather than to *projected* costs based on *uncertain* spot gas price forecasts.[3] Nonetheless, utilities and others often compare the costs of renewable to gas-fired generation using as their fuel price input long-term gas price forecasts that are inherently uncertain, rather than long-term natural gas forward prices that can actually be locked in (Bolinger and Wiser, 2005; Bolinger et al., 2006).

This practice raises the critical question of how these two price streams, i.e. forwards and forecasts, compare. If the two price streams closely match one another, then one might conclude that forecast-based resource acquisition, planning and modelling exercises are implicitly accounting for the price stability benefits of renewable relative to gas-fired generation, approximating an apples-to-apples comparison. If, however, forward prices systematically differ from long-term spot price forecasts, then the use of such forecasts in resource acquisition, planning and modelling exercises will arguably yield results that are either biased in favor of renewable generation (if forwards < forecasts) or biased in favor of natural-gas-fired generation (if forwards > forecasts).

2.2.1 Methodology

The extent to which natural gas forward prices match fundamental price forecasts is investigated by comparing the prices of natural gas futures, swaps, and fixed-price physical supply contracts to 'reference-case' gas price forecasts from the EIA. Although long-term gas price forecasts are easy to come by, e.g. the EIA forecasts are publicly available and updated every fall, long-term forward prices present a greater challenge. The NYMEX natural gas futures strip for delivery to the Henry Hub in Louisiana extends out for between five and six years (but is liquid for less than that), a period that is only about one-third as long as the typical term of a power purchase agreement for renewable energy.[4] Forward gas contracts in excess of six years are traded infrequently and, when traded, are often traded bilaterally 'over the counter' (i.e. not on an organized exchange),

---

[3] Note that this is strictly true only if the consumer benefit of price stability is greater than the incremental cost of achieving that stability through a hedged gas contract. Ideally, one would directly estimate the consumer *value* of a fixed-price (versus variable-price) contract. In the absence of such an estimate, here the focus is on the *cost* of achieving such a fixed-price gas contract. Moreover, although this chapter focuses exclusively on fuel price risk, the cost of fuel (and its impact on total generation costs) is only one of many important considerations involved in any resource comparison. For example, the relative dispatchability of generating resources, regardless of levelized costs, may be of prime importance in some instances.

[4] In February 2008, the NYMEX extended its natural gas futures strip out 12 years, though liquidity is still concentrated in just the first few years of the strip.

and are therefore rarely documented in the public domain. In addition, the sample must further be restricted to those forward prices that were traded or posted at roughly the same time as the generation of a long-term gas price forecast, to ensure an adequate comparison.

Thus, despite efforts to obtain a larger sample, this analysis is limited to comparisons based on publicly available data collected from November 2000 to 2006, and for forward price terms not exceeding ten years. Specifically, the limited sample of natural gas forward contracts includes:

- five- and ten-year natural gas swaps offered by Enron in early November 2000 and 2001;
- a seven-year physical gas supply contract between Williams and the California Department of Water Resources (DWR) signed in early November 2002;
- the 72-month NYMEX natural gas futures strip (averaged each calendar year) from November 2002, October 2003 and October 2004;
- the 60-month NYMEX natural gas futures strip from November 2005 and 2006.

In each case, these forward prices are evaluated against the EIA's then-current reference-case forecast of spot natural gas prices delivered to electricity generators (in nominal dollars[5]), which is generated in the fall of each year and presented in the AEO released toward the end of the year. The locational basis of the EIA's forecast is adjusted to match the stated delivery point of the forward gas price being compared.[6] Although the use of EIA reference-case gas price forecasts for this purpose is somewhat controversial (as discussed later), they are used here because they are publicly available, have been widely vetted and, most importantly, are commonly adopted by the EIA and others as a 'base-case' price scenario in policy evaluations, modelling exercises and even resource acquisition decisions.

2.2.2 Empirical findings of a premium

Each of these comparisons reveals that forward natural gas prices have traded above EIA reference-case price forecasts during this seven-year period, sometimes significantly so. Figure 2.3 consolidates the resulting levelized premiums (in terms of $/MMBtu and ¢/kWh, assuming a heat rate of 7000 Btu/kWh) from each of these comparisons into a single graph.[7] As shown, the magnitude of the

---

[5] Unless otherwise noted, all forecasts are expressed in *nominal* (as opposed to *real* or constant dollar) terms in order to be comparable to forward prices, which are also expressed in nominal terms. For example, the EIA gas price forecasts – which are expressed in real terms – are inflated to nominal terms using the EIA's own inflation projections.
[6] For details on the specific basis adjustments, see Bolinger et al. (2003).
[7] The premiums were derived by levelizing the first five, six, seven and ten years of the EIA forecasts (using a nominal discount rate of 10%) and subtracting the resulting levelized forecast price from the corresponding forward prices. Because in this case levelizing involves taking the present value of a price stream and amortizing it forward *at the same discount rate*, the calculation is relatively insensitive to the level of the discount rate chosen. For example, using a discount rate of 5% barely changed the results.

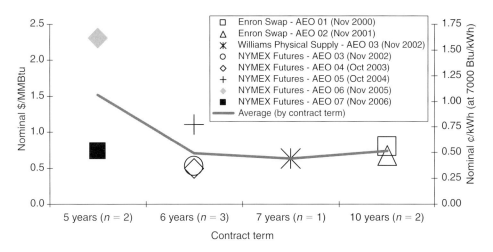

FIGURE 2.3   Levelized premiums (forwards – forecasts).

empirically derived premiums varies somewhat from year to year, contract to contract, and by contract term, ranging from $0.5 to $2.4/MMBtu, or 0.4 to 1.7 ¢/kWh assuming a highly efficient gas-fired power plant.

Although the relatively tight range of premiums (November 2005 excepted) is somewhat remarkable, the limitations in the data sample mean that one cannot easily extrapolate these findings beyond the seven periods for which data are available, or to contract terms longer than those examined (ten years). It is, however, at least apparent that utilities and others who have conducted resource planning and modelling studies based on EIA reference-case gas price forecasts from November 2000 to November 2006 have produced results that, by seemingly ignoring fuel price risk, favor variable-price gas-fired over fixed-price renewable generation (at least over the five- to ten-year terms of the comparisons presented in Figure 2.3).

The EIA's forecasts are by no means the only long-term gas price forecasts available to and used by market participants. In order to assess how the premiums presented in Figure 2.3 would change had natural gas forward prices been compared with some forecast other than the EIA's, a number of other long-term gas price forecasts were reviewed, sourced from the EIA's own forecast comparisons (contained in each year's AEO), as well as from various utility integrated resource plans. As shown in Bolinger et al. (2003), with few exceptions, the EIA reference-case forecast has generally been higher – and often substantially so – than most other forecasts generated from 2000 to 2003 and used by utilities and others. A more recent examination of AEO forecast comparisons reveals that this trend has generally continued since 2003 (EIA, 2004, 2005, 2006a), with the notable exception of the AEO 2007 reference-case forecast, which is in some cases significantly below the other long-term forecasts highlighted in the forecast comparisons (EIA, 2007a). But, in general, these findings suggest that some of the premiums presented in Figure 2.3 would be *even larger* when comparing forward prices with some of the other commonly used gas price forecasts.

2.2.3 Potential explanations for empirical premiums

The differences between natural gas forward prices and spot gas price forecasts revealed in the previous section, and the implications such differences hold for resource comparisons, are significant. The authors are keenly aware, however, that these empirical results are based on limited historical data over a seven-year period characterized by extreme price volatility. In considering whether it is possible to extrapolate these findings, or the implications thereof, into the future, it is important to try to understand some of the causes of the empirical findings.

Two possible explanations may partially or wholly account for the sizeable differences between these limited samples of observed natural gas forwards and forecasts: (1) the forecasts are out of tune with market expectations, so the observed empirical premiums reflect forecast bias; and (2) hedging is not costless, and the observed premiums represent the cost of hedging.[8] Each of these possibilities is assessed below.[9]

2.2.3.1 The forecasts may be out of tune with market expectations

One possible explanation for the empirically derived premiums between gas forwards and forecasts is related to the forecasts themselves, which may have been biased downwards or otherwise inconsistent with broader market expectations of future spot gas prices from November 2000 to November 2006.

Although this explanation cannot be directly supported (or refuted) by the empirical observations, it has gained some traction in the literature. For example, Considine and Clemente (2007) use error decomposition analysis on AEO natural gas price forecasts to identify the proportion of the forecast error that is attributable to randomness, bias or the forecasting model (i.e. the EIA's National Energy Modeling System, or NEMS). Over the period from 1998 to 2006, they find that between 56% and 87% of the forecast errors were attributable to systematic bias, on average (with the range reflecting different forecast periods, from one to four years ahead). In other words, forecast bias could explain a significant portion of the empirical premiums presented earlier in Figure 2.3.

Two considerations potentially call into question the widespread applicability of Considine and Clemente's (2007) results, however. First, given the relatively short term (i.e. one to four years) of the forecasts analyzed, the EIA's *Short-Term Energy Outlook* forecasts (which are *not* generated using NEMS) would arguably

---

[8] Another *possible* explanation for the findings is simply that this analysis is plagued by data issues that prohibit a meaningful comparison between natural gas forward prices and price forecasts. It is possible, for example, that the forward prices are upwardly biased. Or, even if the forward price and price forecast are unbiased, it is possible that potential changes in the market between when the forward prices were sampled and the forecasts were generated could account for some or all of the observed premiums. Both of these possibilities are examined at length in Bolinger et al. (2003), however, and are found to be inconsequential.

[9] More detailed information on these potential explanations is provided in Bolinger et al. (2003).

have been more suitable for analysis than the AEO forecasts, which are longer term in nature. Second, a visual inspection of Figure 2.2 from earlier shows that the sample period chosen (1998–2006) is perhaps itself biased, in that it falls exclusively within the recent period of apparent underforecasts that began in the mid-1990s (as opposed to the earlier period of overforecasts that occurred in the 1980s). Thus, while systematic bias may indeed be a problem over the period 1998–2006 – which, it should be noted, also includes the sample period used in this analysis – these results are somewhat selective.

Another potential problem with the exclusive use of the EIA's AEO forecasts as the point of comparison is that, by definition, the EIA's 'reference-case' forecast may not represent the market's view of future spot gas prices. For example, the EIA itself notes that its 'reference-case' forecast assumes that normal inventories and weather, as well as current laws and regulations, will hold throughout the forecast period, and therefore that the reference-case forecast does not necessarily reflect what the EIA believes to be the 'most likely' outcome.[10] In fact, the EIA does not assign probabilities to any of the forecasts it generates, so the 'high economic growth case' forecast might be considered just as likely as the 'reference-case' or even 'low economic growth case' forecast, for example. Furthermore, by assuming away weather and inventory variability, and possible changes in regulations – all of which have a major impact on prices – the EIA notes that it is not really forecasting *prices* at all, but rather long-term equilibrium *costs*.[11]

Although the EIA reference-case forecast may not be *designed* to represent EIA or market expectations of future gas prices, it deserves note that industry participants and energy analysts regularly adopt the EIA reference-case projection as a 'best estimate' of future energy outcomes; in fact, the EIA itself regularly uses its reference-case forecast as the base-case forecast when evaluating the cost and impacts of

---

[10] Although the EIA apparently does not hold this view about its own forecasts, some have argued that 'reference-case' gas price forecasts can best be thought of as *modes* (i.e. the single scenario that the forecaster believes to be the most likely) rather than *means* (i.e. a probability weighted average of all possible spot prices), and since market expectations are by definition *mean* expectations, reference-case gas price forecasts cannot represent market expectations. Although, again, this argument apparently does not hold for the EIA reference-case forecast, its implications are nonetheless worth noting. Specifically, since gas prices are generally believed to be log-normally distributed (i.e. positively skewed), the mean must lie above the mode, meaning that true market expectations must be higher than reference-case gas price forecasts (if those forecasts do indeed represent the mode). If this argument is accurate, it might explain some or all of the premium observed here between forward prices and price forecasts (although, once again, note that the EIA does not consider its reference-case forecast to be a mode). More importantly, however, this argument calls into question why utilities and others would ever place significant emphasis on the use of reference-case gas price forecasts in modelling and planning exercises. By doing so, they would be *systematically underestimating* the market's expectations of future gas prices (which, by definition, must be a mean estimate), thereby erroneously making gas-fired generation appear to be cheaper than the market expects it to be on average.
[11] Note, however, that *Annual Energy Outlook* uses the term 'price' instead of 'cost' to describe its forecasts.

energy policies. Furthermore, Bolinger et al. (2003) and Bolinger and Wiser (2005) demonstrate that some utilities – one important segment of the energy market – are relying on EIA reference-case forecasts as a 'best estimate' of future gas prices for the purpose of long-term resource planning. Finally, if forecast bias is an issue, it is not restricted to the EIA reference-case forecast: as noted earlier, other private-sector gas price forecasts have often been *even lower* than the EIA's reference-case forecast.

2.2.3.2 Hedging may not be costless

A second potential explanation for the difference between forward and forecast gas prices is that hedging may not be costless, i.e. forward prices may be (for several reasons) biased predictors of future spot prices. This notion has been extensively debated in the literature ever since Keynes first introduced the idea of *normal backwardation* in his 1930 *Treatise on Money*. Specifically, Keynes (1930) argued that *hedgers* who use futures markets to mitigate commodity price risk must compensate *speculators* for the 'insurance' that they provide.[12] A futures market that is dominated by 'short hedgers', i.e. a market with more natural sellers than buyers of future contracts, results in what is known as *positive net hedging pressure*, and requires speculators to step in and purchase the excess futures contracts that are for sale in order to clear the market. To entice speculators to provide this 'insurance' or liquidity, short hedgers must, according to Keynes, be willing to sell futures contracts to speculators at prices that are *lower* than the expected spot price, thereby enabling speculators to earn a positive return simply by buying the contracts and holding them to expiration (when spot and futures prices presumably converge). The opposite occurs in a market dominated by long hedgers, resulting in *negative net hedging pressure*: to clear the market, speculators must *sell* futures contracts, and are compensated in the form of futures prices that are *higher* than the expected spot price, again enabling them to earn a positive return simply by holding the short position to expiration.

Over the years, a number of studies have attempted, with mixed results, empirically to confirm or refute the existence of such 'risk premiums' in the futures prices of a variety of commodities by examining the returns to speculators. Much of this work has been based on a strict application of Keynes' theory, which, among other things, assumes that hedgers – characterized mainly as

---

[12] *Hedgers* can be thought of as commodity producers (e.g. natural gas drillers) or consumers (e.g. electric utilities) who have a natural underlying position in a commodity that can be hedged by selling or buying futures contracts. A 'short' hedge transaction is one in which the hedger currently owns the underlying commodity and will be selling it in the future, and so today sells futures contracts to lock in the future sales price of the commodity and thereby secure a profit margin. Conversely, a 'long' hedge transaction is one in which the hedger does not currently own the commodity but will be buying it in the future, and so today buys futures contracts to lock in the future purchase price of the commodity and thereby secure a profit margin. In contrast to hedgers, *speculators* have no natural underlying position in the commodity, and trade in the futures market purely to make a profit. If a short hedger cannot find a long hedger to trade with, a speculator will step in to buy the short hedger's futures contracts, in the hope of being able to re-sell them at a profit. In this way, speculators provide liquidity to the market.

producers who are natural short hedgers – will be net short futures and that spec-ulators will therefore be net long futures. As noted by Chang (1985), however, a number of researchers have relaxed this constraint in recognition that both pro-ducers and consumers hedge, and that net hedging pressure may therefore not always be positive (i.e. in aggregate, short hedges may not always outnumber long hedges). Although results remain inconclusive, the relaxation of the assump-tion that speculators are always net long has generally led to results, as reported in Chang (1985) and Hull (1999), that are more supportive of Keynes' notion of a risk premium (positive or negative, depending on whether net hedging pressure is negative or positive, respectively) embedded in futures prices.

Still other researchers have searched for risk premiums in commodity futures prices from within the framework of the CAPM. Under CAPM (described in more detail later), it is not the variability of prices per se, but rather the correlation of price variability with changes in total wealth, that determines the risk of the underlying asset. Dusak (1973) examined wheat, corn and soybean futures within the CAPM framework, and found no evidence of either non-zero systematic risk or non-zero futures returns.[13] Hirshleifer (1988), meanwhile, developed a model whereby the risk premium consists of two terms: a systematic risk term (i.e. related to CAPM) and a term due solely to residual risk (i.e. related to net hedging pressure).

Below, the possibility is examined that either net hedging pressure or CAPM-related systematic price risk is responsible for the empirical premiums observed in the natural gas market from November 2000 to November 2006, as presented earlier in Figure 2.3.[14]

---

[13] Note that these findings do not rule out CAPM as a potentially useful tool for this purpose, since, under CAPM, one would expect zero systematic risk to lead to no risk premium.

[14] Although not discussed here, those hedging gas price risk may also incur incremental costs relative to market expectations of future spot prices simply due to the presence of transaction costs. In commodity markets, transaction costs are manifested in the 'bid-offer spread': the spread between the price at which one is willing to buy (bid) and sell (offer) a product. To execute a deal with minimal price risk (i.e. the risk that the market price will either rise or fall while one is trying to buy or sell, respectively), one must typically 'cross' the bid-offer spread (i.e. pay the offer price if buying, or accept the bid price if selling). Since the 'true' market price lies somewhere in between the bid and the offer, crossing the bid-offer spread to execute a deal results in transaction costs being incurred (i.e. paying more, or receiving less, than the 'true' market price). In liquid markets, bid-offer spreads are typically very small, and of little concern. In less-liquid markets, however, bid-offer spreads can be quite wide, and can have a more significant impact on the cost of transac-tions. Bolinger et al. (2003) illustrate that while bid-offer spreads in the NYMEX natural gas futures market are quite small in the front months due to high liquidity, they widen dramatically as liquidity declines along the strip. In fact, beyond 36–48 months, there is not necessarily 'a market' per se, and the cost of executing a deal of any size may depend lar-gely on the ability to locate a willing buyer or seller, which may in turn require significant price concessions at times. In other words, even if futures prices perfectly match market price expectations, those seeking to hedge price risk for lengthy terms may still be required to pay a sizeable premium, in the form of transaction costs, in order to execute the hedge.

Net hedging pressure in the natural gas market   To test whether the positive premiums (natural gas forward prices that are higher than contemporaneous forecasts of spot prices) observed earlier could be the direct result of negative net hedging pressure, Figure 2.4 depicts net hedging pressure in the natural gas market from 2000 to 2006.[15] Net hedging pressure has been largely positive over this period, suggesting that futures prices should generally be *lower* than expected spot prices, although it has clearly varied quite a bit, and has been negative at times. In fact, during three of the seven periods in which forward prices were sampled (dates indicated by shaded vertical lines), net hedging pressure was either neutral or slightly negative, suggesting that futures prices should, at least in those instances, match or be higher than expected spot prices. Nonetheless, because Figure 2.4 does not show systematic negative net hedging pressure over the period of analysis, it does not support the idea that negative net hedging pressure is directly responsible for the positive premiums observed earlier. In fact, if anything, the generally positive net hedging pressure presented in the figure suggests that forward natural gas prices should be lower than expected spot prices, directly contradicting the results presented earlier and clearly

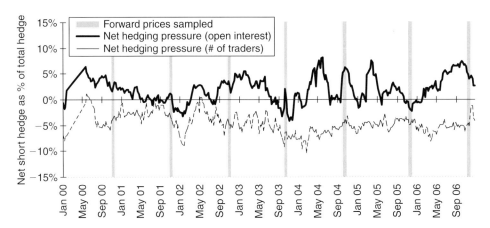

FIGURE 2.4   Net hedging pressure in the natural gas futures market, 2000–2006. Source: Commodity Futures Trading Commission (2007).

[15] The data come from the weekly *Commitments of Traders* reports published by the Commodity Futures Trading Commission (CFTC). Traders are classified as 'commercial' or 'non-commercial' depending on whether their futures positions in a given commodity are used for hedging or speculative purposes, respectively. Net hedging pressure is defined as the difference between the number of outstanding short and long contracts divided by the sum of outstanding short and long contracts (among commercial traders). Commercial traders (i.e. hedgers) currently account for about 50% of the open interest in the natural gas futures market, down sharply from roughly 80% earlier in this decade. The correspondingly sharp increase in the open interest of speculators over the past few years potentially calls into question the continued relevance of commercial net hedging pressure as an indicator of the direction of futures premiums.

demonstrating that net hedging pressure is an unlikely explanation for the earlier findings.

Systematic risk in natural gas prices   Setting aside net hedging pressure and the returns of speculators, what if natural gas producers benefited from price volatility, while natural gas consumers were hurt by it? In this case, producers would require compensation (i.e. a premium) for being locked into long-term fixed-price contracts, and consumers would be willing to pay such compensation. Economic theory provides some support for this very scenario in the form of the CAPM.[16]

Although CAPM was originally derived as a financial tool to be applied to investment portfolios, its basic tenet – that an asset's risk depends on the correlation of its revenue stream with the asset-holder's overall wealth – can be applied much more broadly, for example in evaluating investments in physical assets such as power plants (Awerbuch, 1993, 1994; Kahn and Stoft, 1993). Specifically, in the context of natural-gas-fired generation, one can think about the correlation between a gas consumer's overall wealth (as proxied by the economy or, more specifically, the stock market) and natural gas prices. If gas prices, and therefore consumer expenditures on gas, rise as the stock market declines (e.g. because rising gas prices hurt the economy), then natural gas is said to have a negative 'beta',[17] and is risky to gas consumers and beneficial to gas producers. In other words, at the same time as gas consumers and producers feel the pinch of a weak stock market, expenditures on natural gas also rise, compounding overall wealth depletion among consumers while providing some consolation to producers.

In this specific case, where gas with a negative beta is risky to consumers and beneficial to producers, consumers have an incentive to hedge natural gas price risk, while producers do not. Intuitively, it follows that even if both consumers and producers share identical expectations of future spot gas prices, producers will still require – and consumers will be willing to pay – a premium over expected spot prices in order to lock in those prices today. Both Pindyck (2001) and Hull (1999) mathematically demonstrate this to be the case: when beta is negative, futures prices should, at least theoretically, trade at a premium to expected spot prices. Thus, if the beta of natural gas is indeed negative, this theory might explain the empirical observations of a positive premium embedded in forward prices.

---

[16] For a good introduction to CAPM, see Brealey and Myers (1991).

[17] In its original application to the stock market, beta represents the risk premium of a particular stock, and is related in a linear fashion to that stock's market risk (i.e. beta = expected risk premium on stock/expected risk premium on entire market). Stocks that carry the same market risk as the entire stock market (i.e. stocks whose returns are perfectly correlated with those of the broad market) have a beta of 1, while stocks that are perfectly uncorrelated with the market have a beta of 0. Similarly, stocks that are riskier than the market as a whole have betas > 1, while stocks that are negatively correlated with the market have betas < 0. While *assets* with a negative beta are desirable for diversification purposes, *liabilities* with a negative beta are undesirable for the same reason. In the case of natural gas, the producer holds the asset (and benefits from a negative beta) while the consumer is faced with a liability (and is hurt by a negative beta).

Literature from the early 1990s supports the existence of a negative beta for natural gas. Kahn and Stoft (1993) regressed spot wellhead gas prices against the S&P 500 using annual data from 1980 through the first six months of 1992 and arrived at an estimate of beta of -0.78 (± 0.27 standard error). Awerbuch wrote several papers advocating the use of risk-adjusted discount rates for evaluating investments in generation assets; in them he usually cited a natural gas beta ranging from -1.25 to -0.5 (Awerbuch, 1993, 1994). More recently, Bolinger et al. (2003) regressed annual changes in the price of gas delivered to electricity generators against annual S&P 500 returns, and found a natural gas beta of -0.1 from 1980 to 2002; a 2007 update to that regression yields a beta of -0.2 through 2006 (with a 90% confidence interval ranging from -0.8 to $+0.3$). Both of these more recent estimates are lower in magnitude than those from the 1990s, and could reflect a number of factors, including use of different price series (e.g. the price of gas delivered to electric generators versus wellhead prices), use of average-annual versus year-end prices, different sampling periods, or even potential changes in the relationship between fuel prices and the economy.

It is also possible to back into empirical (rather than regression-based) estimates of the beta of natural gas, by using the sample of forward prices and EIA reference-case gas price forecasts presented earlier. These are the specific betas that would be required to explain fully the empirically derived discrepancy between natural gas forward and forecast prices within the context of CAPM. To do this, one must assume that the forward prices are 'riskless' (i.e. known in advance and able to be locked in), while the price streams represented by the EIA gas forecasts are 'risky' (i.e. merely a forecast and bound to be wrong in one direction or the other). One then calculates the present value of both price streams: the forward market price stream using the known 'riskless' discount rate (i.e. the US Treasury yield at the time), and the EIA forecast price stream using whatever discount rate results in the same present value as the discounted forward market price stream. The difference between the resulting empirically derived risk-adjusted discount rate and the known 'riskless' discount rate is then divided by the 'market risk premium', i.e. the expected outperformance of risky assets (stocks) over riskless assets (Treasuries), to yield the estimated beta.[18]

Performing this exercise using the sample of forward market prices and EIA reference-case forecasts presented earlier, and an assumed market risk premium

[18]Since by definition $R_{\text{risk-adjusted}} = R_{\text{risk-free}} + \beta * \text{Market risk premium}$, then $\beta = (R_{\text{risk-adjusted}} - R_{\text{risk-free}})/\text{Market risk premium}$.

[19]Although the market risk premium has traditionally been estimated based on the long-term historical outperformance of stocks over Treasuries, many analysts think that this backward-looking approach yields a risk premium that is too high (in excess of 6%) and will likely not persist into the future. Although there is considerable debate about the appropriate size of a forward-looking risk premium, estimates in the range of 0–5% can be found. Here, 5% is simply used across the board, so as to acknowledge the concerns of analysts yet at the same time not stray too far from historical estimates. Note that only the magnitude, and not the sign, of the implied beta will change as different estimated market risk premiums are used.

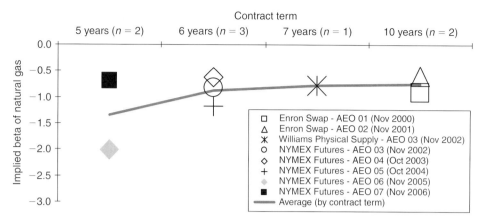

FIGURE 2.5   Implied beta of natural gas, by contract term.

of 5%,[19] gives the various estimates of beta presented in Figure 2.5. These empirical estimates of beta are generally consistent in sign and magnitude with the regression-based estimates from the 1990s' literature presented above, although the more recent regression-based estimates are less negative than the implied betas shown in Figure 2.5. Nonetheless, the consistency in the negative sign of these beta estimates at least suggests that CAPM may partially explain why long-term natural gas forwards have been priced higher than contemporaneous EIA price forecasts from 2000 to 2006.

As theoretically appealing as CAPM may be, however, the authors are nevertheless hesitant to place too much faith in it as the sole explanation for their empirical findings, for several reasons. First, CAPM formally requires that each individual's portfolio be fully diversified so that only market risk remains. Kahn and Stoft (1993) note that this may not be the case for the management of gas-producing companies, whose careers and reputation (if not financial portfolios – witness Enron's retirement plan, which was heavily invested in Enron stock) are closely tied to the profitability of the firm, and who therefore may view gas price volatility as risky even if it is negatively correlated with the stock market. Second, if CAPM were the dominant influence, then one would expect past studies of the efficiency of the natural-gas futures market regularly to demonstrate a systematic positive difference between futures and realized spot prices. As discussed in Bolinger et al. (2006), this is not the case: some, but not all, of these studies do show evidence of a premium, but in some cases the sign of the premium varies with net hedging pressure, rather than always being positive as would be required under CAPM with a negative beta. Finally, recent regression-based estimates of beta as provided in Bolinger et al. (2003) and updated earlier in this chapter are seemingly not consistent in magnitude with the negative betas that would be required to explain fully the observed forward-forecast premiums presented earlier. Thus, while CAPM may be a contributor towards natural gas futures prices that are in excess of expected spot prices, other factors seem likely to be at work as well.

2.2.4 Implications

Unfortunately, neither of the potential explanations presented above is either fully satisfying or easily refutable, although there is some evidence that biased forecasts and the CAPM may play some role in the forward-forecast premiums observed earlier. Regardless of the explanation, however, the basic implication of this study for renewable energy (or any other energy source that is immune to natural gas fuel price risk) remains the same: one should not blindly rely on gas price forecasts when comparing the levelized costs of fixed-price renewable to variable-price gas-fired generation contracts. These forecasts do not inherently allow one to assess the value of renewable energy in hedging uncertain gas prices. If there is a cost to hedging, gas price forecasts do not capture and account for it. Alternatively, if gas price forecasts are at risk of being biased or out of tune with true market expectations, then one clearly would not want to use them as the sole basis for investment decisions or resource comparisons.

A better way of comparing the levelized costs of renewable and gas-fired generation would, arguably, be to use forward natural gas price data as opposed to natural gas price forecasts. In other words, if consumers value price stability (and if the benefits of stability exceed the incremental cost of achieving it through natural gas forwards), then the cost of fixed-price renewable generation should be compared to the *hedged* or *guaranteed* cost of natural-gas-fired generation, rather than to *projected* costs based on *uncertain* gas price forecasts. Such a comparison, over the last seven years at least, would have found an additional hedging benefit of renewable generation that would not have been accounted for if uncertain gas price forecasts were used.

Of course, these findings cannot be easily extrapolated beyond the five- to ten-year terms analyzed here (i.e. we cannot know whether the observed premiums persisted over longer terms), and publicly available forward price information is unlikely to extend much beyond these terms regardless. These shortcomings call into question the practicality of implementing the recommendation to use forwards instead of forecasts in resource comparisons. Although implementation is a challenge, several options, including soliciting information on long-dated forward prices or blending existing forward prices into long-term fundamental forecasts, are arguably superior to doing nothing. These and other implementation issues are discussed further in Bolinger et al. (2003).

2.3 Renewable energy reduces natural gas prices

The previous section explored the notion that, by displacing natural-gas-fired generation, fixed-price renewable generation directly reduces natural gas price risk, providing a benefit to consumers (if not society at large). This section explores a more indirect and subtle, but no less important, consequence of that same displacement: a reduction in demand for natural gas among gas-fired generators will, all else being equal, place downward pressure on natural gas prices for all gas consumers. Many recent modelling studies of increased renewable energy deployment in the USA have demonstrated that this 'secondary' effect of putting downward pressure on natural gas prices could be significant, with the consumer benefits from reduced gas prices in many cases more than offsetting any increase in electricity costs caused by renewables deployment.[20] As a result, this price-suppression effect is increasingly cited as justification for policies promoting renewable generation.[21]

2.3.1 A cursory review of economic theory

2.3.1.1 Natural gas supply and demand curves

Economic theory predicts that a reduction in natural gas demand caused by increased deployment of renewable energy will, by causing an inward shift in the aggregate demand curve for natural gas, generally lead to a reduction in the price of natural gas relative to the price that would have been expected under business-as-usual conditions.[22] The magnitude of the price reduction will depend on the amount of demand reduction, with greater displacement of demand for gas leading to greater drops in the price of the commodity.[23] Equally important, the shape of the natural gas supply curve will have a sizeable impact on the magnitude of the price reduction. The more upwardly sloped the supply curve, the greater the price reduction will be.

The shape of the supply curve for natural gas will, in turn, depend on whether one considers short- or long-term effects. One generally assumes upward, steeply sloping supply curves in the short term when supply constraints exist in the form of fixed inputs like labor, machinery and well capacity (Henning et al., 2003). In the long term, however, the supply curve will presumably flatten because

[20] Awerbuch and Sauter (2005) go one step further and, due to the correlation between natural gas and oil prices, make the case that displacement of gas-fired generation may also reduce oil prices.

[21] Although the impact of renewable generation on natural gas prices is emphasized, it is acknowledged that similar effects would result from greater energy efficiency, as well as increased utilization of other non-gas energy sources whose fuel costs are not highly correlated with the price of natural gas (e.g. coal or nuclear power). In addition, while this analysis focuses on the USA, similar effects might be expected elsewhere.

[22] It is worthy of note that natural gas prices may fall over time even with increasing demand if technological progress allows gas to be extracted at lower prices despite the need to extract resources from increasingly less attractive resource areas. The argument here is simply that a reduction in natural gas demand is expected, *all else being equal*, to result in lower natural gas prices than would be seen under a higher demand scenario.

[23] One would not generally expect any particular threshold of demand reduction to be required to lower the price of gas (unless the supply curve was flat over some of its range). Instead, greater quantities of gas savings should result in higher levels of price reduction. The impact on prices, however, need not be linear over the full range of demand reductions; it will, instead, depend on the exact – as yet unknown – shape of the supply curve in the region in which it intersects the demand curve.

supply will have time to adjust to higher (or lower) demand expectations, for example, through increased (or decreased) exploration and drilling expenditures (Dahl and Duggan, 1998).[24] Therefore, for each unit of gas displacement, a higher price-reduction effect is to be expected in the short term than in the long term.[25]

The present discussion is primarily interested in the long-term impacts of renewable energy investments in the USA as a whole, and thus most of the attention is focused on the shape of the long-term supply curve. This approach is taken for two key reasons. First, renewable energy investments are typically long term in nature, so their most-enduring effects are likely to occur over the long term. Second, the model results reviewed here often do not clearly distinguish between short-term and long-term effects, but appear to focus predominantly on long-term, national-level impacts.

2.3.1.2 Social benefits, consumer benefits and wealth transfers

The case has been made that increased deployment of renewable energy can and should lower the price of natural gas relative to a business-as-usual trajectory. This price reduction will benefit consumers by reducing the price of gas delivered to electricity generators (assumed to be passed through in the form of lower electricity prices), and by reducing the price of gas delivered for direct use in the residential, commercial, industrial and transportation sectors. Before proceeding, however, it is important to address the nature of this 'benefit', because mischaracterizations are common and may lead to unrealistic expectations and policy prescriptions.

In particular, according to economic theory, lower natural gas prices that result from an inward shift in the demand curve do not necessarily lead to a complete gain in economic welfare, but may instead represent, to some degree at least, a shift of resources (i.e. a transfer payment) from natural gas producers to natural gas consumers. As natural gas producers see their profit margins decline (a loss of producer surplus), natural gas consumers benefit through lower gas bills (a gain of consumer surplus). Wealth transfers of this type are not a standard, primary justification for policy intervention on economic grounds.

Reducing gas prices may still be of importance in policy circles, however, where it may be viewed as a positive ancillary effect of renewable energy deployment. Energy programs are frequently assessed using consumer impacts as a key metric. Furthermore, the wealth redistribution effect may, in fact, result in a social welfare gain if economy-wide macroeconomic adjustment costs are expected to be severe in the case of natural gas price spikes and escalation. Such adjustment costs have been found to be significant in the case of oil price shocks and one might expect to discover a similar effect for natural gas, although research

---

[24] Because natural gas is a non-renewable commodity, however, the long-term supply curve must eventually slope upward as the least-expensive resources are exhausted.

[25] Note that the long-term *demand* curve is also expected to be flatter than the short-term *demand* curve (EMF, 2003). This too will moderate the long-term impacts of renewable energy investments on natural gas prices.

has not yet targeted this issue.[26] Moreover, if producers are located outside the country in question – an increasingly likely situation in the USA as the country becomes more reliant on imports of natural gas [especially liquefied natural gas (LNG)] – the wealth redistribution would increase aggregate domestic welfare. Finally, because reduced natural gas demand allows society to postpone more-expensive natural gas exploration and drilling expenses, some fraction of the gain from lower gas prices may be considered a social benefit.

2.3.2 Review of previous studies

A number of recent studies of renewable energy policies have estimated the impact of increased deployment of renewable energy on natural gas prices in the USA. These studies have exclusively evaluated a *renewables portfolio standard* (RPS), a policy that requires electricity suppliers to source an increasing percent-age of their supply from renewable generation over time. In most cases, national-level policies have been the focus of attention, but state- or regional-level policies have also been evaluated.

Information from 13 such studies was compiled and evaluated: (1) six stud-ies by the EIA focusing on US national RPS policies, two of which model multiple RPS scenarios; (2) six studies of national RPS policies by the Union of Concerned Scientists (UCS), three of which model multiple RPS scenarios; and (3) one study by the Tellus Institute that evaluates three different standards of a state-level RPS in Rhode Island (combined with RPS policies in Massachusetts and Connecticut).[27] All relevant studies for which comprehensive data could be obtained were included.

Each of the studies reviewed here relies on the EIA's National Energy Modeling System (NEMS), which is a national integrated general equilibrium energy model.[28] NEMS does not exogenously define a simple, transparent, long-term natural gas supply curve; instead, a variety of modelling assumptions and inputs are made that, when combined, implicitly define the long-term supply curve. For this reason, the long-term impact of increased renewable energy deployment on natural gas prices must be evaluated in an implicit fashion, i.e. by reviewing modelling results.

---

[26] Although the literature on the macroeconomic impacts of oil price escalation is broad, the authors are aware of no research that has explored the impact of natural gas price escalation. Extrapolating from studies that have looked at oil price shocks, Brown (2003) estimates that a sustained doubling of natural gas prices might reduce US gross domestic product (GDP) by 0.6–2.1% below what it otherwise would be.

[27] In some instances, the studies included in the analysis actually incorporated multiple sensitivity cases in addition to different RPS standard levels (e.g. different cost caps or policy sunset provisions). In these instances, just one of the sensitivity cases was selected to be reported here.

[28] Because NEMS is revised annually and many of these studies were conducted in dif-ferent years, they used different versions of NEMS. In addition, some of the studies used modified versions of NEMS with, for example, different renewable energy potential and cost assumptions.

2.3.2.1 National gas consumption and price impacts

Table 2.1 summarizes some of the key results of these studies.[29] As shown, some of the studies predict that increased renewable generation will modestly increase retail electricity prices on a national average basis, although more-recent studies have sometimes found small price reductions (due to improved renewable energy economics relative to gas-fired generation). Increased renewable generation also causes a reduction in US natural gas consumption, ranging from less than 1% to nearly 11%, depending on the study. This reduced gas consumption suppresses natural gas prices, with expected price reductions ranging from virtually no change in the US average wellhead price to an 18% reduction in that price.[30]

These wellhead price reductions translate into lower gas bills for natural gas consumers and, by reducing the price of gas delivered to the electricity sector, moderate the expected renewable-energy-induced increase in electricity prices predicted by many of the studies. Although not shown here, Wiser et al. (2005) demonstrate that the absolute reduction in delivered natural gas prices for the electricity and non-electricity sectors largely mirrors the reduction in wellhead gas prices shown in Table 2.1. This suggests that changes in wellhead prices flow through to delivered prices for all US consumers – even those consumers located in regions that do not experience significant renewable energy development – on an approximate one-for-one basis.

Figure 2.6 graphically presents the expected impact of increased renewable generation on the displacement of US gas consumption in 2020 (relative to the 'base-case' forecast) among the 13 studies in the sample. Figure 2.7, meanwhile, shows the expected impact of increased renewable generation on the average wellhead price of natural gas in the USA. As expected, increased levels of renewable energy deployment generally lead to higher levels of gas displacement and greater price reductions. Variations in the level of gas displacement and wellhead price reductions are also caused by different assumptions for the degree to which renewable energy offsets natural gas generation (versus other forms of generation), and the shape of the natural gas supply curve.

---

[29] Table 2.1 presents the projected impacts of increased renewable energy deployment in each study relative to some baseline. The baselines differ from study to study, which partially explains why, for example, a 10% RPS in two studies can lead to different impacts on renewable generation (in TWh and in percentage increase in renewable generation, above the baseline). The impact on renewable generation also varies because of assumed cost caps used in some studies or sunset provisions that in some studies terminate the RPS in a certain year, leading to fewer modelled renewable capacity additions in later years of the study because there are fewer years under the RPS in which to recoup investment costs. Additional variations among model runs include renewable technology and cost assumptions and the treatment of the federal production tax credit for wind power.

[30] Note that NEMS captures the secondary or 'rebound' effects of reduced prices on natural gas consumption and exploration (i.e. lower prices cause demand to rebound somewhat, and gas exploration and drilling to drop). The modelling results presented here represent projected impacts after such rebound effects have taken place.

Table 2.1 Summary of results from past renewable energy deployment studies

Study	RPS	Increase in US renewable generation TWh (% of total generation)	Reduction in US gas consumption Quads (%)	Gas wellhead price reduction $/MMBtu (%)	Retail electric price increase Cents/kWh (%)
EIA (1998)	10%-2010 (US)	336 (6.7%)	1.12 (3.4%)	0.34 (12.9%)	0.21 (3.6%)
EIA (1999)	7.5%-2020 (US)	186 (3.7%)	0.41 (1.3%)	0.19 (6.6%)	0.10 (1.7%)
EIA (2001b)	10%-2020 (US)	335 (6.7%)	1.45 (4.0%)	0.27 (8.4%)	0.01 (0.2%)
EIA (2001b)	20%-2020 (US)	800 (16.0%)	3.89 (10.8%)	0.56 (17.4%)	0.27 (4.3%)
EIA (2002a)	10%-2020 (US)	256 (5.1%)	0.72 (2.1%)	0.12 (3.7%)	0.09 (1.4%)
EIA (2002a)	20%-2020 (US)	372 (7.4%)	1.32 (3.8%)	0.22 (6.7%)	0.19 (2.9%)
EIA (2003b)	10%-2020 (US)	135 (2.7%)	0.48 (1.4%)	0.00 (0.0%)	0.04 (0.6%)
EIA (2007b)	15%-2020 (US)	242 (5.1%)	0.35 (1.3%)	0.08 (1.7%)	0.10 (1.4%)
UCS (2002a)	10%-2020 (US)	355 (7.1%)	1.28 (3.6%)	0.32 (10.4%)	−0.18 (−2.9%)
UCS (2002a)	20%-2020 (US)	836 (16.7%)	3.21 (9.0%)	0.55 (17.9%)	0.19 (3.0%)
UCS (2002b)	10%-2020 (US)	165 (3.3%)	0.72 (2.1%)	0.05 (1.5%)	−0.07 (−1.1%)
UCS (2003)	10%-2020 (US)	185 (3.7%)	0.10 (0.3%)	0.14 (3.2%)	−0.14 (−2.0%)
UCS (2004a)	10%-2020 (US)	181 (3.6%)	0.49 (1.6%)	0.12 (3.1%)	−0.12 (−1.8%)
UCS (2004a)	20%-2020 (US)	653 (13.0%)	1.80 (5.8%)	0.07 (1.9%)	0.09 (1.3%)
UCS (2004b)	10%-2020 (US)	277 (5.5%)	0.62 (2.0%)	0.11 (2.6%)	−0.16 (−2.4%)
UCS (2004b)	20%-2020 (US)	647 (12.9%)	1.45 (4.7%)	0.27 (6.7%)	−0.19(−2.9%)
UCS (2007)	20%-2020 (US)	310 (6.6%)	1.07 (3.8%)	0.25 (5.3%)	−0.13 (−1.7%)
Tellus (2002)	10%-2020 (RI)	31 (0.6%)	0.13 (0.4%)	0.00 (0.0%)	0.02 (0.1%)
Tellus (2002)	15%-2020 (RI)	89 (1.8%)	0.23 (0.7%)	0.01 (0.4%)	−0.05 (−0.3%)
Tellus (2002)	20%-2020 (RI)	98 (2.0%)	0.28 (0.8%)	0.02 (0.8%)	−0.07 (−0.4%)

All dollar figures are in constant 2000 $.

The increase in US renewable energy generation reflects the TWh and percentage increase *relative* to the reference case scenario for the year 2020. The percentage figures do not equate to the size of the renewables portfolio standard (RPS) for a variety of reasons: (1) existing renewable generation and new renewable generation that comes on line in the reference case may also be eligible for the RPS, and (2) the RPS is not always achieved, given assumed cost caps in some studies.

The reference case in most studies reflects an EIA Annual Energy Outlook (AEO) reference case, with some studies making adjustments based on more-recent gas prices or altered renewable-technology assumptions. The one exception is UCS (2003), in which the reference case reflects a substantially higher gas-price environment than in the relevant AEO reference case.

The Tellus study models an RPS for Rhode Island, also including the impacts of the Massachusetts and Connecticut RPS policies. All the figures shown in this table for the Tellus study are for the predicted national-level impacts of the regional policies that were evaluated.

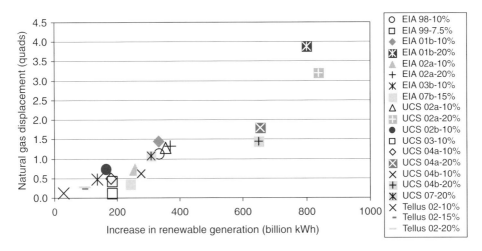

FIGURE 2.6   Forecast natural-gas displacement in 2020.

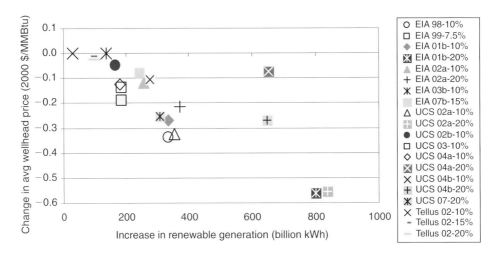

FIGURE 2.7   Forecast natural-gas wellhead price reduction in 2020.

2.3.2.2 Induced natural gas price reductions, in context

The previously presented results show that increased renewable generation is predicted to reduce natural gas consumption and prices while retail electricity prices are predicted to rise in at least some instances. The net predicted effect on US consumer energy bills could be positive or negative, depending on the relative magnitude of changes in electricity and natural gas bills.

Figure 2.8 presents these offsetting effects for a subset of the studies that were reviewed.[31] Although there are variations among the different studies, the present

---

[31] Figure 2.8 shows the energy bill impacts only for the national RPS studies (i.e. it excludes Tellus, 2002).

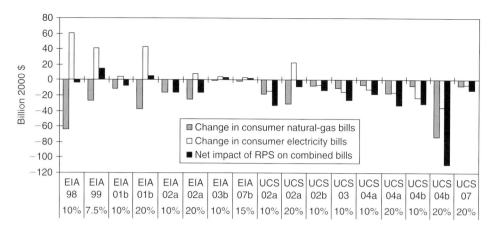

FIGURE 2.8   Present value of renewables portfolio standard (RPS) impacts on natural gas and electricity bills (through 2020, 7% real discount rate).

value cost of the cumulative (through 2020) predicted increase in consumer electricity bills (if any) in the RPS cases compared with the reference case is often on the same order of magnitude as the present value benefit of the predicted decrease in consumer natural gas bills. From an aggregate *consumer* perspective, therefore, the net consumer cost of these policies is typically predicted to be rather small, with 13 of 17 RPS analyses even showing net consumer savings (i.e. negative cumulative bill impacts).[32] Clearly, the expected consumer benefit of renewable energy in reducing natural gas prices can be substantial, and ignoring this benefit may cause a substantial bias in any estimate of the net consumer costs of increased renewable energy deployment.

To put this possible price-reduction benefit in another context, Figure 2.9 shows the range of consumer benefits delivered by increased renewable energy, by study, expressed in terms of $ per MWh of renewable generation. Here, the focus is only on the consumer benefits that derive from induced reductions in natural gas prices, and the positive and negative impacts and costs of renewable energy on other segments of the energy-economy are not considered.

Results from these studies suggest that each MWh of renewable energy provides, in aggregate, US consumer gas price reduction benefits that range from $5 to $35/MWh, with most studies showing a range of $7.50–20/MWh. Variations in this value are caused by different implied inverse price elasticities of natural gas supply (defined and discussed in the next section), and by differences in the

---

[32] In several of these studies, RPS cost caps are reached, ensuring that consumers pay a capped price for some number of *proxy* renewable energy credits (and leading to increased electricity prices) while not obtaining the benefits of increased renewable generation on natural gas prices. Accordingly, if anything, Figure 2.8 underestimates the possible consumer benefits of a well-designed renewable energy program with less-binding cost caps.

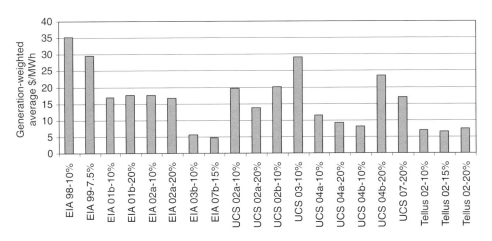

FIGURE 2.9   Consumer gas-savings benefits of increased renewable production (in $/MWh).

amount of gas displacement caused by renewable energy. Even at the low end of the range, however, these consumer benefits are sizeable.

Note that the annual gas bill savings per MWh of renewable generation are weighted by the amount of yearly renewable generation to derive this weighted average figure. In doing so, early-year 'noise' in the data is ignored, and the average is extended to the last year of the forecast period (2020–2030, depending on the study).

2.3.3 Summary of implied inverse price elasticities of supply

The *price elasticity of natural gas supply* is a measure of the responsiveness of natural gas supply to changes in the price of the commodity at a specific point on the supply curve, and is calculated by dividing the percentage change in quantity supplied by the percentage change in price. In the case of renewable-energy-induced shifts in the demand for natural gas, however, the interest lies in understanding the change in price that will result from a given change in quantity demanded, or the *inverse price elasticity of supply* ('inverse elasticity').

The inverse elasticity provides a convenient, normalized measure by which to compare the price responses predicted by the 13 modelling studies described earlier. The calculation requires annual data on the predicted average US wellhead price of natural gas and total gas consumption in the USA for both the business-as-usual scenario and the policy scenario of increased renewable energy deployment.[33] Because relying on the implied inverse elasticity for any *single* year could be misleading, Figure 2.10 compares the *average* value of the long-term implicit

---

[33]The specific calculation is: E^{-1} = (Wellhead price$_{business-as-usual}$/Wellhead price$_{policy}$ − 1)/ (Gas demand$_{business-as-usual}$/Gas demand$_{policy}$ − 1)

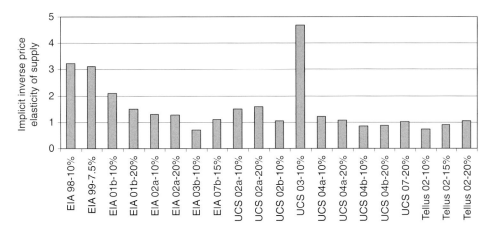

FIGURE 2.10   Average inverse price elasticities of supply.

inverse elasticities among studies.[34] Despite substantial variations among studies and results for individual years (Wiser et al., 2005), there is some consistency in the *average* long-term inverse elasticities; the overall range is between 0.7 and 4.7, with elasticities from 14 of 20 analyses falling between 0.8 and 2.0.[35] This means that each 1% reduction in US natural gas demand is expected to lead to something in the order of a 0.8% to 2% reduction in wellhead gas prices on average.

Natural gas demand is equivalent to natural gas supply in this instance. The inverse elasticity calculations presented here use US price and quantity data under the assumption that the current market for natural gas is more regional than worldwide in nature (Henning et al., 2003). Of course, the market for natural gas consumed in the USA is arguably a North American market, including Canada and Mexico, with LNG expected to play an increasing role in the future. Trade with Mexico is relatively small, however, and Canadian demand for gas is relatively small compared to US demand. LNG, meanwhile, remains a modest contributor to total US consumption.

2.3.4 Benchmarking elasticities against other models and empirical estimates

The energy-modelling results reviewed previously tell a consistent, basic story: reducing the demand for natural gas through the use of renewable energy is

---

[34] Average inverse elasticities are calculated as the average of each year's inverse elasticity, ignoring early-year 'noise' in the data and extending to the last year of the forecast period, 2020–2030, depending on the study.

[35] The average inverse elasticity from UCS (2003) is substantially higher than that from most of the other studies. As noted earlier, UCS (2003) evaluated the potential impact of an RPS under a scenario of higher gas prices than in a typical AEO reference case, so that study is not strictly comparable to the others covered in this chapter (specifically, the UCS study includes a more-constrained gas supply than most of the other analyses, especially in the later years, and so is likely to be measuring changes along a steeper portion of the supply curve).

expected to lead to lower natural gas prices than would be the case in a business-as-usual scenario. These are modelling predictions, however, which are based on an estimated shape of a natural gas supply curve that is not known with any precision. Furthermore, all of the studies reviewed so far used the same basic model – the EIA's National Energy Modeling System – which raises questions about the degree of confidence one should place in these modelling results. These questions can be at least partly addressed by comparing the NEMS-based implied inverse elasticities presented earlier to those from other energy models, and by benchmarking them against empirical estimates of historical inverse elasticities.

As shown in Wiser and Bolinger (2007), a survey of studies conducted using models other than NEMS found that most use long-term inverse elasticity assumptions that are broadly consistent with, and in some cases higher than, the range shown in Figure 2.10. For example, studies from the American Council for an Energy-Efficient Economy (ACEEE, 2003), the National Petroleum Council (NPC, 2003a, b) and the National Commission on Energy Policy (NCEP, 2003) used a model from Energy and Environmental Analysis, Inc. Although not directly comparable to those in the present sample, these studies exhibit more-aggressive long-term inverse elasticities, in the range of 4.0. More recently, the California Energy Commission used a model from Global Energy Decisions to measure the gas price suppression impact; it found even-more-aggressive long-term inverse elasticities in the neighborhood of 5.0 (CEC, 2007). Other models used in a Stanford Energy Modeling Forum study were more consistent with NEMS (EMF, 2003). This general level of cross-model consistency provides at least some comfort that the renewable energy studies within this sample are modelling this effect within reason, and perhaps even conservatively.

Empirical research on energy elasticities has focused almost exclusively on the impact of supply shocks on energy *demand* (demand elasticity) rather than the impact of demand shocks on energy *supply* (supply elasticity). As a result, there are few empirical estimates of supply elasticities against which to benchmark the modelling output described earlier. Nonetheless, as discussed in Wiser and Bolinger (2007), the few published empirical estimates of historical long-term inverse elasticities for gas, coal and oil are positive, and are not wildly out of line with the long-term inverse elasticity of natural gas implied by the modelling output presented earlier. Although more work is certainly warranted in this area, the extant literature at least suggests that the models are evaluating price responsiveness within reason and within the limits of current knowledge.

2.4 Conclusions

Renewable energy has historically been supported primarily because of its perceived environmental, economic development and national security benefits. More recently, sharp price escalations in wholesale electricity and natural gas markets have led to renewed discussions about the potential risk mitigation value of renewable resources in the USA and elsewhere. Specifically, by displacing variable-priced natural-gas-fired generation, fixed-price renewable generation not only directly mitigates fuel price risk, but also reduces demand for natural gas, which in turn places downward pressure on natural gas prices – a benefit that flows through to all gas-consuming sectors of the economy.

This chapter has explored both of these possible risk mitigation benefits. It was found that the direct risk mitigation benefit of renewable generation is not always recognized in quantitative resource evaluations, which often use uncertain natural gas price forecasts to compare the levelized cost of gas-fired generation to the cost of fixed-price renewables, and therefore do not consider the consumer benefits of price stability. The analysis shows that, over the past seven years at least, substituting natural gas forward prices, i.e. prices that can be locked in to create price certainty, for the uncertain price forecasts would have significantly improved the relative economics of fixed-price renewables.

In addition, a review of results from 13 studies that model the economy-wide impact of increased renewables penetration suggests that each 1% reduction in nationwide demand for natural gas will result in a long-term wellhead price reduction somewhere in the order of 0.8% to 2.0% on average. Price reductions are likely to be even larger in the short term (and perhaps also in the long term, according to some models). In many cases, this gas price suppression is predicted to be sizeable enough to outweigh any incremental costs associated with increased renewables penetration, even without considering the direct risk mitigation benefit discussed above.

Both of these findings support the notion that, on an economic basis alone, renewable generation should play a larger role in generating portfolios. This, of course, is the same message that our late colleague, Shimon Awerbuch, so tenaciously espoused throughout his distinguished career. Unfortunately, this message and its underlying concepts have only recently spread beyond the academic community and taken root among energy policy and decision makers, who have become more-receptive in this era of sharply higher and volatile fuel prices, energy insecurity, more competitive renewable generation technologies and growing concern over global climate change. In turn, the composition of generation portfolios around the world has only recently begun to shift more towards renewables. We regret that Shimon, a thinker before his time (and well ahead of the rest of us), will not be here to witness the full transformation to this more-sustainable, and less-risky, energy future.

References

American Council for an Energy-Efficient Economy: Elliot, R., Shipley, A., Nadel, S. and Brown, E. (2003). *Natural Gas Price Effects of Energy Efficiency and Renewable Energy Practices and Policies.* Report No. E032. Washington, DC: ACEEE.

Awerbuch, S. (1993). The surprising role of risk in utility integrated resource planning. *The Electricity Journal*, 6(3), 20–33.

Awerbuch, S. (1994). Risk-adjusted IRP: it's easy!!! In Proceedings of the NARUC-DOE Fifth National Conference on Integrated Resource Planning, Kalispell, MT, pp. 228–269.

Awerbuch, S. (2003). Determining the real cost: why renewable power is more cost-competitive than previously believed. *Renewable Energy World*, 6(2).

Awerbuch, S. and Sauter, R. (2005). *Exploiting the Oil–GDP Effect to Support Renewables Deployment.* SPRU Working Paper 129. Brighton: University of Sussex.

Bolinger, M. and Wiser, R. (2005). *Balancing Cost and Risk: The Treatment of Renewable Energy in Western Utility Resource Plans.* LBNL-58450. Berkeley, CA: Lawrence Berkeley National Laboratory.

Bolinger, M., Wiser, R. and Golove, W. (2003). *Accounting for Fuel Price Risk: Using Forward Natural Gas Prices Instead of Gas Price Forecasts to Compare Renewable to Natural Gas-Fired Generation.* LBNL-53587. Berkeley, CA: Lawrence Berkeley National Laboratory.

Bolinger, M., Wiser, R. and Golove, W. (2006). Accounting for fuel price risk when comparing renewable to gas-fired generation: the role of forward natural gas prices. *Energy Policy*, 34(6), 706–720.

Brealey, R. A. and Myers, S. C. (1991). *Principles of Corporate Finance.* San Francisco, CA: McGraw-Hill.

Brown, S. (2003). *US Natural Gas Markets in Turmoil.* Testimony prepared for a hearing on The Scientific Inventory of Oil and Gas Resources on Federal Lands, US House of Representatives (19 June).

California Energy Commission (2007). Scenario analyses of California's electricity system: preliminary results for the *2007 Integrated Energy Policy Report*, Second Addendum Appendices. CEC-200-2007-010-AD2-AP.

Chang, E. C. (1985). Returns to speculators and the theory of normal backwardation. *Journal of Finance*, 40(1), 193–208.

Commodity Futures Trading Commission (2007). *Commitments of Traders Reports.* Available at www.cftc.gov (accessed 16 August 2007).

Considine, T. J. and Clemente, F. A. (2007). Betting on bad numbers. *Public Utilities Fortnightly*, (July), 53–59.

Dahl, C. and Duggan, T. (1998). Survey of price elasticities from economic exploration models of US oil and gas supply. *Journal of Energy Finance and Development*, 3(2), 129–169.

Dusak, K. (1973). Futures trading and investor returns: an investigation of commodity market risk premiums. *Journal of Political Economy*, 81(6), 1387–1406.

Energy Information Administration (1998). *Analysis of S. 687, the Electric System Public Benefits Protection Act of 1997.* SR/OIAF/98-01. Washington, DC: EIA.

Energy Information Administration (1999). *Annual Energy Outlook 2000.* DOE/EIA-0383(2000). Washington, DC: EIA.

Energy Information Administration (2000). *Annual Energy Outlook 2001.* DOE/EIA-0383(2001). Washington, DC: EIA (December).

Energy Information Administration (2001a). *Annual Energy Outlook 2002.* DOE/EIA-0383(2002). Washington, DC: EIA (December).

Energy Information Administration (2001b). *Analysis of Strategies for Reducing Multiple Emissions from Electric Power Plants: Sulfur Dioxide, Nitrogen Oxides, Carbon Dioxide, and Mercury and a Renewable Portfolio Standard.* SR/OIAF/2001-03. Washington, DC: EIA.

Energy Information Administration (2002). *Impacts of a 10-Percent Renewable Portfolio Standard.* SR/OIAF/2002-03. Washington, DC: EIA.

Energy Information Administration (2003a). *Annual Energy Outlook 2003.* DOE/EIA-0383(2003). Washington, DC: EIA (January).

Energy Information Administration (2003b). *Analysis of a 10-Percent Renewable Portfolio Standard.* SR/OIAF/2003-01. Washington, DC: EIA.

Energy Information Administration (2004). *Annual Energy Outlook 2004.* DOE/EIA-0383(2004). Washington, DC: EIA (January).

Energy Information Administration (2005). *Annual Energy Outlook 2005.* DOE/EIA-0383(2005). Washington, DC: EIA (February).

Energy Information Administration (2006a). *Annual Energy Outlook 2006.* DOE/EIA-0383(2006). Washington, DC: EIA (February).

Energy Information Administration (2006b). *Electric Power Annual 2005.* DOE/EIA-0348(2005). Washington, DC: EIA (November).

Energy Information Administration (2007a). *Annual Energy Outlook 2007.* DOE/EIA-0383(2007). Washington, DC: EIA (February).

Energy Information Administration (2007b). *Impacts of a 15-Percent Renewable Portfolio Standard.* SR/OIAF/2007-03. Washington, DC: EIA.

Energy Modeling Forum (2003). *Natural Gas, Fuel Diversity and North American Energy Markets*. EMF Report 20, Stanford, CA: Stanford University.

Henning, B., Sloan, M. and de Leon, M.(2003). *Natural Gas and Energy Price Volatility*. Arlington, VA: Energy and Environmental Analysis.

Hirshleifer, D. (1988). Residual risk, trading costs, and commodity futures risk premia. *Review of Financial Studies* (Summer).

Hull, J. C. (1999). *Options, Futures, and Other Derivatives*. Upper Saddle River, NJ: Prentice Hall.

Humphreys, H. B. and McClain, K. T. (1998). Reducing the impacts of energy price volatility through dynamic portfolio selection. *Energy Journal*, 19(3), 107–131.

Kahn, E. and Stoft, S. (1993) (unpublished draft). *Analyzing Fuel Price Risks Under Competitive Bidding*. Berkeley, CA: Lawrence Berkeley National Laboratory.

Keynes, J. M. (1930). *A Treatise on Money*. London: Macmillan.

National Commission on Energy Policy (2003). *Increasing US Natural Gas Supplies: A Discussion Paper and Recommendations from the National Commission on Energy Policy*. Washington, DC: NCEP.

National Petroleum Council (2003a). *Balancing Natural Gas Policy – Fueling the Demands of a Growing Economy. Volume I, Summary of Findings and Recommendations*. Washington, DC: NPC.

National Petroleum Council (2003b). *Balancing Natural Gas Policy – Fueling the Demands of a Growing Economy. Volume II. Integrated Report*. Washington, DC: NPC.

Pindyck, R. S. (2001). *The Dynamics of Commodity Spot and Futures Markets: A Primer*. Cambridge, MA: Massachusetts Institute of Technology.

Tellus (2002). *Modeling Analysis: Renewable Portfolio Standards for the Rhode Island GHG Action Plan*. Boston, MA: Tellus Institute.

Union of Concerned Scientists (2002a). *Renewing Where We Live (February 2002 edn)*. Cambridge, MA: UCS.

Union of Concerned Scientists (2002b). *Renewing Where We Live (September 2002 edn)*. Cambridge, MA: UCS.

Union of Concerned Scientists (2003). *Renewing Where We Live (September 2003 edn)*. Cambridge, MA: UCS.

Union of Concerned Scientists (2004a). *Renewable Energy Can Help Ease the Natural Gas Crunch*. Cambridge, MA: UCS.

Union of Concerned Scientists (2004b). *Renewing America's Economy: A 20 Percent National Renewable Energy Standard Will Create Jobs and Save Consumers Money*. Cambridge, MA: UCS.

Union of Concerned Scientists (2007). *Cashing in on Clean Energy*. Cambridge, MA: UCS.

Wiser, R. and Bolinger, M. (2007). Can deployment of renewable energy put downward pressure on natural gas prices? *Energy Policy*, 35(1), 295–306.

Wiser, R., Bolinger, M. and St Clair, M. (2005). *Easing the Natural Gas Crisis: Reducing Natural Gas Prices through Increased Deployment of Renewable Energy and Energy Efficiency*. LBNL-56756. Berkeley, CA: Lawrence Berkeley National Laboratory.

CHAPTER 3

Using Portfolio Theory to Value Power Generation Investments

The Editors
(based on work from **Shimon Awerbuch** and **Spencer Yang**[1]*)

3.1 Introduction

Traditional valuation approaches such the "levelised cost" methodology value generation technologies on a stand alone basis. But the various generation technologies have different risk return profiles, such that there are potentially great advantages in operating a diversified portfolio of plants. Mean-Variance Portfolio theory can usefully complement traditional valuation approaches to inform a country or utility on the optimal generation portfolio to minimize the impact of some critical risks in liberalized power markets. This chapter introduces the intuition and theory underpinning the application of Mean-Variance Portfolio theory to the power sector. The second section of the book provides a number of case studies illustrating the different types of insights that can be learned from this new valuation approach, while the third section of the book concentrates on some frontier applications and identifies issues for further research.

---

[1] The second section is based on Awerbuch and Yang (2007) which is the last published work of Shimon Awerbuch.

---

*Sussex Energy Group, SPRU, University of Sussex, Brighton, UK

Analytical Methods for Energy Diversity and Security © 2008 Elsevier Ltd.
978-0-08-056887-4 All rights reserved.

3.2 Capturing risk in power investment valuation techniques

While many of the risks facing power producers in liberalized electricity markets existed in the fully regulated and vertically integrated industry, investors can no longer pass these costs on to consumers or taxpayers automatically. Investors now have additional risks to consider and manage in the liberalized industry, chief among which is electricity price risk. Investment in power generation comprises a large and diverse set of risks, as classified by IEA/NEA (2005):

- economy-wide factors that affect the demand for electricity and availability of labor and capital;
- factors under the control of the policy makers, such as regulatory (economic and non-economic) and political risks, with possible implications for costs and financing conditions and on earnings; an example of such risk is the cost of additional emissions controls;[2]
- factors under the control of the company, such as the size and diversity of its investment program, the choice and diversity of generation technologies, and control of costs during construction and operation;
- the price and volume risks in the electricity market;
- fuel price and, to a lesser extent, availability risks;
- financial risks arise from the financing of investment; they can to some extent be mitigated by the capital structure of the company.

These risks will affect competing technologies differently. Some risks are inherent to the technology involved; others involve the interaction of technology and the environment in which the generating company operates. Table 3.1, reproduced from IEA/NEA (2005), provides a qualitative assessment of how the various types of risk in liberalized electricity markets affect the three main base load generation technologies (gas, coal and nuclear power plants).

The traditional 'levelized cost' valuation approach was well adapted to assess power investments prior to liberalization. Fraser (2003) points out that it, 'reflected the reality of long-term financing, passing on costs to the customers, known technology paradigms, a predictable place in the merit order, a steady increase in consumption, and, in the presence of steady technical progress, no problem in securing a favourable position in the merit order for new plant'. The levelized cost methodology remains widely used in the liberalized industry, both by energy planners and by electric companies (IEA/NEA, 2005). Power companies will apply this methodology based on an internal target for return on equity (the 'hurdle rate') to make a decision whether to invest or not and to decide between different projects.

However, in the liberalized electricity industry, what matters to the investor is the profitability of the investment against the risk to the capital employed. The

---

[2] In the European Union (EU), one of the greatest uncertainties for investors in new power plants is the future evolution of the European Carbon Trading Scheme. Worldwide, the uncertainty over controls on future carbon dioxide (CO_2) emissions will grow in the future, particularly as future restrictions on levels of CO_2 emissions beyond the first commitment period of the Kyoto Protocol are unknown.

Table 3.1   Qualitative comparison of generic features of generation technology

Technology	Unit size	Lead time	Capital cost/kW	Operating cost	Fuel prices	CO_2 emissions	Regulatory risk
CCGT	Medium	Short	Low	Low	High	Medium	Low
Coal	Large	Long	High	Medium	Medium	High	High
Nuclear	Very large	Long	High	Medium	Low	Nil	High
Hydro	Large	Long	Very high	Very low	Nil	Nil	High
Wind	Small	Short	High	Very low	Nil	Nil	Medium
Reciprocating engine	Small	Very short	Low	Low	High	Medium	Medium
Fuel cells	Small	Very short	Very high	Medium	High	Medium	Low
Photovoltaics	Very small	Very short	Very high	Very low	Nil	Nil	Low

Source: IEA/NEA (2005).
CCGT: combined cycle gas turbine.

level of risk anticipated by an investor in a power plant will be reflected in the level of return expected on that investment. The greater the business and financial risks, the higher the return that will be demanded. It is difficult for the levelized cost methodology to incorporate risks and uncertainty effectively (Roques, 2006). In order to assess various risks, different scenarios or sensitivities are usually calculated, which often give only a limited assessment of the risks involved. IEA/NEA (2005) reckons for instance that '[the levelized cost] methodology for calculating generation costs does not take business risks in competitive markets adequately into account' and that 'it needs to be complemented by approaches that account for risks in future costs and revenues'.

Most importantly, the levelized cost methodology values investment projects on a stand-alone basis. The various generation technologies have different risk–return profiles, and there are potentially great advantages in operating a diversified portfolio of plants for a utility. Because it does not take into account the complementarity in the risk–return profiles of different plants that a utility operates, the levelized cost methodology cannot inform a utility or country on the optimal technological choice for an additional power plant, given the current portfolio that a utility or country operates. In other words, the best technology choice depends on the portfolio of plants that the utility/country already operates, and this interdependency needs to be captured by power valuation methodologies using a mean-variance portfolio (MVP) theory approach.

3.3 Applying portfolio optimization to power investment choices

Given the uncertain environment in which utilities make their investment decisions, it makes sense to shift electricity planning from its current emphasis on evaluating alternative technologies to evaluating alternative electricity generating portfolios and strategies. The techniques for doing this are rooted in modern finance theory – in particular, MVP theory.[3] Portfolio analysis is widely used by financial investors to create low-risk, high-return portfolios under various economic conditions. In essence, investors have learned that an efficient portfolio takes no unnecessary risk to its expected return. In short, these investors define efficient portfolios as those that maximize the expected return for any given level of risk, while minimizing risk for every level of expected return.

Portfolio theory is highly suited to the problem of planning and evaluating electricity portfolios and strategies because energy planning is not unlike investing in financial securities where financial portfolios are widely used by investors to manage risk and to maximize performance under a variety of unpredictable outcomes. Similarly, it is important to conceive of electricity generation not in terms of the cost of a particular technology today, but in terms of its portfolio cost. At any given time, some alternatives in the portfolio may have high costs while others have lower costs, yet over time, an astute combination of alternatives can serve to minimize overall generation cost relative to the risk. In sum, when portfolio theory is applied to electricity generation planning, conventional and renewable alternatives are evaluated not on the basis of their stand-alone cost, but on the basis of their portfolio cost – that is, their contribution to overall portfolio generating cost relative to their contribution to overall portfolio risk. Portfolio-based electricity planning techniques – pioneered by Awerbuch and Berger (2003), Berger (2003), Awerbuch (1995, 2000), Humphreys and McLain (1998) and Bar-Lev and Katz (1976) – thus suggest ways to develop diversified generating portfolios with known risk levels that are commensurate with their overall electricity generating costs. Simply put, these techniques help to identify generating portfolios that can minimize a society's or a utility's energy price cost and risk.

Portfolio theory was developed for financial analysis to locate portfolios with maximum expected return at every level of expected portfolio risk. Box 3.1 reviews the basics of this theory in the context of electricity generation mixes from the perspective of a social planner, where the generating cost is the relevant measure, as defined in Berger (2003) and Awerbuch and Berger (2003). Note that other measures of costs and returns can be used to reflect different objective functions, for instance from the perspective of a private investor, as described in Roques et al. (this volume, chapter 11). For instance, the net cash flow or the net present value is a better suited measure to capture the impact of both costs and revenues risks (electricity price risk).

---

[3] MVP theory, an established part of modern finance theory, is based on the pioneering work of Nobel Laureate Harry Markowitz 50 years ago. For a recent contribution see Fabozzi et al. (2002).

Box 3.1 Electricity generating costs, risks, and correlations

Electricity generating cost and returns

Portfolio theory was initially conceived in the context of financial portfolios, where it relates expected portfolio return to expected portfolio risk, defined as the year-to-year variation of portfolio returns. This box illustrates portfolio theory as it applies to a two-asset generating portfolio from the perspective of a social planner, where the generating cost is the relevant measure, as it was originally defined in Awerbuch and Berger (2003) and Berger (2003). Generating cost (€/kWh) is the inverse of a return (kWh/€); that is, a return in terms of physical output per unit of monetary input.

Expected portfolio cost

Expected portfolio cost is the weighted average of the individual expected generating costs for the two technologies:

$$\text{Expected portfolio cost} = X_1 E(C_1) + X_2 E(C_2) \qquad (3.1)$$

where X_1 and X_2 are the fractional shares of the two technologies in the mix, and $E(C_1)$ and $E(C_2)$ are their expected levelized generating costs per kWh.

Expected portfolio risk

Expected portfolio risk, $E(\sigma_p)$, is the expected year-to-year variation in generating cost. It is also a weighted average of the individual technology cost variances, as tempered by their covariances:

$$\text{Expected portfolio risk} = E(\sigma_p) = \sqrt{X_1^2\sigma_1^2 + X_2^2\sigma_2^2 + 2X_1 X_2 \rho_{12}\sigma_1\sigma_2} \qquad (3.2)$$

where X_1 and X_2 are the fractional shares of the two technologies in the mix, σ_1 and σ_2 are the standard deviations of the holding period returns (HPRs) of the annual costs of technologies 1 and 2 as further discussed, and ρ_{12} is their correlation coefficient

Portfolio risk is here estimated as the standard deviation of the HPRs of future generating cost streams. The HPR is defined as: $\text{HPR} = (EV - BV)/BV$, where EV is the ending value and BV the beginning value (see Brealey et al., 2004, for a discussion on HPRs). For fuel and other cost streams with annual reported values, EV can be taken as the cost in year $t + 1$ and BV as the cost in year t. HPRs measure the rate of change in the cost stream from one year to the next. A detailed discussion of its relevance to portfolios is given in Berger (2003).

Each individual technology actually consists of a portfolio of cost streams (capital, operating and maintenance, fuel, CO_2 costs, and so on).

Total risk for an individual technology (i.e. the portfolio risk for those cost streams) is σ_T. In this case, the weights, X_1, X_2, and so on, are the fractional share of total levelized cost represented by each individual cost stream. For example, total levelized generating costs for a coal plant might consist of {1/4} capital, {1/4} fuel, {1/4} operating costs and {1/4} CO_2 costs, in which case each weight $X_j = 0.25$.

Correlation, diversity and risk

The correlation coefficient, ρ, is a measure of diversity. Lower ρ among portfolio components creates greater diversity, which reduces portfolio risk σ_p. More generally, portfolio risk falls with increasing diversity, as measured by an absence of correlation between portfolio components. Adding a fuelless (i.e. fixed-cost, riskless) technology to a risky generating mix lowers expected portfolio cost at any level of risk, even if this technology costs more (Awerbuch, 2006). A pure fuelless, fixed-cost technology has $\sigma_i = 0$, or nearly so. This lowers σ_p, since two of the three terms in Equation (3.2) reduce to zero. This, in turn, allows higher risk/lower cost technologies into the optimal mix. Finally, it is easy to see that σ_p declines as $\rho_{i,j}$ falls below 1.0. In the case of fuelless renewable technologies, fuel risk is zero and its correlation with fossil fuel costs is zero too.

Expected portfolio generating cost is the weighted average of the individual technology costs. The expected risk of an electricity portfolio (i.e. the expected year-to-year fluctuation in portfolio generating cost) is a weighted average of the risks of the individual technology costs, tempered by their correlations or covariances. Each technology itself is characterized by a portfolio of cost streams, comprising capital outlays, fuel expenditures, operation and maintenance (O&M) expenditure, and CO_2 costs. It follows that for each technology, risk is the standard deviation of the year-to-year changes of these cost inputs.

Portfolio theory improves decision making in the following way. First, since the investor only needs to consider the portfolios on the so-called efficient frontier, rather than the entire universe of possible portfolios, it simplifies the portfolio selection problem, Second, it quantifies the notion that diversification reduces risk. For electricity planning, portfolio optimization exploits the interrelationships (i.e. correlations) among the various technology generating cost components. Take for example fossil fuel prices. Because they are correlated with each other, a fossil-dominated portfolio is undiversified and exposed to fuel price risk. Conversely, renewables, nuclear and other non-fossil options diversify the mix and reduce its expected risk because their costs are not correlated with fossil prices.

The portfolio diversification effect is illustrated in Figure 3.1, which shows the costs and risks for various possible two-technology portfolios. Technology A is representative of a generating alternative with higher cost and lower risk, such as photovoltaics (PV). It has an expected (illustrative) cost around €0.10/kWh with an expected year-to-year risk of 8%. Technology B is a lower cost/higher

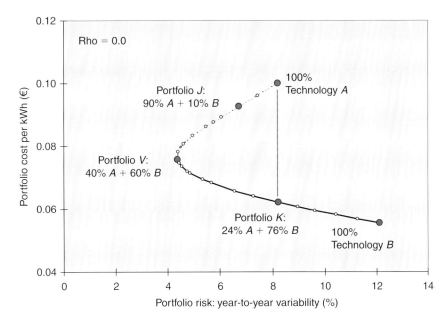

FIGURE 3.1   Portfolio effect for illustrative two-technology portfolio.

risk alternative, such as gas-fired generation. Its expected cost and risk are about €0.055/kWh and 12%, respectively. The correlation factor between the total cost streams of the two technologies is assumed to be zero. This is a simplification since in reality the capital and variable cost of PV will exhibit some non-zero correlation with the capital and variable cost of gas generation.

As a consequence of the portfolio effect, total portfolio risk decreases when the riskier technology B is added to a portfolio consisting of 100% A. For example, portfolio J, which comprises 90% of technology A plus 10% B, exhibits a lower expected risk than a portfolio comprising 100% A. This is counter-intuitive since technology B is riskier than A. Portfolio V, the minimum variance portfolio, has a risk of around 4%, which is one-half the risk of A and one-third the risk of B. This, however, illustrates the point of diversification.

Investors would not hold any mix above portfolio V because mixes exhibiting the equivalent risk can be obtained at lower cost on the solid portion of the line. Portfolio K is therefore superior to 100% A. It has the same risk, but lower expected cost. Investors would not hold a portfolio consisting only of technology A, but rather would hold the mix represented by K. Taken on a stand-alone basis, technology A is more costly, yet properly combined with B, as in portfolio K, it has attractive cost and risk properties. Not only is the mix K superior to 100% A, most investors would also consider it superior to 100% technology B. Compared to B, mix K reduces risk by one-third while increasing cost by just 10% (€0.005/kWh), which gives it a higher Sharpe ratio than other mixes.[4] Mix K

---

[4] Developed by Nobel Laureate William F. Sharpe, this ratio is a risk-adjusted performance of an asset and is used to characterize how well the return of an asset compensates the investor for the risk taken.

illustrates that astute portfolio combinations of diversified alternatives produce efficient results, which cannot be measured using stand-alone cost concepts. To summarize, portfolio optimization locates minimum-cost generating portfolios at every level of portfolio risk, represented by the solid part of the line in Figure 3.1, that is, the stretch between V and B.

3.4 Conclusion

Electric utilities operating in liberalised markets are faced with a wide range of risks and uncertainties when evaluating different generation technologies. This chapter introduced Mean-Variance Portfolio theory as a new valuation approach to identify optimal generation portfolio to minimize the impact of some critical risks. The second section of the book provides a number of case studies illustrating the different types of insights that can be learned from this new valuation approach, while the third section of the book concentrates on some frontier applications and identifies issues for further research.

References

Awerbuch, S. (1995). New economic cost perspectives for valuing solar technologies. In Böer, K. W. (Ed.), *Advances in Solar Energy: An Annual Review of Research and Development*. Vl. 10. Boulder, CO: ASES.

Awerbuch, S. (2000). Getting it right: the real cost impacts of a renewables portfolio standard. *Public Utilities Fortnightly* (15 February).

Awerbuch, S. (2005). Portfolio-based electricity generation planning: policy implications for renewables and energy security. *Mitigation and Adaptation Strategies for Global Change*, in press.

Awerbuch, S. and Berger, M. (2003). *Energy Security and Diversity in the EU: A Mean-Variance Portfolio Approach*. IEA Report No. EET/2003/03 (February). Paris: IEA. http://library.iea.org/dbtw-wpd/textbase/papers/2003/port.pdf

Awerbuch and Yang (2007). Efficient electricity generating portfolios, for Europe: maximising energy security and climate change mitigation. *European Investment bank papers*, Vl. 12 No. 2.

Awerbuch, S. (2006). Portfolio-Based Electricity Generation Planning: Policy Implications for Renewables and Energy Security, Mitigation and Adaptation Strategies for Global Change, Vl. 11, No. 3, May 2006, pp. 693–710(18).

Bar-Lev, D. and Katz, S. (1976). A portfolio approach to fossil fuel procurement in the electric utility industry. *Journal of Finance*, 31(3), 933–947.

Berger, M. (2003). *Portfolio Analysis of EU Electricity Generating Mixes and Its Implications for Renewables*. Ph.D. Dissertation, Technische Universität Wien, Vienna (March).

Brealey, R. A., Myers, S. C. and Allen, F. (2004). *Principles of Corporate Finance*.

Franklin Allen, Richard and Stewart, C. Myers (2004). Principles of Corporate Finance. Mcgraw-hill Series in Finance, 7th edition, 2004. ISBN: 9780073130828.

Fabozzi, F., Gupta, F. and Markowitz, H. (2002). The legacy of modern portfolio theory. *Journal of Investing, Institutional Investor* (Fall), 7–22.

Fraser, P. (2003). *Power Generation Investment in Electricity Markets*. Paris: OECD/IEA.

Humphreys, H. and McLain, K. T. (1998). Reducing the impacts of energy price volatility through dynamic portfolio selection. *Energy Journal*, 19(3), 107–134.

International Energy Agency/Nuclear Energy Agency (2005). *Projected Costs of Generating Electricity, 2005 Update*. Paris: OECD.

Roques, F. (2006). Power Generation Investments in Liberalised Markets: Methodologies to Capture Risk, Flexibility, and Portfolio Diversity, special issue of *Economies et Sociétés*, "Risks and Uncertainties in the Energy Industry", edited by J.-M. Chevalier, Nb. 10 (Oct./Nov. 2006)

CHAPTER 4

Use of Real Options as a Policy-Analysis Tool

William Blyth*

4.1 Relationship between portfolio theory and real options theory

There are many different ways to assess the financial implications of risk. The suitability of different approaches depends on the type of decision being analyzed, who is making that decision, and the context in which the decision is being made. This book focuses mainly on portfolio theory, but also discusses other approaches, including the real options analysis presented in this chapter. Before launching into an exposition of real options theory, it may be helpful to take a step back and look at the interrelationship between these different approaches.

In both portfolio theory and real options theory, risk is characterized as a variance in returns from a particular asset. Portfolio theory is concerned with minimizing risk for a given return (or maximizing return for a given risk) through combining assets with different risk characteristics into a diversified portfolio, whereas real options theory is concerned with optimizing investments in the face of uncertain future states of the world taking account of managerial flexibility in decision making. Both theories are partial equilibrium models, and their credibility depends on their relationship with a broader general equilibrium model for the pricing of risk.

This general equilibrium model is provided by the Capital Asset Pricing Model (CAPM), which provides an overarching theory of the pricing of risk in a perfect market under equilibrium. The key principle of portfolio theory is that when considering the acquisition of a new asset, the relevant risk to be considered is not the absolute variance of that asset's returns, but its contribution

---

*Associate Fellow, Chatham House London, UK

Analytical Methods for Energy Diversity and Security © 2008 Elsevier Ltd.
978-0-08-056887-4 All rights reserved.

to the portfolio's total variance. This contribution is not simply additive, but depends on the correlation between the variances of the asset and the portfolio. The CAPM utilizes the same principle as portfolio theory, but extends it to cover the whole asset class r_m, usually interpreted to mean the entire market for equity stocks.

Under the CAPM, the expected returns $E(r_j)$ on asset j are related to the risk-free rate of return r, the expected returns on the overall market portfolio $E(r_m)$, and β_j, the asset's volatility relative to the market:

$$E\left(r_j\right) = r + \beta_j \left[E\left(r_m\right) - r\right]$$

where:

$$\beta_j \equiv \frac{\mathrm{cov}\left(r_j, r_m\right)}{\mathrm{Var}\left(r_m\right)}$$

An asset's risk may be divided into two distinct components: that which can be attributed to variations in return in the aggregate market portfolio (*non-diversifiable risks*), and variations that are firm specific, driven by factors other than the market (*diversifiable risks*). The important insight behind the CAPM is that investors will only require a risk premium to cover the additional marginal contribution of the asset to the overall market portfolio risk. In other words, it is only the *non-diversifiable* risks, as measured by β, that matter to an investor, since the CAPM assumes that the investor is able to eliminate all *diversifiable* risks through spreading their investments sufficiently widely. It can be shown that this valuation of risk applies in principle at the project level as well as at the company level. In principle, companies should consider a project's β (relative to the entire market portfolio) when deciding the expected returns they would require from that project.

Portfolio theory and real options theory both take the CAPM as their point of departure, but they depart from different points, based on deviations from different underlying assumptions made in the CAPM.

As pointed out above, portfolio theory uses essentially the same insight as the CAPM, namely that an investor should be concerned with the contribution of an asset's risk–return profile to the total portfolio risk–return. But portfolio theory applies this insight to a particular market sector (in this case the electricity sector) rather than to the market as a whole. A rationale for doing so is that there are stakeholders other than equity investors who are affected by electricity price risk. If markets were complete, this would not matter, as individual preferences and strategic concerns such as security of supply would be priced into the market. But arguably, markets are not complete, and equity markets do not necessarily bear the full costs of price volatility. The use of portfolio theory for a given sector therefore departs from the general equilibrium model over the assumption of perfect markets, making the case that the risk–return relationship should be optimized at the sector level because of incomplete representation of the interests of all stakeholders in the CAPM model.

Justification of the use of portfolio theory is further elaborated in the other chapters of this book. However, this initial perspective indicates that portfolio theory is designed to identify those portfolios of generation mix at the sector level that are in some sense 'optimal' (i.e. that satisfy the condition of the efficient frontier, namely maximizing expected return given a particular risk exposure, or minimizing risk given a particular expected return). The key policy insight that this approach brings is to identify situations where the generation mix is suboptimal from a risk–return perspective, or to identify where incentives for new additions to the power generation portfolio do not improve the risk–return characteristics of the portfolio. These deviations from an 'optimal' generation mix may then be considered as an externality that is not being adequately addressed by the market design, pointing to areas for policy intervention.

Real options theory departs from the CAPM model over the assumption of market equilibrium. Under the CAPM, the present value of an investment opportunity can be calculated by deriving an appropriate risk-adjusted discount rate, using the project's β to calculate the necessary risk adjustment. In general, a project's β may vary over time giving a time-dependent discount rate to reflect expectations about when various sources of uncertainty will be resolved. But the calculation of present value can only be done if these variations in discount rate are deterministic (i.e. based on initial expectations of project risks). For projects where the risks are reasonably predictable, and where the market has experience of pricing risk for many of these projects types in that asset class, such assumptions are valid, and lead in the simple case of constant discount rates to the normal investment rule requiring a positive net present value (NPV). However, the use of deterministic discount rates may be a poor assumption if there is managerial flexibility to adapt investment behavior in response to actual outcomes of uncertain states of the world. Real options theory explicitly incorporates a valuation of such flexibility into the investment decision-making analysis. This gives a rather different set of policy insights. The issue here is concerned not so much with a 'top–down' definition of an optimal generation mix, but with a 'bottom–up' understanding of the incentives facing an individual investment decision in the face of uncertainty and flexibility. This will be of interest to policy makers who wish to understand how companies may respond to uncertain market conditions or uncertain regulatory drivers.

This chapter describes an application of real options theory to the case of investment in power generation plant when faced with uncertain climate change regulation. The power sector is one of the largest emitters of greenhouse gases in most economies, and is likely to be among the first targets of climate change policy. However, in most countries, climate policy is still in an emergent state, and uncertainty is high. It is important to understand the way in which this public policy uncertainty feeds through into private investment decision making because actual investment outcomes in terms of investment timing and/or choice of technology may differ from expectations if managerial flexibility is not taken into account. Analyzing the potential impacts of policy uncertainty on investment risk using the CAPM through postulating the effects on a firm's values of β could quickly become intractable. It is one of the strengths of the real options analysis approach that individual sources of risk can be treated explicitly using an assumed probability distribution for unknown future events, although the limitations of this approach also need to be recognized. One of the aims of this chapter is to make the case for why a real options approach is an appropriate tool for analyzing this problem. But as for any partial equilibrium model, it is always worth reflecting back to see how the results relate to a general equilibrium view of the world in order to understand what has been added by the analysis, and what may be missing. This is discussed in the final section of this chapter.

4.2 Electricity price risk

The act of investment involves exchanging a lump sum of money now in return for an income stream in the future. Investors will make this exchange if the expected project returns are high enough to cover the initial lump sum as well as compensating them for taking on the project risks.

Risks arise from many different sources, and affect the finances of the project principally through uncertainty in either the project's costs or its revenues. Cost uncertainties include capital cost uncertainty and operating cost uncertainty, including fuel costs, environmental costs, and so on. Revenue uncertainty stems primarily from uncertainty over the price of electricity, and uncertainty over the utilization levels of the new plant.

Centralized system planning models will tend to focus on only the cost element of these risks, as they try to optimize the electricity system with cost-minimization as their objective. Under public ownership, the objective of energy policy was to identify least-cost investment options, with much of the risk analysis focused on the costs and performance of different technologies and the potential cost of fuel.

In competitive electricity markets, revenue risk can be at least as important as cost risks. This can be illustrated by comparing levelized cost estimates with overall NPV estimates for investment in new gas, coal and nuclear plants based on assumptions in the UK's July 2006 Energy Review (DTI, 2006).[1] The levelized cost (£/kWh) is the average cost per unit of electricity generated over the lifetime of the plant.

Figure 4.1 reproduces the levelized cost figures from the Energy Review for gas (combined cycle gas turbine), coal (pulverized fuel coal plus flue gas desulfurization) and nuclear (pressurized water reactor). The low and high cases for coal and nuclear refer to more favorable and less favorable technology assumptions (regarding capital costs, operating costs, efficiency, etc.) as reported in the Energy Review. The ranges for gas and coal relate to the maximum and

---

[1] The following analysis draws on work presented in UKERC (2007).

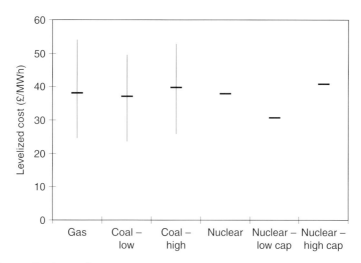

FIGURE 4.1 Levelized costs from UK Energy Review.

minimum levelized costs for the different fuel price and carbon price scenarios used in the Energy Review. The fuel price scenarios include two central scenarios (one favorable to coal, one favorable to gas), plus a high fuel price and a low fuel price scenario. There are four CO_2 price scenarios: £0/tCO_2, £10/tCO_2, £17/tCO_2 and £25/tCO_2

Figure 4.2 takes the same technology data, and constructs a simple cash-flow analysis assuming that electricity prices are determined in a competitive market where the short-run marginal cost (SRMC) of generation of the last plant dispatched in the system sets the electricity price. The last dispatched plant is assumed to be either coal or gas depending on the relative price of gas and coal, and also depending on the price of CO_2 allowances. Again, these prices are simply taken from the same Energy Review scenarios as used before for the cost calculations. The assumption is made here that the full price of CO_2 allowances is passed through to the price of electricity at a rate determined by the emission rate of the marginal plant. The efficiency of the marginal gas plant was taken to be 40%, and the efficiency of the marginal coal plant was taken to be 30%. Standard emission factors for each type of fuel were applied to calculate the rate at which a given CO_2 price would be passed through to the price of 1 kWh of electricity.

The contrast with the levelized cost representation in Figure 4.1 is immediately apparent. The range of financial outcomes is much greater in the case of nuclear than for the other technologies, even if capital cost uncertainty is not taken into account, whereas gas plant has the smallest range of financial outcomes.

This reflects the fact that under the assumptions made about technology performance and the Energy Review's price scenarios, gas plant is mostly on the margin of the merit-order. Coal plant becomes the marginal plant in the case of six out of the 16 scenarios (those which combine higher CO_2 price with lower gas price), whereas for the other scenarios, gas is assumed to be on the margin.

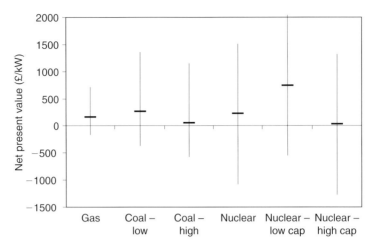

FIGURE 4.2   Net present value representation.

This means that gas price variations and risk are to a large extent incorporated into the electricity price. This provides a natural hedge for gas plants, since the uncertain gas price appears in both the cost and revenue sides of the financial equation, leaving the gross margin (difference between revenue and cost) relatively stable to gas price changes. Likewise, sensitivity of gas plant to CO_2 price risk is somewhat dampened when gas plant is predominantly on the margin. For coal plant, and especially for nuclear plant, however, the situation is the reverse. The pass-through of these price uncertainties into the electricity price leaves the revenue of the plant quite exposed to fuel price and CO_2 price risk, and leads to a substantially greater spread in possible financial outcomes depending on the scenario.

It is these spreads, or variance, in NPV that usually form the basis for analysis of project risk. Although the NPV spreads give some indication of the range of financial outcomes, they do not in themselves provide a quantification of the risk in a way which could inform decision making. There are a number of different ways that such risk analysis can be done, and the following sections explores one of these approaches (real options) in some detail.

4.3 Evaluating risk using real options

Real options theory is an extension of standard financial appraisal methods, adding the ability to model explicitly the effect of individual sources of uncertainty, and accounting for the flexibility that managers often have over the timing of their investment when faced with uncertain future cash flows. Originally developed for valuing financial options in the 1970s (Black and Scholes, 1973; Merton, 1973), economists soon realized that option pricing also provided considerable insight into decision making concerning capital investment. Hence the term 'real' options. Early frameworks were developed by McDonald and Siegel (1986) and

Pindyck (1988, 1991, 1993), and comprehensive textbook treatments are given by Dixit and Pindyck (1994) and Trigeorgis (1996).

Investment in the electricity sector has been analyzed within real options frameworks. Work was carried out by the Electric Power Research Institute (EPRI, 1999) to provide a framework for managing the effects of regulatory uncertainty, and Ishii and Yan (2004) looked at the overall effects of regulatory uncertainty on investment rates in the USA. In the area of short-term planning, real options were used by, among others, Tseng and Barz (2002) and Hlouskova et al. (2005). At the same time, several long-term planning frameworks have emerged. Recent examples include Fleten et al. (2007), who found that investment in power plants requires greater returns than the traditional NPV break-even point when a real options approach with stochastic prices is used.

Real options have also been quite widely used to model the effects of uncertain climate change policy. Rothwell (2006) found that returns on investment in nuclear plant need to be higher in a scenario with uncertain carbon prices than in a world with certain prices. Laurikka (2004), Laurikka and Klojonen (2006) and Kiriyama and Suzuki (2004) deal with the influence of future uncertain emissions trading and with CO_2 penalties within a real options setup. In these models the design of emissions trading schemes and the number of allowances that are freely distributed are main features of the overall model. Another example of the application of real options to the problem of uncertain climate policy is Reedman et al. (2006), who show that uptake of various electricity generation technologies varies significantly depending on the investor's view of carbon price uncertainty.

4.3.1 Toward an intuitive understanding of real options

For a company faced with uncertain future costs or revenues, there may be financial benefit to reducing the range of these uncertainties. If new information can be acquired before investment, then there may be an opportunity to avoid the worst financial outcomes, e.g. by investing in a different type of technology, or avoiding investment altogether. Since the expected value of the NPV is a probability-weighted mean, the ability to avoid some of the worst outcomes leads to an increase in the expected value of the project. It would therefore be rational for a company to pay some money to acquire this information in exchange for an improved (expected) financial performance of the project. Management flexibility on the timing of investment with respect to uncertain future events is an important way in which companies may acquire such information – in other words, waiting until the uncertainty is resolved.

It follows that a company which already possesses this flexibility on timing of investment would need to be financially compensated in order to give up this flexibility and go ahead with an investment immediately rather than waiting for new information (Blyth et al., 2007). Figure 4.3 follows a recent International Energy Agency (IEA) publication on the effects of policy uncertainty (IEA, 2007), and illustrates the economic rationale for waiting to gain information about an expected regulatory uncertainty at time T_p. This could be, for example, the introduction of a

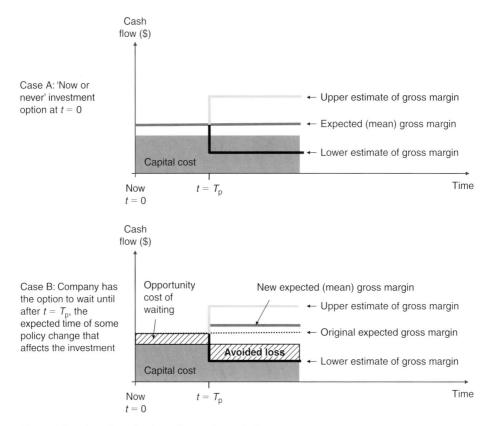

FIGURE 4.3   The value of waiting for regulatory information.

new policy, or a new phase of an existing policy that could affect the project's financial outcome either positively or negatively. For simplicity, the diagram assumes an equal probability of an increase or decrease in gross margin, so that the expected value is unaffected by the introduction of the new policy.

In case A, the company has to choose whether to invest immediately, or not invest at all. In this case, since the expected NPV is positive (gross margin is greater than capital costs), the company would choose to invest despite the future uncertainty, assuming the company is not risk averse.

In case B, the company has flexibility over the timing of its investment. In this case, there is a financial benefit to waiting until after T_p when information is available on how the new policy will affect the project. This gives the company the option to avoid investing in a loss-making project, which increases the expected gross margin of the project. The company will pay for this option by forgoing income from the project in the period up to T_p. The value of the option to wait therefore has to take into account the opportunity cost of waiting as well as the possible rise in expected project value.

The greater the range of uncertainty, and the less time available until T_p, the greater the option value of waiting will be. In order for the company to go ahead with the investment immediately rather than waiting, this option value would have to be recouped by the project, i.e. there would need to be an increase in the expected value of investing immediately to overcome the value of waiting. Another way to think about this is that if the expected gross margin in case A is sufficiently high compared to the capital costs, then the opportunity costs of waiting will outweigh the value of waiting. The option value of waiting therefore creates an additional financial threshold that the project must exceed in order to justify immediate investment.

4.3.2 Mathematical formulation

The approach taken here is an optimization of investment decision making under uncertainty using dynamic programming, as described for example in Dixit and Pindyck (1994). Dynamic programming compares the expected outcome of investing in a project in the current year with an alternative ('continuation value') which delays investment until the timing is optimal. The calculation of the continuation value requires solving the problem from the final year of the scenario, working backward to the first year in order to deduce the optimal investment rule over the whole possible investment horizon.

This can be described mathematically as follows. Consider a project with lifetime L which can be irreversibly initiated in any year t ($0 < t < T$) for a total capital outlay of K. The cash-flow in year t without the investment is A_t, and the annual cash-flow with the investment is B_t. Since these values are uncertain, the project value will depend on the expectation $E[.]$ of these values. The total NPV V_t^{inv} of the project if investment goes ahead in year t is:

$$V_t^{inv} = \left(\sum_{n-t}^{L} d(t,n) E[B_n] \right) - K \tag{4.1}$$

where $d(t, n)$ denotes the discount factor applied at time t to cash-flows occurring at time n. The continuation value which is the NPV of the project if one chooses not to invest in the project at period t (but assuming optimal investment opportunity depending on future conditions) is given by:

$$V_t^{cont} = A_t + d(t, t+1) E\left[V_{(t+1)}^*\right] \tag{4.2}$$

where V^* is the optimal NPV of the project cash-flows from year $t + 1$ until the end of the project lifetime under the assumption of optimal investment behavior. The assumption of optimal investment behavior in future years requires the comparison of V^{inv} with V^{cont} in every future year. Since the continuation value always depends on the total expected value of the project in the following year, this procedure needs to be solved from the end of the project, working backward. The final possible year for investing in the project is year T, at which it is assumed that the decision is a 'now or never' investment choice. In year T, V^{inv} becomes the expected value of the project over its lifetime, and V^{cont} equals zero (since there is no further opportunity to invest beyond this date). Investment will therefore go ahead in year T if $V_T^{inv} > 0$.

From the perspective of year $T - 1$, the decision in year T will depend on the random changes in variables in the intervening year. Therefore the continuation value will be based on expectations in year $T - 1$, denoted $E_T^{-1}[.]$. In year $T - 1$, the continuation value becomes the current year's income A_T^{-1} plus the discounted value of the expected project value given the expected outcome of the decision in year T, and the current state of information in year $T - 1$.

$$V_{T-1}^{cont} = A_{T-1} + d(T - 1, T) \max\left\{ E_{T-1}\left[V_T^{inv}\right] 0 \right\} \tag{4.3}$$

Once the continuation value in year $T-1$ has been calculated, this provides a minimum value which V_T^{-1inv} must exceed in order for investment to proceed in that year. This provides an optimal investment rule for year $T - 1$ given expectations about how prices will evolve in the intervening period. Once the optimal investment behavior for period $T - 1$ has been calculated, the same procedure can be used to derive the optimal investment rule for period $T - 2$, $T - 3$ and so on. Working backward, an optimal investment rule can be derived for each year in the period $0 < t < T$.

As discussed in the next section, the stochastic price processes are chosen in such a way that the expected future values for the variables can be derived from their current values. This means that the expected total revenues over the lifetime of the project can also be derived from the current values of the stochastic variables. The optimal investment rule in a given year expresses a threshold level that the annual returns have to exceed (given the prices in that year and the implication of these prices for the expected future returns) in order for investment to proceed. In other words, the optimal investment rule sets out the minimum project returns required to justify immediate investment rather than waiting. As will be shown, these minimum project returns can significantly exceed the normal positive NPV rule that would be derived from the same cash-flow calculations under certainty. The degree to which the optimal investment rule exceeds the standard positive NPV rule is a measure of the option value of optimizing investment timing. It can also be interpreted as a measure of the risk premium that could be incorporated into investment decision making as a result of price uncertainty. This risk premium can further be expressed as \$/kW at an additional cost of construction for investors and/or as surcharges in cents/kWh for electricity end-users.

4.4 Case study: CO_2 and fuel price risks

This section describes the results of an application of dynamic programming to the question of power generation investment decisions in the face of uncertain fuel prices and uncertainty in climate change regulation, drawing from analysis in Yang et al. (2008). Uncertainty in climate change regulation is modelled as an expected event at some point in time $t = \tau$, which creates a 'shock' to the carbon price C:

$$C_t = \begin{cases} \alpha_c C_{t-1}dt + \sigma_c C_{t-1}dz + \eta_c C_{t-1}dy & \text{when } t = \tau \\ \alpha_c C_{t-1}dt + \sigma_c C_{t-1}dz & \text{when } t \neq \tau \end{cases}$$

where α_c is the expected growth rate of the carbon price for the period dt, σ_c is the standard deviation of a geometric Brownian motion random walk process dz, and η_c is a parameter determining the size of the shock to carbon prices associated with a jump process dy. In the results shown here, the expected (mean) carbon price is $25/tCO_2$ with a zero growth rate ($\alpha_c = 0$); σ_c is taken to be ±7.75% per year, leading to a standard deviation over 15 years of ±30% as a result of general price uncertainties such as technology costs and uncertain balance of supply and demand. In addition, the jump process introduces a step change in carbon price at time τ, assumed to be in the range ±100%, i.e. anywhere between a doubling or a drop to zero in the carbon price. In the results shown here, this carbon price jump is expected to occur in year 11. The model can choose to build at any time between year 1 and year 20, and calculates a risk premium required to go ahead with the investment decision in that year rather than waiting until resolution of the regulatory uncertainty in year 11.

Fuel prices are modelled as a geometric Brownian motion process:

$$dF = \alpha_f F dt + \sigma_f F dx$$

where α_f and σ_f are the growth rate and standard deviation of fuel price, respectively, and dx is a random walk process that is separate from but moderately correlated with the dy process for carbon prices with a correlation factor of around 50%. Gas prices are modelled with an annual standard deviation σ of ±7.75%, and coal prices with a ±1.8% standard deviation. This gives a standard deviation from the expected mean after 15 years of ±30% for gas prices, and ±7% for coal prices after 15 years, approximately in line with the IEA's high and low price scenarios (IEA, 2004). The expected (mean) price levels are $5.2/GJ (55 US cents per therm) for gas and $1.9/GJ for coal price throughout the modelling period.

Electricity prices are assumed to be formed in a competitive market, following the SRMC of generation of the marginal plant in the system. Different assumptions can be made in the model about which type of plant is at the margin: it can be assumed that either coal or gas is *always* on the margin, or the model can choose between coal and gas plant depending on which has the higher SRMC at any given point in time. The results for the mixed coal/gas case tend to be partway between the two extreme cases of either *always coal* or *always gas*, so it is instructive to look at these two extremes to see what drives the investment risks for different technologies under different market conditions.

Figures 4.4 and 4.5 show the risk premium required to overcome the value of waiting until after regulatory uncertainty is resolved in year 11, combined with the risk premium associated with the annual random walks in fuel and carbon

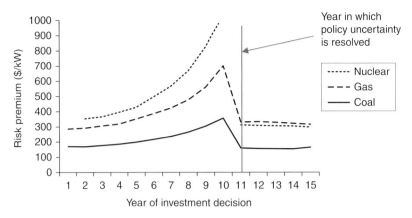

FIGURE 4.4   Risk premiums when coal is assumed to be the electricity price maker.

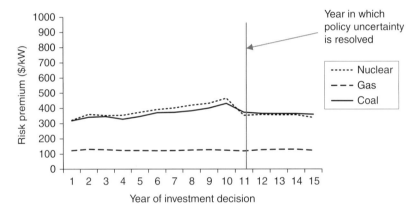

FIGURE 4.5   Risk premiums when gas is assumed to be the electricity price maker.

prices. The results show that as the regulatory event is approached, the risk pre-mium rises, i.e. the risk premium for making an investment decision in year 10 (only one year away from the regulatory event) is greater than the risk premium for making an investment decision in year 1 (ten years away from the regula-tory event). This concurs with the intuition that companies are unlikely to make a major investment decision a short time before a regulatory announcement that significantly affects their business case. The details of the results in Figures 4 and 5 give an indication of which technologies are more sensitive to these risks.

Figure 4.4 shows the results when it is assumed that coal plant are always on the margin of the electricity system (i.e. representing a coal-dominated electricity market). Since coal prices are assumed to be relatively stable, there is relatively little fuel price risk feeding through to the electricity price. However, coal plants are the most carbon-intensive plans, so the feed-through of carbon price varia-tions to the electricity price is strong. When considering investment in new coal plant (solid line in Fig. 4.4), this carbon price risk is relatively weak, since cost fluctuations associated with carbon price uncertainty will be positively correlated with revenue fluctuations, i.e. although under a high carbon price scenario costs go up, revenues also go up to compensate because carbon prices feed through to the electricity price.

For gas and nuclear plants, this strong variability in the electricity price in relation to carbon price uncertainty represents a strong risk factor, since cost and revenue fluctuations are no longer correlated, and it can be seen that the risk premiums for these two technologies are higher than the risk premium for coal. It should be noted that in this example, the risk premium for a new gas plant also contains a significant element of fuel price risk, since gas price fluctuations would affect operating costs without the possibility of recouping these variations through the electricity price.

Figure 4.5 shows the alternative case where gas is assumed to always be on the margin (i.e. in a gas-dominated electricity market). In this situation, gas price uncertainty is a key risk factor feeding through to the electricity price, whereas carbon price uncertainty is a weaker risk because of the lower emission factor associated with gas generation compared to coal generation. In this case, the risks for investment in new gas plant are relatively low because although gas price uncertainty affects generation costs, it also affects revenues to a similar extent.

The gas price uncertainty does, however, create a significant risk premium for investment in coal and nuclear plant. In this example, fuel price risk dominates carbon price uncertainty, as can be seen from the relatively minor increase in risk premium as the regulatory event in year 11 is approached.

The key message from these results is that carbon price uncertainty can be important under some market conditions, particularly if coal plays the role of electricity price maker. The results indicate that in this case, if the regulatory event is sufficiently far in the future (e.g. greater than ten years), then the risk premiums will be significantly reduced, as the value of waiting for such a long time becomes eroded.

In situations where gas plant acts as price maker, gas prices may be a much more significant risk factor than carbon price uncertainty, which helps to put the problem in some kind of context. In any case it is interesting to note that the results reinforce the results seen in Section 4.2, namely that the risk premiums for nuclear plant are relatively high compared to the other technologies, showing sensitivity to both carbon price risk and gas price risk, even though the *costs* of generation from nuclear are unaffected by gas and carbon prices.

4.5 Conclusions on the benefits and limitations of real options

The case study in Section 4.4 illustrates the benefits of the real options approach as a tool for policy analysis. It can be used to calculate investment risks posed by the introduction of hypothetical new policies in a way that would be difficult using general equilibrium models of risk such as the CAPM. Investment risks are important to understand when designing new policies, as they may affect investment timing and technology choice. In addition, an increase in the risk premium implies that profit expectations of investors will increase, which could raise electricity prices or carbon prices depending on what drives investment.

However, as discussed in the introduction, it is important to recognize the limitations of any partial equilibrium approach, and to think about what is missing from such analyses so as to inform future research developments. In the case of real options, such issues include the following:

- Climate policy risks are exogenous to the model, allowing any assumption to be made about the probability distribution of unknown future policy events. This is very flexible, but rather arbitrary. In the absence of 'real' data on this probability distribution, a relevant factor to consider is the attitudes and expectations of investors regarding this risk profile, since they will be the ones ultimately taking the decisions. Gaining better information about how companies regard these risks should therefore help to improve the analysis.
- It may also be important to build a more rational view of the policy-making process into such analysis. The case study above effectively assumes that there is no knowledge whatsoever about the direction of future carbon prices prior to the carbon price 'jump'. In reality, information will evolve in a rather smoother fashion. Finding ways for policy makers and companies to come to a shared understanding of the price-setting mechanism may be an important step. It could allow the design of policies that reduce risks for investors while allowing governments the flexibility to respond to their own learning in relation to technology costs, developments in international negotiations and the expected benefits of greenhouse gas mitigation.
- Understanding how the option to wait interacts with the potential for first-mover advantage in a competitive market is another issue that would be fruitful ground for further work. This requires development of agent-based decision-making analysis tools combined with real options analysis, a relatively new area of research.

More generally, the case for taking a real options approach may be supported if it can be shown that similar effects are likely to be replicated in a general equilibrium model. Real options analysis generally considers individual investment options in isolation, and does not take account of portfolio effects, nor does it make assumptions about changes to β that would be relevant in a CAPM framework. Nevertheless, it seems intuitively clear that the risk premiums calculated using the real options approach would be likely to survive a translation into an equilibrium view of the world. The climate change policy risks analyzed here affect all sectors of the economy, most notably the energy sector. Other economic shocks to the energy sector have long been suspected of contributing to economy-wide market effects, and it seems highly likely that climate policy 'shocks' would also be expected to shift the β value of projects in the energy sector. Exploring ways of combining optionality effects within an equilibrium analysis of electricity and broader financial markets could be a useful way of taking this type of intuition further.

References

Black, F. and Scholes, M. (1973). The pricing of options and corporate liabilities. *Journal of Political Economy*, 81, 637–659.

Blyth, W., Bradley, R., Bunn, D., Clarke, C., Wilson, T. and Yang, M. (2007). Investment risks under uncertainty. *Energy Policy*, 35, 5766–5773.

Dixit, A. K. and Pindyck, R. S. (1994). *Investment under Uncertainty*. Princeton University Press.

Department of Trade and Industry (2006). *The Energy Challenge: Energy Review*. Report No. Cm 6887 (July).

Electric Power Research Institute (1999). *A Framework for Hedging the Risk of Greenhouse Gas Regulations*. Palo Alto, CA: EPRI. TR-113642.

Fleten, S., Maribu, K. and Wangensteen, I. (2007). Optimal investment strategies in decentralized renewable power generation under uncertainty. *Energy*, 32, 803–15.

Hlouskova, J., Kossmeier, S., Obersteiner, M. and Schnabl, A. (2005). Real options and the value of generation capacity in the German electricity market. *Review of Financial Economics*, 14(3–4), 297–310.

International Energy Agency (2004). *World Energy Outlook*. Paris: IEA.

International Energy Agency (2007). *Climate Policy Uncertainty and Investment Risk*. Paris: IEA.

Ishii, J. and Yan, J. (2004). *Investment under Regulatory Uncertainty: US Electricity Generation Investment Since 1996*. CSEM Working Paper 127 (March). Center for the Study of Energy Markets, University of California Energy Institute.

Kiriyama, E. and Suzuki, A. (2004). Use of real options in nuclear power plant valuation in the presence of uncertainty with CO_2 emission credit. *Journal of Nuclear Science and Technology*, 41(7), 756–764.

Laurikka, H. (2004). The impact of climate policy on heat and power capacity investment decisions. In Hansjürgens, B. (Ed.), *Proceedings of the Workshop 'Business and Emissions Trading'*, Wittenberg, Germany.

Laurikka, H. and Koljonen, T. (2006). Emissions trading and investment decisions in the power sector – a case study of Finland. *Energy Policy*, 34, 1063–1074.

McDonald, R. and Siegel, D. (1986). The value of waiting to invest. *Quarterly Journal of Economics*, 101, 707–723.

Merton, R. (1973). The theory of rational option pricing. *Journal of Economic Management Science*, 4, 141–183.

Pindyck, R. (1988). Irreversible investment, capacity choice, and the value of the firm. *American Economic Review*, 79, 969–985.

Pindyck, R. (1991). Irreversibility, uncertainty and investment. *Journal of Economic Literature*, 29, 1110–1152.

Pindyck, R. (1993). Investments of uncertain cost. *Journal of Financial Economics*, 34, 53–76.

Reedman, L., Graham, P. and Coombes, P. (2006). Using a real options approach to model technology adoption under carbon price uncertainty: an application to the Australian electricity generation sector. *The Economic Record*, 82 (Special Issue), 64–73.

Rothwell, G. (2006). A real options approach to evaluating new nuclear power plants. *The Energy Journal*, 27(1), 37.

Trigeorgis, L. (1996). *Real Options: Managerial Flexibility and Strategy in Resource Allocation*. Cambridge, MA: MIT Press.

Tseng, C. and Barz, G. (2002). Short-term generation asset valuation: a real options approach. *Operations Research*, 50(2), 297–310.

UKERC (2007). Investment in electricity generation: the role of costs, incentives and risks. Technology and Policy Assessment Report (May).

Yang, M., Blyth, W., Bradley, R., Bunn, D., Clarke, C. and Wilson, T. (2008). Evaluating the power investment option with uncertainty in climate policy. *Energy Economics Journal*, 30(4), 1933–1950.

PART II

Applying Portfolio Theory to Identify Optimal Power Generation Portfolios

CHAPTER 5

Efficient Electricity Generating Portfolios for Europe

Maximizing Energy Security and Climate Change Mitigation[*]

Shimon Awerbuch[†*] and **Spencer Yang**[**]

Abstract

This chapter applies portfolio theory optimization concepts from the field of finance to produce an expository evaluation of the 2020 projected European Union business-as-usual (EU-BAU) electricity generating mix. Optimal generating portfolios are located that reduce cost and market risk as well as carbon dioxide emissions relative to the BAU mix. Optimal generating portfolio mixes generally include greater shares of wind, nuclear and other non-fossil technologies that often cost more on a stand-alone engineering basis, but overall costs and risks are reduced because of the effect of portfolio diversification. They also enhance energy security. The benefit streams created by these optimal mixes warrant current investments in the order of €250–500 billion. The analysis further suggests that the optimal 2020 generating mix is constrained by shortages of wind, especially offshore, and possibly nuclear power, so that even small incremental additions of these two technologies will provide sizeable cost and risk reductions.

[†]Shimon Awerbuch (http://www.awerbuch.com) was Senior Fellow, Sussex Energy Group, SPRU, University of Sussex until his sudden death in a plane crash on 10 February 2007, only a few days after presenting this paper at the 2007 EIB Conference on Economics and Finance. He was a financial economist specializing in electric utilities, energy and technology, and had previously served as Senior Advisor for Energy Economics, Finance and Technology with the International Energy Agency in Paris. EIB staff who worked with Shimon will remember him as a charming, friendly and inspiring person, and as a dedicated economist enthusiastically presenting his ideas and convictions. We are very grateful to Dr Spencer Yang, Dr Awerbuch's co-author, for finishing their joint work.

[*]Sussex Energy Group, SPRU, University of Sussex, Brighton, UK
[**]Visiting Fellow, Sussex Energy Group, SPRU, University of Sussex, Brighton, UK; Manager, Energy Practice, Bates White, LLC, Washington DC, USA

5.1 Introduction

Traditional energy planning in Europe and the USA focuses on finding the least cost generating alternative. This approach worked sufficiently well in a techno-logical era, marked by relative cost certainty, low rates of technological progress, and technologically homogeneous generating alternatives and stable energy prices (Awerbuch, 1993, 1995). However, today's electricity planner faces a broadly diverse range of resource options and a dynamic, complex and uncertain future. Given the uncertain environment, it makes sense to shift electricity plan-ning from its current emphasis on evaluating alternative technologies to evalu-ating alternative electricity generating portfolios and strategies. The techniques for doing this are rooted in modern finance theory, in particular mean-variance portfolio theory. Portfolio analysis is widely used by financial investors to create low-risk, high-return portfolios under various economic conditions.

This also has important implications for security of energy supply. Although energy security considerations are generally focused on the threat of abrupt sup-ply disruptions (e.g. European Commission, 2001), a case can also be made for the inclusion of a second aspect: the risk of unexpected electricity cost increases. This is a subtler, but equally crucial, aspect of energy security. Energy security is reduced when countries (and individual firms) hold inefficient portfolios that are needlessly exposed to the volatile fossil fuel cost risk.

The purpose of this chapter is to describe a portfolio optimization analysis that develops and evaluates optimal and efficient European Union (EU) electricity generating mixes for 2020, in an environment of uncertain carbon dioxide (CO_2) prices. These optimal portfolio mixes are designed to minimize expected generat-ing cost and risk, while simultaneously enhancing energy security, and they can be used as a benchmark for evaluating electricity generating strategies aimed at minimizing CO_2 emissions. A key finding of the analysis is that compared with the projected 2020 EU business-as-usual (BAU) electricity generating portfolio, there exist optimal and efficient portfolios that are less risky and less expensive, and that substantially reduce CO_2 emissions and energy import dependency.

In developing these results, this chapter proceeds as follows. Section 5.2 describes the data needed for applying a portfolio planning approach and how they have been compiled and estimated. Using these data, Section 5.3 identifies optimal EU elec-tricity generating portfolios for 2020 and presents key features of these portfolios. Section 5.4 probes more deeply into some of the findings, highlighting the role of nuclear energy, the scope for minimizing CO_2 emissions, the economic consequences of real-world technology constraints and the effects of carbon pricing. Section 5.5 summarizes, concludes, and stresses the potential and limitations of this analysis.

5.2 Data needed for computing optimal electricity generating portfolios

Applying portfolio optimization to the EU generating mix requires the follow-ing inputs: (1) capital, fuel, operating and CO_2 costs per unit of output (kWh) for each generating technology; (2) the risk or standard deviation of each cost component; and (3) the correlation factors between all cost components. The following subsections will address each input and how they are used to identify optimal portfolios. A more detailed presentation of the data and estimation can be found in Awerbuch and Yang (2007).

5.2.1 Technology generating cost

Figure 5.1 shows levelized 2020 generating cost for various technologies based on TECHPOLE performance and cost data.[1] Fossil fuel costs reflect the most recent projections of the European Commission (2006) and the International Energy Agency (IEA, 2006).

As for the cost of CO_2, a value of €35/t CO_2 has been used. This can be interpreted as an expected market price of CO_2, assuming that economic policies aimed at internalizing the economic cost of CO_2 emissions yield a market price of CO_2, for example under the European Union Emissions Trading Scheme. Alternatively, in the absence of such policies, the cost of CO_2 can be interpreted as the shadow price of CO_2, estimated on the basis of the economic cost of CO_2 emissions and of CO_2 abatement cost.[2] As for capital cost, this study assumes full

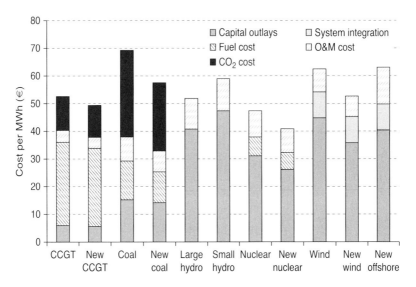

FIGURE 5.1 2020 generating costs (€/MWh) for various technologies. Economic costs of carbon dioxide (CO_2) assumed to be €35/t CO_2. Sources: based on TECHPOLE database, LEPII, University of Grenoble; European Commission (2006) and IEA (2006). O&M: operating and maintenance.

[1]TECHPOLE database, LEPII, University of Grenoble, CNRS. Fuel input costs, reflecting the most recent projections of the European Commission and IEA, are shown in the Appendix (Table 5.A1).
[2]For example, in its cost–benefit analyses of energy sector projects, the European Investment Bank currently uses a baseline shadow price that rises from €25/t CO_2 in 2007 to €45/t CO_2 in 2030.

capital cost recovery for new and already installed generating capacity. Although capital costs are sunk from an economic perspective, it is assumed that electricity producers set prices so as to recover their sunk costs. This assumption may not hold in day-to-day decision making, but over time, producers cannot remain viable unless they recover their capital costs. Thus, a full-cost recovery approach is implemented for both existing and new plant.

As Figure 5.1 shows, a system integration charge is added to wind generation to compensate for 'intermittency costs'. This adjustment is necessary because wind is a variable-output technology. System integration is a complex issue. Many think of wind as intermittent, although there are very few times when wind output is actually zero (Sinden, 2005; Grubb et al., 2006). The existing electricity network organization and protocols do require wind integration to have some extra level of backup capacity to balance the system when wind electricity output is reduced.[3] The costs have been quantified in multiple studies with similar results (e.g. Dale et al., 2004; DENA Grid Study, 2005; UKERC, 2006). The present analysis follows the results of the UKERC (2006) survey, which estimates the aggregate intermittency costs in the range of €7.5–12/MWh (£5–8/MWh) for 20% wind penetrations. Because intermittency cost estimates in Europe are somewhat lower (DENA Grid Study, 2005), for instance, estimated cost at or under €10/MWh, a system integration charge of €10/MWh is applied. This analysis, however, does not include possible associated systematic risks that may become more significant for wind penetrations in excess of 20–30%.

5.2.2 Technology risk estimates

Table 5.1 summarizes the technology risk estimates, expressed as the standard deviations of the holding-period returns based on historical data for each cost component.

Let us start with capital, or construction, cost risk. This varies by technology type and is generally related to the complexity and length of the construction period. A World Bank analysis covering a large number of projects estimates the standard deviation of construction period outlays for thermal plants (e.g. coal-fired power stations) at 23% and 38% for large hydro plants (Bacon et al., 1996). For the purpose of this analysis, the thermal plant value is applied to the construction period risk of nuclear plant (but alternative values will be considered in Section 5.4.1). To some extent, this is an arbitrary simplification. Many believe that these risks are significantly higher. Others, however, believe that such risks will resolve themselves with experience. The estimates for wind, gas, geothermal and solar risk were determined from developer interviews as reported in Awerbuch et al. (2005: Sandia Report). Construction cost risk of existing capacity was estimated

---

[3]This being said, new electricity network protocols and information systems have even been proposed to exploit wind variability and obviate the need for standby reserve capacity (e.g. Awerbuch, 2004; Fox and Flynn, 2005). These proposals generally involve matching variable output wind to interruptible load applications to prevent system balancing and/or backup generation.

Table 5.1 Holding-period returns standard deviations for generating technology cost streams (%)

	Construction	Fuel	O&M	CO_2
Coal	23.0	14.0	5.4	26.0
Oil	23.0	25.0	24.2	26.0
Gas – CC turbine	15.0	19.0	10.5	26.0
Nuclear	23.0	24.0	5.5	–
Hydro – large	38.0	0.0	15.3	–
Hydro – small	10.0	0.0	15.3	–
Wind	5.0	0.0	8.0	–
Wind – offshore	10.0	0.0	8.0	–
Biomass	20.0	18.0	10.8	–
PV	5.0	0.0	3.4	–
Geothermal	15.0	0.0	15.3	–

Source: authors' own calculations.
Holding-period returns (HPRs) measure the year-to-year fluctuation in the underlying cost stream; as a result, the standard deviation is expressed as a percentage, while the cost stream itself is measured in €/kWh; construction cost HPRs for existing capacities are not shown as they are estimated at about zero.
CO_2: carbon dioxide; O&M: operation and maintenance; CC: combined cycle; PV: photovoltaics.

at around 0%. This suggests that 'new' vintage assets are riskier than old ones; for example, risks for a new, not yet constructed coal plant are greater than those for an existing coal plant.

Fuel cost risks have been estimated on the basis of historical (1980–2005) European fossil fuel import prices taken from an IEA database. Annual price observations were used because they eliminate seasonal variations that could potentially bias the results. In practice, electricity producers buy fuel through spot and contract purchases so that the cost of fuel in any calendar period is best measured as the total fuel outlays divided by total fuel delivered. The holding-period return (HPR) standard deviations of fuel cost range from 0.14 for coal to 0.24 for oil. Obviously, renewable technologies and geothermal require no fuel outlays and there is thus no fuel cost risk.

The risks of operating and maintenance outlays are difficult to estimate. Typically, estimates can be found in corporate records, but often these records are not publicly available. Even if they were, maintenance policies may not keep the records in a format suitable for the analysis carried out here. In addition, companies design these records to promote overall corporate objectives, which can result in biased numbers. For example, during periods of poor financial performance, corporate managers may choose to defer maintenance in order to meet specified corporate objectives, such as reducing operation and maintenance

(O&M) expense. Thus, maintenance outlays might be arbitrarily recorded as capital improvements and be depreciated over time. In the case of rate-regulated utilities, there is a significant incentive to charge these outlays to capital improvements because they earn a regulated rate of return.

The US Energy Information Agency and the Federal Energy Regulatory Commission databases maintain records covering every generator operated by a regulated utility. These data were used to estimate the HPR standard deviations for O&M costs (along with the correlations between these costs discussed in the next subsection). By using these data, it is implicitly assumed that the maintenance volatility for a large portfolio of generating assets in the USA will not differ materially from those that would be found for a similar European portfolio. As Table 5.1 shows, different technologies show different year-to-year fluctuations in maintenance outlays, ranging from 3.4% for photovoltaics to 24.2% for oil.[4]

This takes us to the risk associated with last cost category, that is, the cost of CO_2 emissions, which is relevant for fossil fuel technologies. As Table 5.1 indicates, the HPR standard deviation for CO_2 has been estimated at 0.26. The approach underlying this estimate will be presented next in the context of discussing the correlation between the cost of different fuels, the correlation between O&M costs of different technologies, and the correlation between the cost of fossil fuels, on the one hand, and CO_2 cost on the other. A more comprehensive presentation of the technology cost and risk estimation can be found in Awerbuch and Yang (2007).

5.2.3 Correlation coefficients

This section starts with a brief description of the authors' approach to estimating the HPR standard deviation for CO_2 and the correlation between CO_2 cost and fuel prices. The estimates are derived using both analytical techniques and Monte Carlo simulation. The analytical approach to estimating CO_2 risk and correlation follows the spirit of Green (2006), who expresses CO_2 price in terms of gas and coal prices. This relationship is used to derive the HPR standard deviation of CO_2 as well as its correlation with fossil fuels. The Monte Carlo approach uses a series of simulations that provide a second set of CO_2 risk and fossil fuel correlation estimates. In the Monte Carlo analyses, the volatility and other trends from 18 months of actual historical data were used to simulate 20 years of trading. This, and its correlation to coal, gas and oil, provides an estimate of annual risk factors for CO_2.

The two methods provide a range of estimates of CO_2 risk and correlations. The analytical and Monte Carlo results were compared and various sensitivity analyses performed to test the reasonableness and robustness of these estimates. The HPR standard deviation for CO_2 used in the portfolio optimization model

---

[4] In principle, the O&M cost category should include outlays for property taxes, insurance and other non-maintenance categories. These would be likely to exhibit lower risk and potentially dampen the results in Table 5.1. Because the focus in this chapter is on CO_2 risk, this O&M issue is not pursued further.

Table 5.2   Fuel and carbon dioxide (CO_2) holding-period return correlation coefficients

	Coal	Oil	Gas	Uranium	CO_2	Biomass
Coal	1.00	0.27	0.47	0.12	−0.49	−0.38
Oil	0.27	1.00	0.49	0.08	0.19	−0.17
Gas	0.47	0.49	1.00	0.06	0.68	−0.44
Uranium	0.12	0.08	0.06	1.00	0.00	−0.22
CO_2	−0.49	0.19	0.68	0.00	1.00	0.00
Biomass	−0.38	−0.17	−0.44	−0.22	0.00	1.00

Source: authors' own calculations.

(0.26) is shown in the last column of Table 5.1.[5] The CO_2 cost/fuel cost correlation coefficient used in the portfolio optimization is shown in the penultimate column (or row) of Table 5.2.

As can be seen from these correlation coefficients, there is a negative correlation between CO_2 and coal prices and a positive correlation between CO_2 and gas. This is the expected result. Intuitively, as gas becomes more expensive, electricity generation shifts to coal, putting upward pressure on CO_2 prices, be they market determined or shadow prices. Conversely, rising coal prices shift generation to gas, which emits about half as much CO_2. As a result, the price of CO_2 falls with rising coal prices.

Table 5.2 also shows the correlation coefficients for the various fuels, indicating a positive correlation between fuels, with the notable exception of biomass. Although the data used for this analysis do not obtain a negative fuel correlation for nuclear, a number of researchers (Awerbuch and Berger, 2003; Roques, 2005) find a negative correlation between nuclear and fossil fuels, suggesting a greater diversification potential than that resulting from this analysis.

The estimated O&M correlation coefficients are shown in Table 5.A2 in the Appendix.

5.2.4 Total technology cost and risk

The previous subsections described the cost and risk inputs for the various generating technologies.

Figure 5.2 shows the costs per kWh for each of the generating technologies in 2020 along with its risk, with the added assumption that CO_2 costs €35/tonne. For comparison, Figure 5.2 also shows the cost–risk combination of the projected

---

[5]While these CO_2 risk estimates are statistically robust, it is important to note that they are based on just 18 months of CO_2 trading. Because the results of the CO_2 risk and correlation estimates were relatively consistent over various unrelated estimation procedures, the authors are relatively confident in applying them to the analysis.

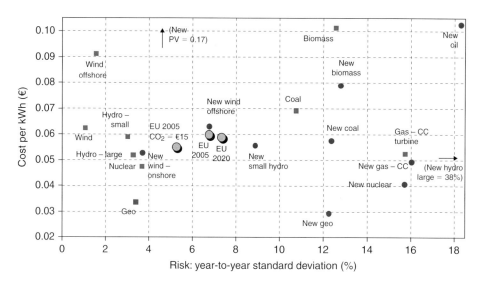

FIGURE 5.2   Cost and risk of existing and new EU generating alternatives in 2020. Estimates for individual technologies and the EU 2020 BAU mix are based on a carbon dioxide (CO_2) emission cost of €35/t CO_2. For comparison, the figure shows the actual EU 2005 generation mix for €35/t CO_2 and for €15/t CO_2. See text for details. PV: photovoltaics. Source: authors' own calculations.

2020 EU-BAU mix; in addition, it shows two variants of the EU 2005 mix: one assuming a CO_2 cost of €15/tonne and the other €35/tonne. The former reflects the approximate price of CO_2 in 2005 and the latter enables a direct comparison between the 2005 mix and the 2020 BAU mix. This comparison shows that relative to the 2005 mix, the 2020 BAU mix slightly reduces electricity generating cost from 5.98 to 5.87 €-cents/kWh. This cost reduction is attained by increasing expected risk from 6.8% to 7.3%. Compared with the 2005 EU mix, the 2020 BAU mix represents a cost–risk tradeoff that few investors would make: a cost reduction of less than 2% would come with an increase in risk of almost 9%.

The results also show that compared with existing vintages, new vintages exhibit lower cost and larger risk (in Figure 5.2, new vintages lie to the southeast of existing vintages). The cost decline is because new-vintage technologies increase energy efficiency and, thus, lower cost. For example, electricity produced by new coal plants cost 5.8 €-cents/kWh, which is 1.2 €-cents less than for existing coal plants. Risk for new vintages increases because the construction period risk of existing vintages is sunk or zero, while new generating assets yet to be constructed are exposed to construction period risk. The largest differences between the new and existing vintages show up in capital-intensive technologies such as nuclear, wind (especially offshore) and geothermal.

Not unexpectedly, the inclusion of CO_2 charges increases the generating cost of fossil alternatives relative to non-fossil technologies. CO_2 prices also increase the risk of the fossil alternatives to the extent that the HPR risk of the CO_2

Table 5.3 Effect of carbon dioxide (CO_2) costs on coal and gas generating cost-risk

| | CO_2 cost per tonne | | | | | |
| | €0.00 | | €15.00 | | €35.00 | |
	Cost (€/MWh)	Risk (%)	Cost (€/MWh)	Risk (%)	Cost (€/MWh)	Risk (%)
Coal	3.8	5.6	5.1	6.2	6.9	10.7
Coal – new	3.3	11.7	4.3	10.3	5.8	12.3
Gas – CC	4.0	14.3	4.6	14.9	5.3	15.7
Gas – CC New	3.8	14.7	4.3	15.2	4.9	16.0
Oil	8.2	20.2	9.2	18.8	10.6	17.8
Oil – new	8.0	20.8	8.7	19.3	10.2	18.3

Source: authors' own calculations.
CC: combined cycle.

exceeds the HPR risk of the fossil fuel. As shown in Table 5.1, the HPR standard deviation for CO_2 is 0.26 as compared to 0.14 for coal fuel and 0.19 for natural gas fuel. Observe that with €35/t CO_2, the standard deviation of existing coal technology rises from 5.6% to 10.7%, while the risk of existing gas generation increases much less, from 14.3% to 15.7% (Table 5.3). The increase for new coal is also smaller than for existing coal because the risk of new coal includes the construction period risks, reducing the fractional share of CO_2 outlays.[6]

In the case of oil-fired electricity generation, the HPR fuel price risk is 25% (slightly lower than for CO_2). Because of the low correlation between CO_2 and oil (0.19 as shown in Table 5.2), the inclusion of CO_2 charges reduces overall risk of this technology as the proportional weight of CO_2 outlays rises as a share of total costs.

The general outcome is that the 26% estimate for the CO_2 HPR risk and the estimated CO_2–fossil fuel correlations, along with the addition of CO_2 charges, do not significantly raise total HPR risks of new fossil fuel generating assets, and in some cases lowers them. This is contrary to widely held beliefs. Of course, higher CO_2 risk estimates (or higher correlation with fossil fuels) will affect even new assets to a greater extent.

---

[6] Note that the risk for new coal decreases slightly as CO_2 costs move from €0 to €15/t; this is no doubt caused by the negative correlation between CO_2 cost and coal prices. As CO_2 cost rise to €35, however, the magnitude of the price overwhelms the negative correlation, and overall risk again rises.

5.3 Portfolio optimization of EU electricity generating mix

5.3.1 Efficient multitechnology electricity portfolios: an illustration

As previously stated, the aim of this study is to evaluate whether there exist feasible 2020 generating mixes that are 'superior' to the 2020 EU-BAU mix by virtue of reducing risk or the cost of producing electricity. To prepare for the interpretation of the results of this portfolio optimization model, it is useful to offer a general illustration of possible results.

Figure 5.3 shows an infinite number of different generating mixes that could meet the 2020 electricity needs with a unique mix of the various technology options. The different portfolios all have different cost–risk, as represented by the small dots. Interestingly, technology shares do not change monotonically in any direction in Figure 5.3, so that two mixes with virtually identical cost–risk characteristics (i.e. two mixes located close to each other in cost–risk space) can have radically different technology generating shares. Indeed, Awerbuch and Berger (2003) show that costs and risks of the entire EU projected 2010 generating mix are virtually identical to a mix consisting of 100% coal. Likewise, radically different mixes can have nearly identical cost–risk characteristics; that is, they could be virtually co-located in the risk–cost space. The intuition for this is straightforward: there are many ways to combine ingredients to produce a given quantity of salad at a given price.

The curve (*PNSQ*) is the so-called efficient frontier (EF), the locus of all optimal mixes. There are no feasible mixes below the efficient frontier, and along it, only accepting greater risk can reduce cost. The small-dot mixes in Figure 5.3 are suboptimal or inefficient because it is still possible to reduce both cost and risk by finding mixes on the efficient frontier by moving below or to the left. As shown below, the 2020 EU-BAU mix lies above the efficient frontier.

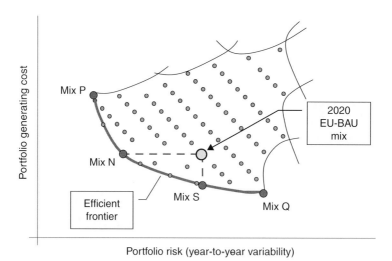

FIGURE 5.3   Feasible region and efficient frontier for multitechnology electricity portfolios.

Although an infinite number of possible generating portfolios lies on the efficient frontier, the focus here is on four typical optimal mixes, *P*, *N*, *S* and *Q*. Taking the 2020 EU-BAU mix as the benchmark, they are defined as follows:

- Mix *P* is a high-cost/low-risk portfolio. It is usually the most diverse (e.g. Stirling, 1996; Awerbuch et al., 2006).
- Mix *N* is an equal-cost/low-risk portfolio, i.e. it is the mix with the lowest risk for costs equal to that of the 2020 EU-BAU mix.
- Mix *S* is an equal-risk/low-cost portfolio, i.e. it is the mix with the lowest costs for a risk equal to that of the 2020 EU-BAU mix.
- Mix *Q* is a low-cost/high-risk portfolio. It is usually the least diverse portfolio.

The portfolio analysis does not advocate for any particular generating mixes, but rather displays the risk–cost tradeoffs across many mixes. Although it may turn out that solutions in the region of the 2020 EU-BAU mix, e.g. solutions between portfolios *N* and *S*, may be the most practical, it is not claimed that these optimization results help to set technology targets for 2020. Rather, the idea is to highlight and quantify the tradeoffs between generating mixes.

5.3.2 Efficient multitechnology electricity portfolios for 2020: results

The portfolio optimization evaluates the 2020 EU-BAU mix shown in Figure 5.4 against two cases: 'baseline' and 'realizable'. These cases differ in the extent to which future technology choices are constrained because of upper (and lower) bounds, representing either maximum attainable deployment levels for each technology or maximum resource limits, as in the case of renewables such as

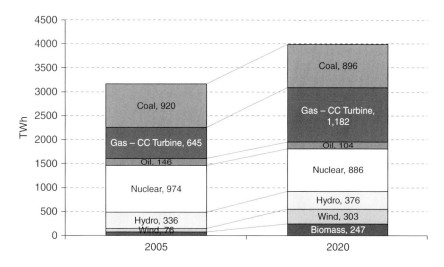

FIGURE 5.4   2005 and 2020 EU-BAU generation mix (in TWh). Source: European Commission (2005).

wind or hydro (see Awerbuch and Yang, 2007, for a more detailed discussion). The baseline case represents aggressive technology deployment levels that would probably be difficult to attain in practice. Its purpose is to help to explore practical policy limits and identify policies that may be worth pursuing. The realizable case, however, represents a set of upper technology limits that could be attained in practice given sufficiently focused policies and accelerated resource deployments. Table 5.4 shows baseline and realizable case lower and upper limits for the share of alternative technologies in the overall generation mix. For each set of constraints, efficient electricity generation mixes are computed and the level of CO_2 emissions associated with them is analyzed.

5.3.2.1 Efficient portfolios: baseline case
This section discusses the 2020 baseline optimization results and compares their risk–return characteristics and CO_2 emissions with those of the projected 2020 EU-BAU mix. The results indicate that the optimal baseline portfolios minimize cost and risk and reduce CO_2 emissions. This is shown in Figure 5.5, which illustrates the risk and return for the projected 2020 EU-BAU and for the typical optimized mixes under baseline assumptions. The efficient frontier $PNSQ$ shows the location of all optimal mixes.

The EU-BAU mix has an overall generating cost of 5.9 €-cents/kWh and a risk of 7.6%. By comparison, mix N, the equal-cost/low-risk mix, cuts risk nearly in half, to 3.4%. Alternatively, mix S, has the same risk as the BAU but reduces

Table 5.4   Lower and upper technology limits (% of electricity mix)

	Baseline case		Realizable case	
	Lower limit (%)	Upper limit (%)	Lower limit (%)	Upper limit (%)
Coal	3	52	5	35
Gas – CC old	5	16	10	16
Gas – CC new	0	50	0	20
Oil	2	8	2	5
Nuclear	15	52	15	33
Hydro	8	13	8	11
Biomass	2	22	2	13
PV	0	5	0	1
Geothermal	0	1/2	0	0
Wind – onshore	2	32	2	7
Wind – offshore	0	40	0	7

CC: combined cycle; PV: photovoltaics.

generating costs by 0.9 €-cents/kWh, which equates to an EU-wide cut in annual electricity outlays of €36 billion.[7]

Mix P, the minimum-risk mix, reduces risk slightly relative to mix N. But this seems to represent an unattractive cost–risk tradeoff over mix N. Similarly, mix Q, the minimum-cost mix, hardly reduces cost relative to mix S, but comes with a noticeable increase in risk. It thus seems that in cost–risk terms, the practical range of policy interest generally runs from mix N down to mix S.

Policy makers tend to view climate change mitigation as an objective that competes with cost and, indeed, it is widely believed that low-carbon electricity generation will increase cost. But such beliefs are based on stand-alone cost concepts. The baseline results show that in addition to reducing cost and/or risk relative to the EU-BAU mix, the optimal mixes also reduce CO_2 emissions, in contradiction to widely held beliefs that climate change mitigation policies inevitably increase cost.[8] This is illustrated in Figure 5.6, which shows technology shares and portfolio risk on the left vertical axis, CO_2 emissions on the right axis, and portfolio generating cost along the top of the graph. The low-risk mixes, P and N, reduce annual CO_2 to 199 million tonnes, which is 85% lower than emissions in the BAU mix (1273 million tonnes of CO_2). They accomplish this primarily by

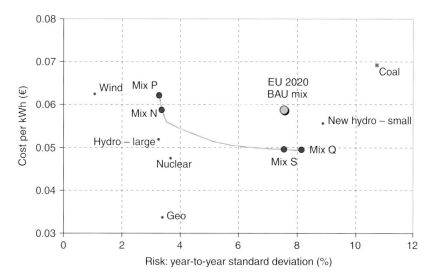

FIGURE 5.5 Efficient frontier for 2020 electricity generation mix: baseline case. Results for €35/t carbon dioxide (CO_2). Source: authors' own calculations.

[7]Based on an annual consumption in 2020 of 4006 TWh (€0.009/kWh $\times$ 4006 $\times$ 10^9 kWh = €36 billion).

[8]This is true only to the extent that the underlying generating costs shown in Figure 5.5 reflect all economic cost. However, since the costs shown in Figure 5.5 do not fully incorporate some economic costs such as investment grants that benefited some of these technologies (e.g. wind and nuclear), the resulting climate change mitigation may cost more than suggested by Figure 5.5.

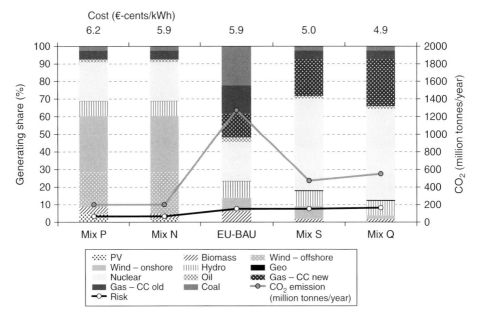

FIGURE 5.6 Technology shares, portfolio risk and cost and carbon dioxide (CO_2) emissions: baseline case. Results for €35/t CO_2. PV: photovoltaics; CC: combined cycle. Source: authors' own calculations.

substituting wind for gas and coal. Indeed, the share of onshore wind is 32%, its permissible upper limit (Table 5.5). Mixes S and Q, the low-cost mixes, reduce CO_2 emissions to 472 and 549 million tonnes, respectively, by incorporating larger shares of nuclear generation, which reaches its 52% upper limit in both mixes. This result – that is, that optimal low-risk mixes increase wind shares relative to the BAU while optimal low-cost mixes increase nuclear – tends to hold for the realizable case, too, as shown next.

5.3.2.2 Efficient portfolios: realizable case

Compared to the baseline case, the realizable case considers technology deployment levels that can be attained by 2020, assuming focused policy efforts. This case incorporates upper bounds for renewables based on the 'realizable' scenarios developed by Ragwitz et al. (2005), who estimate the realizable market potential for renewable energy technologies as 'the maximal achievable potential, assuming that all existing barriers can be overcome and all driving forces are active' (Ragwitz, M., personal communication, 2006). Compared to the baseline case, the realizable case has less latitude to search for optimal solutions because it is limited to a smaller feasible region. As a consequence, optimal realizable mixes are costlier and riskier, and they emit more CO_2 than optimal baseline mixes.

Figure 5.7 shows the cost and risk results for the realizable case (solid line). There are mixes on the efficient frontier that exhibit lower cost–risk than the projected EU-BAU mix. However, as the realizable case is more constrained, the

Table 5.5 Optimal portfolio shares and carbon dioxide (CO_2) emissions in 2020: baseline case

	EU-BAU	Mix P	Mix N	Mix S	Mix Q	Technology bounds	
						Lower (%)	Upper (%)
Share in electricity generating (%)							
Coal	22	3 [L]	3 [L]	3 [L]	3 [L]	3	52
Gas – CC old	16	5 [L]	5 [L]	5 [L]	5 [L]	5	16
Gas – CC new	13	0 [L]	0 [L]	19	27	0	50
Oil	3	2 [L]	2 [L]	2 [L]	2 [L]	2	8
Nuclear	22	22	22	52 [U]	52 [U]	15	52
Hydro	9	8 [L]	8 [L]	8 [L]	8 [L]	8	13
Biomass	6	4	3	2 [L]	2 [L]	2	22
PV	0	5 [U]	2	0 [L]	0 [L]	0	5
Geothermal	0	0	0	0	0	0	1/2
Wind – onshore	6	32 [U]	32 [U]	9	2 [L]	2	32
Wind – offshore	1	19	23	0 [L]	0 [L]	0	40
CO_2 emissions (million tonnes/year)							
	1273	199	199	472	549		

Source: authors' own calculations.
EU-BAU: European Union – business-as-usual; CC: combined cycle; PV: photovoltaics; [L, U]: technology share is at lower or upper bound, respectively; results for €35/t CO_2.

efficient frontier is shorter, riskier and more costly relative to the baseline. The tighter resource limits, particularly the penetration levels for onshore wind and nuclear, increase the cost of mixes S and Q and the risk of mixes P and N.

For example, the cost of mix S rises by 0.3 €-cents/kWh (6%) relative to the baseline case. This increase in cost equates to an increase in total annual outlays by EU electricity consumers of €12 billion. This figure represents about 0.1% of the current gross domestic product (GDP) of the EU. To illustrate the impact of tighter technology deployment limits on risk: with less wind resource available, the optimization cannot reach the low risk levels of the baseline. For example, in mix N, lower limits for wind (in particular) increase coal and nuclear shares, thereby raising risk by some 60%, from 3.3% in the baseline case to 5.4% in the realizable case.

Compared to the baseline case, the realizable case is characterized by significantly lower shares of nuclear, wind and, in some cases, new and existing gas-fired power plants (Table 5.6 and Figure 5.8). This is driven by the lower upper technology bounds for most technologies, as can be seen by comparing the right-hand column of Table 5.6 with that of Table 5.5. Further, as Table 5.6

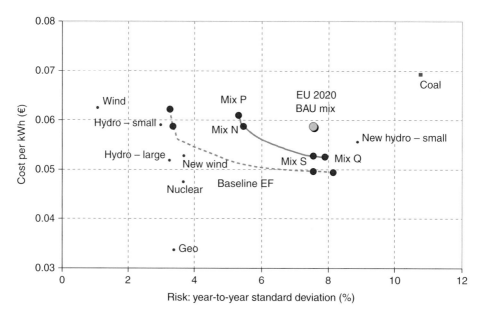

FIGURE 5.7 Efficient frontier (EF) for 2020 electricity generation mix: realizable case. Results for €35/t carbon dioxide (CO_2). Source: authors' own calculation.

indicates, wind hits its upper limit in all the optimal realizable mixes, while offshore wind hits the upper limit in the low-risk mixes P and N, where offshore wind is required to balance and complete the mix. Nuclear is at its upper limit in all except for mix P. The results of Table 5.6 suggest that additional deployment of these technologies could lower cost, risk and CO_2 emissions. As a comparison of the last row in Table 5.6 to the last row in Table 5.5 shows, the realizable case reduces annual CO_2 emissions at best by 548 million tonnes (mix S), while they may fall by as much as 1074 million tonnes under baseline assumptions (mixes P and N).

Table 5.7 summarizes the changes in technology generating shares and CO_2 emissions for the typical optimal mixes relative to the 2020 EU-BAU. The low-risk mixes P and N show large percentage increases for nuclear, biomass and wind, coupled with significant percentage reductions for gas, oil and coal (in mix N only). The low-cost mixes S and Q show large percentage rises for gas, nuclear and wind (in mix S), coupled with large reductions for coal, oil and biomass.

In practice, the move from the 2020 BAU mix to the realizable mix S is probably the most attractive of the realizable possibilities. If new policies were to redirect investment so that mix S is achieved, this would have the highly desirable effect of cutting annual electricity costs by €24 billion[9] and CO_2 emissions by 548 million tonnes without changing risk.

However, other moves involving alternative risk choices are possible. For example, to the left of the 2020 BAU mix in Figure 5.9 (i.e. EF for 2020, realizable case)

---

[9](5.9 − 5.3) €-cents/kWh × 4006 TWh.

Table 5.6 Optimal portfolio shares and carbon dioxide (CO_2) emissions in 2020: realizable case

	EU-BAU	Mix P	Mix N	Mix S	Mix Q	Technology bounds	
						Lower (%)	Upper (%)
Share in electricity generating (%)							
Coal	22	22	17	5 [L]	10	5	35
Gas – CC Old	16	10 [L]	11	15	16 [U]	10	16
Gas – CC New	13	0 [L]	0 [L]	20 [U]	20 [U]	0	20
Oil	3	2 [L]	2 [L]	2 [L]	2 [L]	2	5
Nuclear	22	29	33 [U]	33 [U]	33 [U]	15	33
Hydro	9	9	9	10	10	8	11
Biomass	6	13 [U]	13 [U]	2 [L]	2 [L]	2	13
PV	0	1 [U]	0 [L]	0 [L]	0 [L]	0	1
Geothermal	0	0 [U]	0 [U]	0 [U]	0 [U]	0	0
Wind – onshore	6	7 [U]	7 [U]	7 [U]	7 [U]	2	7
Wind – offshore	1	7 [U]	7 [U]	5	0 [L]	0	7
CO_2 emissions (million tonnes/year)							
	1273	981	825	725	836		

Source: authors' own calculations.
EU-BAU: European Union – business-as-usual; CC: combined cycle; PV: photovoltaics; [L, U]: technology share is at lower or upper bound, respectively; results for €35/t CO_2.

lies mix N. Compared to the BAU mix, mix N cuts the portfolio risk by about one-third while simultaneously reducing annual CO_2 emission by 448 million tonnes, or 35%. This move produces no cost reductions and while CO_2 reductions are not as big as when moving from the BAU to mix S, risk is significantly reduced. Obviously, comparing the risk–cost and CO_2 combinations of N against S requires knowledge of societal preference functions.

Over the long run, further technology deployment may make it possible to move closer to baseline mix S from the BAU mix (or the realizable mix S). The decline in CO_2 emissions would be 46% higher (801 versus 548 million tonnes per year), accompanied by a 33% greater cut in the EU's electricity bill (€36 compared to €24 billion).

5.3.3 A summary of key results

The results in this section highlight the importance of focused technology deployment policies designed to move the EU generating mix away from the BAU mix

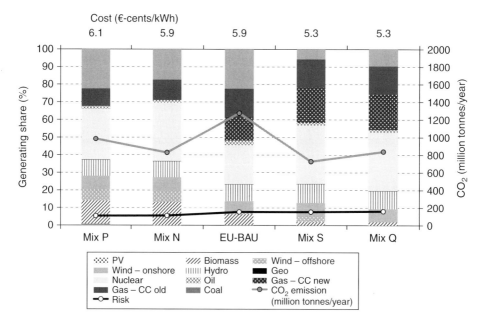

FIGURE 5.8   Technology shares, portfolio risk and cost and carbon dioxide (CO_2) emissions: realizable case. Results for €35/t CO_2. PV: photovoltaics; CC: combined cycle. Source: authors' own calculations.

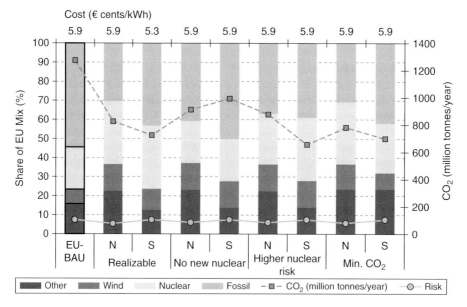

FIGURE 5.9   Technology shares, portfolio risk and cost and carbon dioxide (CO_2) emissions: sensitivity analyses. Results for €35/t CO_2. Source: authors' own calculations.

Table 5.7   2020 EU-BAU electricity generation mix versus optimal realizable mixes

	Mix P	Mix N	EU-BAU	Mix S	Mix Q
Portfolio risk	5.3%	5.5%	7.6%	7.6%	7.9%
Portfolio cost (€/MWh)	61	59	59	53	53
	% change from EU-BAU			% change from EU-BAU	
Annual CO_2	−22%	−35%	1273 m tonnes	−45%	−34%
Coal	0%	−22%	897 TWh	−78%	−57%
Gas – CC	−66%	−61%	1182 TWh	+19%	+22%
Oil	−31%	−42%	104 TWh	−42%	−42%
Nuclear	+29%	+50%	886 TWh	+50%	+50%
Hydro	−4%	−4%	376 TWh	+12%	+8%
Biomass	+115%	+115%	247 TWh	−70%	−70%
Wind	+85%	+85%	303 TWh	+67%	+0%
Other	265%	−7%	12 TWh	−7%	−7%
Total			4006 TWh		

Source: authors' own calculations.
Results for €35/tCO_2.
EU-BAU: European Union – business-as-usual; CC: combined cycle.

and closer to electricity generating portfolios such as the realizable mix S. This mix would reduce annual EU electricity cost by around €24 billion and annual CO_2 emissions by more than 500 million tonnes. Taking annual electricity cost saving as perpetual and assuming an interest rate of 5–10% would justify investment today to the tune of €240–480 billion.

A key finding is that the low-risk mixes (P and N) generally reduce fossil shares and increase wind and other non-fossil shares relative to the BAU mix, while the higher risk/lower cost mixes (S and Q) increase primarily nuclear along with gas, wind and hydro electricity at the expense of coal and oil. There thus seems to be a dichotomy between wind and nuclear, suggesting that this analysis reinforces rather than solves the wide-ranging debate between pro-nuclear and pro-wind forces. However, this debate incorporates numerous additional considerations that are not reflected in the optimization, including highly uncertain waste disposal management costs. The next section tries to shed more light on the role of nuclear power and other factors influencing the results of the portfolio analysis.

5.4 An eclectic view on factors influencing optimal electricity mixes

5.4.1 The role of nuclear power

The nuclear cost estimates used for identifying efficient electricity portfolios do not account for the costs and risks of storing nuclear waste, which are essentially incalculable. The Committee on Radioactive Waste Management (CORWM, 2006) recommends a lengthy, potentially decades-long process, involving interim waste storage in preparation for ultimate geological disposal. Although much of what is risky about nuclear seems to be a matter of expectations and is not necessarily always rational, countries may decide not to build new nuclear power stations, as is currently the case in Germany, for instance. Against this background, it is useful to test a policy of a nuclear moratorium, that is, no new nuclear, to see its effects on cost and risk of the EU portfolio mix. In principle, this can be done for the baseline case and the realizable case, but in what follows the focus is on the latter (for ease of comparison, it is called the 'benchmark' realizable case). In addition, this assessment concentrates on generating mixes N that minimize portfolio risk for the cost of the 2020 EU-BAU mix and on mixes S that minimize portfolio cost for the risk of the 2020 EU-BAU mix.

As Figure 5.9 shows, for mix S cost rises from 5.3 €-cents to 5.5 €-cents/kWh. For mix N, risk stays approximately unchanged. The big change is in terms of additional CO_2 emissions, where CO_2 emissions rise from 725 to 993 million tonnes (mix S) and from 825 to 912 million tonnes (mix N). This is because for these portfolios, a good part (in mix N virtually all) of the drop in the share of nuclear is compensated for by fossil fuel-fired electricity generation.

Another sensitivity test examined the impact of a change in risk of constructing and decommissioning nuclear power plants. To recall from Section 5.3, total generating costs of new nuclear power stations have been estimated at 4.1 €-cents/kWh, including decommissioning costs equivalent to 70% of the overnight plant construction cost of €1710/kW. This makes nuclear attractive relative to other alternatives. It can be argued, however, that nuclear risk is understated because construction period risk was arbitrarily set to the World Bank estimate for the construction period risk of coal at 23% (Bacon et al., 1996). To account for this, the scenarios were rerun several times, gradually increasing nuclear construction risk from 0.23 to 0.38. This raises total technology risk for nuclear from about 16% to 26%.

As can be seen from Figure 5.9, for the realizable case ('Higher nuclear risk'), a higher risk level for nuclear capital costs has a relatively small effect on the optimal cost–risk combination; that is, mix N comes with only a marginal increase in risk relative to the benchmark realizable case, while mix S is associated with only a small increase in portfolio generating costs (5.4 €-cents/kWh instead of 5.3 €-cents/kWh). As expected, both portfolios have a lower share of nuclear, but the change is small because of the already tight upper and lower bounds for most technologies. It is interesting to observe that for the low-risk mix N, the share of renewables is virtually constant, with an increase in fossils making up for the drop in nuclear. As a result, CO_2 emissions rise. As for the low-cost mix S, the decline in nuclear is associated with a decline in fossils and an increase in renewables, all in all resulting in lower CO_2 emissions. The main reason why renewables become more important in mix S, but not in N, is that in the benchmark realizable mix S, renewables, biomass in particular, are not close to their technology upper bounds, whereas they are in the benchmark realizable mix N.[10]

5.4.2 Efficient electricity portfolios that minimize CO_2 emissions

We now turn to something that is not so much a sensitivity analysis but, rather, a change in perspective: to identify the combinations of portfolio risk and portfolio generating cost (and the associated technology shares) that minimize CO_2 emissions. For the realizable case, the results are shown on the very right-hand side of Figure 5.9. Comparing them to the benchmark realizable case suggests only a moderate decline in CO_2 emissions: from 825 million tonnes/year to 782 million tonnes for mix N and from 725 million tonnes to 700 million tonnes for mix S. It is straightforward to illustrate that minimizing CO_2 emissions is most likely to be economically inefficient. As Figure 5.9 shows, for mix S, portfolio generating costs increase by 0.3 €-cents/kWh, implying an increase in annual electricity cost of €12 billion and, thus, carbon reduction cost of €480/tCO_2, a value way above current estimates of global warming damages.

Although not shown in Figure 5.9, results are very different when taking the baseline case rather than the realizable case as a benchmark. As shown in Awerbuch and Yang (2007), moving to the carbon-minimizing mix S would cut CO_2 emissions by 273 million tonnes, implying carbon reduction cost of €44/t CO_2. Awerbuch and Yang (2007) also show that the risk–cost characteristics of the baseline carbon-minimizing portfolios are very similar to – in fact, slightly better than – those of the realizable case shown in Figure 5.9. Although it is unlikely that baseline technology penetration levels could be attained by 2020, this illustrates the significant benefits that could be achieved over a longer period by pursuing deeper penetrations of these technologies.

5.4.3 The effect of upper limits on technology shares

Awerbuch and Yang (2007) investigate in a more rigorous way the economic cost of the constraints that prevent the share of wind, nuclear and gas to be larger than the upper limit of the realizable case. Linear programming techniques show that easing these constraints, and thus allowing technology shares to move toward the baseline case, has considerable economic value. More specifically, for the realizable mix S, increasing the upper limit for the share of nuclear energy by 1 percentage point would result in portfolio cost savings equivalent to 46% of the lifetime generating costs of additional nuclear power stations. The comparable results for wind and

---

[10] A word of caution at the end: the sensitivity of results to changes in underlying assumptions about nuclear energy do not and are not intended to resolve the nuclear–renewables debate. Rather, they are meant to quantify and highlight some of the important factors.

gas are 21% and 8%, respectively. The results for wind could significantly and positively impact the current debate regarding development of an EU offshore 'super-grid' to connect diverse offshore wind sites. They also impact on the nuclear debate in a similar fashion. All in all, they indicate that failure to exploit fully the EU energy resource potentials needlessly raises generating costs and CO_2 emissions.

5.4.4 The effect of pricing CO_2 emissions

So far, this analysis has assumed a charge of €35/tonne of CO_2 emitted, which was interpreted as either a market price or a shadow price for carbon emissions. This section will investigate the effect of pricing CO_2 emissions on the cost–risk characteristics of the 2020 EU-BAU mix and of efficient generating portfolios. In addition, the impact of carbon pricing on CO_2 emissions is discussed. To keep things simple, only the effect of moving from a carbon price of zero to one of €35/tCO_2 is considered, concentrating on the BAU mix and mixes N and S in the realizable case.[11]

As Figure 5.10 illustrates, portfolio risks and costs rise with rising CO_2 prices. This is true for the BAU mix and the efficient electricity generating portfolios. For instance, the cost of the BAU mix increases by 23% or 1.1 €-cent/kWh (from 4.8 €-cents to 5.9 €-cents/kWh). The risk of that mix, however, rises by a whopping 40% (from 5.4% to 7.6%), illustrating its considerable sensitivity to changing CO_2 (and fossil fuel) prices. By definition, the share of each technology in the BAU mix and, thus, CO_2 emissions do not change with a rise in CO_2 prices. Clearly, it makes little sense to keep technology shares constant when CO_2 prices rise.

On the contrary, with rising CO_2 prices it is optimal to reduce the share of fossil fuels in electricity generation, as indicated by the amount of CO_2 emissions, which is shown by parenthetical values next to the mixes in Figure 5.10. Since mixes P and N have lower shares of fossils than mixes S and Q, they have lower emissions at any given CO_2 price. In the absence of CO_2 charges, the realizable mix N emits 1358 million tonnes of CO_2 per year.[12] As the CO_2 price increases, optimal mixes are reshuffled to minimize portfolios costs and risks. For a carbon price of €35/tCO_2, emissions fall by almost 40% to 825 million tonnes per year.

Let us take a closer look at the effect of carbon pricing by considering mix S. In general, the change in portfolio costs and CO_2 emissions is the result of two interrelated changes: a rise in CO_2 charges and the reoptimization of portfolio mixes in response to this rise. Considered in isolation, the increase in the CO_2 price raises the cost of electricity from 4.4 €-cents/kWh (see Figure 5.10) by about 1.3 €-cents/kWh. This increase reflects the cost of carbon (€35/tCO_2

---

[11] Results for carbon prices between zero and €35/tCO_2 and for other efficient generating mixes (in both the realizable case and the baseline case) are discussed in Awerbuch and Yang (2007).

[12] It is worth pointing out that without carbon pricing, efficient portfolios that generate electricity at the same or lower cost than the BAU mix are more carbon intensive than the BAU mix (see the points that lie to the south-east of mix N on the '$CO_2 = €0$' efficient frontier in Figure 5.10).

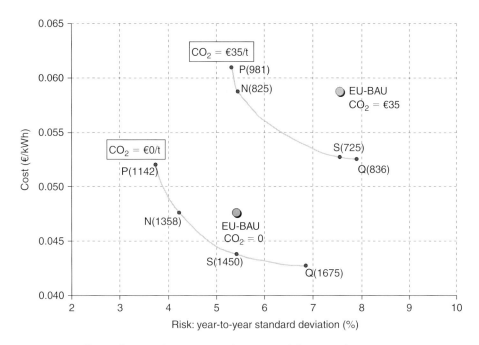

FIGURE 5.10   Efficient frontiers (€0/t CO_2 and €35/t CO_2) for 2020 electricity generation mix: realizable case. Values in parentheses next to the mixes show annual CO_2 emissions in million tonnes. The 2020 EU-BAU emits 1273 million tonnes per year. Source: authors' own calculation.

multiplied by 1450 million tonnes of CO_2) for total electricity production of around 4000 TWh. But as pictured in Figure 5.10, portfolio generating costs increase only by around by 0.9 €-cents/kWh to a total of 5.3 €-cents/kWh. The cost savings of around 0.4 €-cents/kWh are due to the portfolio reoptimization triggered by the pricing of carbon. But the associated decline in the share of fossil fuels in mix S not only offsets, in part, the increase in electricity costs resulting from the pricing of carbon; it also lowers CO_2 emissions from 1450 million tonnes to 725 million tonnes.

5.5 Summary and conclusions

This chapter has presented a mean-variance portfolio optimization analysis that develops and evaluates optimal (i.e. efficient) EU electricity generating mixes for 2020. The results suggest that greater shares of non-fossil technologies, primarily nuclear or wind, can help to reduce the cost and risk of the EU generating portfolio as well as its CO_2 emissions. To illustrate, an efficient generating mix that the authors consider to be achievable by 2020 is estimated to cut annual EU electricity generating cost by €24.0 billion and CO_2 emissions by 548 million tonnes. This mix thus produces perpetual annual benefits sufficient to justify current investments of up to €500 billion, which compares to an estimated EU investment of €900 billion in new electricity generation capacity needed by 2030. It is also shown that easing constraints on investment in nuclear and wind energy capacity would lower overall generating cost enough to offset 46% and 21% of the kWh costs of nuclear and wind generation. Against this background, policies designed to accelerate the deployment of key non-fossil technologies appear to be highly cost-effective.

Perhaps the single most important lesson of the portfolio optimization analysis is that adding a fuelless, fixed-cost technology (such as wind energy) to a risky generating mix lowers expected portfolio cost at any level of risk, even if the fuelless technology costs more when assessed on a stand-alone basis. This underscores the importance of policy-making approaches grounded in portfolio concepts as opposed to stand-alone engineering concepts.

This is a tall order, since quantitative indicators in energy markets are primarily focused on stand-alone performance. In contrast, financial markets provide a beta measure to help investors to think in terms of portfolio performance. The lack of a similar measure in energy markets prevents some from embracing the energy planning portfolio optimization approach.

Ironically, this issue is akin to the practical problems that initially confronted Markowitz's (1952) portfolio approach. The new technique required massive analytical efforts (without computers) to estimate the covariance of returns to each stock in the US market against every other stock. It was not until Sharpe and Lintner developed the Capital Asset Pricing Model (CAPM) that a single covariance with the market portfolio was shown to be sufficient (Varian, 1993). Perhaps with further research, it may be possible to develop energy analogs that will enable a beta type measure to index the risk of particular generating technologies against a large generating mix such as the EU mix. This would provide a simple and expedient method for evaluating the costs and risks of individual technologies and their CO_2 emissions.

Today's dynamic and uncertain energy environment requires portfolio-based planning procedures that reflect market risk and de-emphasize stand-alone generating costs. Portfolio theory is well tested and ideally suited to evaluating electricity expansion strategies.[13] It identifies solutions that enhance energy diversity and security and are therefore considerably more robust than arbitrarily mixing technology alternatives. Portfolio analysis reflects the cost–risk relationship (covariances) among generating alternatives. Although crucial for correctly estimating overall cost, electricity planning models universally ignore this fundamental statistical relationship and instead resort to sensitivity analysis and other ill-suited techniques to deal with risk. Sensitivity analysis cannot replicate the important cost interrelationships that dramatically affect estimated portfolio costs and risks (Awerbuch, 1993), and it is no substitute for portfolio-based approaches. The mean-variance portfolio framework offers solutions that enhance energy

---

[13]Other techniques have also been applied. For instance, Stirling (1994, 1996) develops maximum-diversity portfolios based on a considerably broader uncertainty spectrum. Although radically different in its approach, his diversity model yields qualitatively similar results.

diversity and security and are therefore considerably more robust than arbitrarily mixing technology alternatives.

This being said, we must be clear about the purpose and the limitations of the portfolio approach to electricity sector planning. The portfolio optimization presented in this chapter does not point to a specific capacity-expansion plan. Such outputs would require considerably more detailed models. The results presented here are largely expositional, but they demonstrate the value of portfolio optimization approaches and suggest that capacity planning made on the basis of stand-alone technology costs is likely to lead to economically inefficient outcomes.

Moreover, in deregulated markets, individual power producers who evaluate only their own direct costs and risks make investment decisions. These decisions do not reflect the effects that the producers' technologies may have on overall generating portfolio performance. Wind investors, for example, cannot capture the risk-mitigation benefits they produce for the overall portfolio, which leads to underinvestment in wind relative to levels that are optimal from society's perspective. Similarly, some investors may prefer the risk menu offered by fuel-intensive technologies such as combined cycle gas turbines, which have low initial costs. Given sufficient market power, gas generators may be able to externalize fuel risks onto customers. In effect, these investors do not bear the full risk effects they impose onto the generating mix, which may lead to overinvestment in gas relative to what is optimal from a total portfolio perspective. (A quantitative treatment of this issue is given in Roques, 2006.) All this suggests a rationale for economic policies in favor of technologies that bring diversification benefits.

References

Awerbuch, S. (1993). The surprising role of risk and discount rates in utility integrated-resource planning. *The Electricity Journal*, 6(3), 20–33.

Awerbuch, S. (1995). Market-based IRP: it's easy!. *Electricity Journal*, 8(3), 50–67.

Awerbuch, S. (2004). *Restructuring our Electricity Networks to Promote Decarbonization: Decentralization, Mass Customization and Intermittent Renewables in the 21st Century*. Tyndall Centre Working Paper No. 49 (March). www.tyndall.ac.uk/publications/working_papers/wp49.pdf

Awerbuch, S. and Berger, M. (2003). *Energy Security and Diversity in the EU: A Mean-Variance Portfolio Approach*. Report No. EET/2003/03 (February). Paris: IEA. http://library.iea.org/dbtw-wpd/textbase/papers/2003/port.pdf

Awerbuch, S. and Yang, S. (2007). *Efficient Electricity Generating Portfolios for Europe: Maximizing Energy Security and Climate Change Mitigation*. SPRU Working Paper. forthcoming.

Awerbuch, S., Jansen, J. C. and Drennen, T. (2005). *The Cost of Geothermal Energy in the Western US Region: A Portfolio-Based Approach: A Mean-Variance Portfolio Optimization of the Region's Generating Mix to 2013*. SAND 2005–5173 (Sandia Report).

Awerbuch, S., Stirling, A. and Jansen, J. (2006). Portfolio and diversity analysis of energy technologies using full-spectrum risk measures. In Leggio, K. B., Bodde, D. L. and Taylor, M. L. (Eds), *Managing Enterprise Risk: What the Electric Industry Experience Implies for Contemporary Business*. Oxford: Elsevier.

Bacon, R. H., Besant-Jones, J. F. and Heidurian, J. (1996). *Estimating Construction Costs and Schedules: Experience with Power Generation Projects in Developing Countries*. World Bank Technical Paper No. 325 (August).

Committee on Radioactive Waste Management (2006). *Managing our Radioactive Waste Safely*. London: CORWM (July).

Dale, L., Milborrow, D., Slark, R. and Strbac, G. (2004). Total cost estimates for large-scale wind scenarios in UK. *Energy Policy*, 32(17), 1949–1956.

DENA Grid Study (2005). Presented in detail at the International Conference on the Integration of Wind Energy into the German Electricity Supply, Berlin, 10 May 2005. http://www.ewea.org/fileadmin/ewea_documents/documents/publications/briefings/Dena_Study.pdf

European Commission, Commission of the European Communities (2001). *Towards a European Strategy for the Security of Energy Supply*. Green Paper. Brussels: European Commission.

European Commission (2006). Staff Working Document. *Annex to the Green Paper A European Strategy for Sustainable, Competitive and Secure Energy. What is at Stake – Background Document*. SEC(2006) 317/2. Brussels: European Commission.

European Union (2005). *European Energy and Transport: Trends to 2030 – Update 2005*. European Commission, EREC (undated), SHP Potential. http://www.erec-renewables.org/sources/hydro.htm

Fox, B. and Flynn, D. (2005). *Managing Intermittency of Wind Generation with Heating Load Control*. In collaboration with P. O'Kane, The Queen's University of Belfast.

Green, R. (2006). *Carbon Tax or Carbon Permits: The Impact on Generators' Risks*. Institute for Energy Research and Policy, University of Birmingham (September).

Grubb, M., Butler, L. and Twomey, P. (2006). Diversity and security in UK electricity generation: the influence of low carbon objectives. *Energy Policy*, 34(18), 4050–4062.

International Energy Agency (2006). *World Energy Outlook 2006*. Paris: IEA.

Markowitz, H. M. (1952). Portfolio selection. *Journal of Finance*, 7, 77–91.

Ragwitz, M., Resch, G., Huber, C. and Faber, T. (2005). *Potential of Renewable Energy Technologies*. Fraunhofer Institute and Vienna University of Technology. Presented at Renewable Energy for Europe – Research in Action, Brussels, November 2005.

Roques, F. A. (2006). The value of diversity. Technology choices and nuclear investment in liberalised electricity markets. http://zzz.cerna.ensmpofr/Documents/Enseignement/CoursEnergie/FRoques.pdf

Sinden, G. (2005). *Renewable Resources Characteristics*: Wind power, marine renewables and integration. http://www.electricitypolicy.org.uk/events/seminars/sinden.pdf

Stirling, A. C. (1994). Diversity and ignorance in electricity supply – addressing the solution rather than the problem. *Energy Policy*, 22(3), 195–216.

Stirling, A. C. (1996). *On the Economics and Analysis of Diversity*. Paper No. 28. Science Policy Research Unit (SPRU), University of Sussex. www.sussex.ac.uk/spru

UK Energy Research Centre (2006). *The Costs and Impacts of Intermittency: An Assessment of the Evidence*. London: UKERC, Imperial College (March).

Varian, H. R. (1993). A portfolio of Nobel laureates: Markowitz, Miller and Sharpe. *Journal of Economic Perspectives*, 7(1), 159–169.

Further reading

American Wind Energy Association (1996). *Wind Energy Weekly*, 15 (680; 15 January). http://ces.iisc.ernet.in/hpg/envis/wwdoc1.html (last accessed August 2002).

Awerbuch, S. (1995). New economic cost perspectives for valuing solar technologies. In Böer, K. W. (Ed.), *Advances in Solar Energy: An Annual Review of Research and Development*. Vol. 10. Boulder, CO: ASES.

Awerbuch, S. (2000a). Getting it right: the real cost impacts of a renewables portfolio standard. *Public Utilities Fortnightly* (15 February).

Awerbuch, S. (2000b). Investing in photovoltaics: risk, accounting and the value of new technology. *Energy Policy*, 28 (Special Issue 14).

Awerbuch, S. (2005). Portfolio-based electricity generation planning: policy implications for renewables and energy security. http://www.sussex.ac.uk/spru/documents/portfolio-based_planning-dec-26-04-miti-final-for_distribution.doc/

Awerbuch, S. (2006). *Briefing Paper: Wind intermittency*. www.awerbuch.com

Awerbuch, S. and Sauter, R. (2006). Exploiting the oil–GDP effect to support renewables deployment. *Energy Policy*, 34(17), 2805–2819.

Bar-Lev, D. and Katz, S. (1976). A portfolio approach to fossil fuel procurement in the electric utility industry. *Journal of Finance*, 31(3), 933–947.

Benz, E. and Trück, S. (2006). Modelling CO2 Emission Allowance Prices. Bonn Graduate School of Economics, University of Bonn.

Berger, M. (2003). Portfolio Analysis of EU Electricity Generating Mixes and Its Implications for Renewables. Ph.D. Dissertation, Technische Universität Wien, Vienna (March).

Bolinger, M., Wiser, R. and Golove, W. (2002). *Quantifying the Value that Wind Power Provides as a Hedge Against Volatile Natural Gas Prices*. LBNL-50484 (June).

Bolinger, M., Wiser, R. and Golove, W. (2006). Accounting for fuel price risk when comparing renewable to gas-fired generation: the role of forward natural gas prices. *Energy Policy*, 34(6), 706–720.

Borak, S., Hardle, W., Trück, S. and Weron, R. (2005). *Risk Premiums for CO$_2$ Emission Allowances in the EEX Market*. Center for Applied Statistics and Econometrics (CASE).

Brealey, R. A. and Myers, S. C. (2004). *Principles of Corporate Finance*. New York: McGraw Hill.

Brower, M. C., Bell, K., Spinney, P., Bernow, S. and Duckworth, M. (1997). *Evaluating the Risk Reduction Benefits of Wind Energy*. Report to the US Department of Energy, Wind Technology Division (January).

Bureau of Labor Cost Statistics Data (2002). *US Employment Cost Index*. http://www.bls.gov/data/2002

Copeland, T. E. and Weston, J. F. (1988). *Financial Theory and Corporate Policy*, 3rd edn. Reading, MA: Addison-Wesley.

Dantzig, G. B., Orden, A. and Wolfe, P. (1955). The generalized simplex method for minimizing a linear form under linear inequality constraints. *Pacific Journal of Mathematics*, 5, 183–195.

DeLaquil, P. and Larson, E. (2002). *Global Renewable Energy Resource Estimates for the SAGE Model*. Final Report Prepared for the Energy Information Administration. Washington, DC: US Department of Energy.

Doherty, R., Outhred, H. and O'Malley, M. (2005). *Generation Portfolio Analysis for a Carbon Constrained and Uncertain Future*. Electricity Research Centre, University College Dublin.

Duke, R. and Kammen, D. M. (1999). The economics of energy market transformation programs. *Energy Journal*, 20(4), 15–50.

Dunlop, J. (1996). Wind power project returns – what should equity investors expect? *Journal of Structured Finance* (Spring).

Electric Power Research Institute (1983). *Analysis of Risky Investments for Utilities*. Final Report. Palo Alto, CA: EPRI (September).

ERU (1995) *ERU 11: Development and Implementation of an Advanced Control System for the Optimal Operation and Management of Medium Size Power Systems with a Large Penetration from Renewable Power Sources*. http://www.eru.rl.ac.uk/broch94/eru11.html (last accessed August 2002).

EURATOM Supply Agency (2000). *Annual Report 2000*.

Eurelectric (undated). *Study on the Importance of Harnessing the Hydropower Resources of the World*. Brussels: Hydro Power and Other Renewable Energies Study Committee. http://www.eurelectric.org

European Commission (2007). *An Energy Policy for Europe: Communication from the Commission to the European Council and the European Parliament*. Brussels, 10.1.2007. Com(2007) 1 Final {Sec(2007) 12}.

European Renewable Energy Council (2004). *Renewable Energy Share by 2020 – The RE Industry Point of View*. Christine Lins, Secretary General, European Renewable Energy Council, GREEN-X Conference, Brussels, 23 September 2004.

Fabozzi, F., Gupta, F. and Markowitz, H. (2002). The legacy of modern portfolio theory. *Journal of Investing*, 11 (Fall), 7–22.

Fama, E. F. and French, K. R. (1998). Value versus growth: the international evidence. *Journal of Finance*, 53, 1975–1999.

Felder, F. (1994). *Hedging Natural Gas Price Risk by Electric Utilities*. Master's Dissertation, MIT Center for Energy Policy Research-Program in Technology and Policy (May).

Ferderer, J. P. (1996). Oil price volatility and the macroeconomy. *Journal of Macroeconomics*, 18(1), 1–26.

Garrad, H. and Germanischer, L. (1995). Study of offshore wind energy in the EC. http://www.garradhassan.com/

Gawell, K., Reed, M. and Wright, M. (2002). *Geothermal Energy: The Potential for Clean Power from the Earth*. Excerpt from preliminary report provided by Gawell in September 2002. US Geothermal Energy Association, US Department of Energy, Energy and Geosciences Institute, University of Utah.

Glynn, P. and Manne, A. (1988). On the valuation of payoffs from a geometric random walk of oil prices. *Pacific and Asian Journal of Energy*, 2(1), 47–48.

Grubb, M. J. and Meyer, N. I. (1993). Wind energy: resources, systems, and regional strategies. In Johansson, T. B. et al. (Eds), *Renewable Energy: Sources for Fuels and Electricity*. Washington, DC: Island Press.

Harlow, W. V. (1991). Asset allocation in a downside-risk framework. *Financial Analysts Journal*, 47(5), 29–40.

Hassett, K. and Metcalf, G. (1993). Energy conservation investment: do consumers discount the future correctly? *Energy Policy*, 21(6), 710–716.

Helfat, C. E. (1988). *Investment Choices in Industry*. Cambridge, MA: MIT Press.

Herbst, A. (1990). *The Handbook of Capital Investing*. New York: Harper-Business.

Hoff, T. E. (1997). *Integrating Renewable Energy Technologies in the Electric Supply Industry: A Risk Management Approach*. National Renewable Energy Laboratory (NREL) (March).

Holt, R. (1988). *Boom and Bust: Chaos in Oil Prices, 1901–1987 – A Statistical Analysis*. Washington, DC: US Department of Energy, Office of Technology Policy.

Hoyos, I., Barquín, J., Centeno, E. and Sánchez, J. (2005). Risk analysis of fuel and CO_2 prices volatility in electricity generation expansion planning. Instituto de Investigación Tecnológica ICAI, Universidad Pontificia Comillas, Madrid. Presented at the IX Spanish–Portuguese Electrical Engineering Congress, Marbella, Spain, June. http://www.aedie.org/9CHLIE-paper-send/310-hoyos.pdf

Huber, C., Haas, R., Faber, T., Resch, G., Green, J., Twidell, J., Ruijgrok, W. and Erge, T. (2001). *Final Report of the Project ElGreen*. Energy Economics Group (EEG). Vienna University of Technology.

Humphreys, H. B. and McLain, K. T. (1998). Reducing the impacts of energy price volatility through dynamic portfolio selection. *Energy Journal*, 19(3), 107–131.

Ibbotson, R. and Rex, S. (1982). *Stocks, Bonds, Bills, and Inflation: The Past and the Future*. Charlottesville, VA: Financial Analysts Research Foundation.

Ibbotson Associates (1998). *Stocks, Bonds Bills and Inflation 1998 Yearbook*. Chicago, IL: Ibbotson Associates.

International Energy Agency (2000). *Experience Curves for Energy Technology Policy. World Energy Outlook 2000*. Paris: IEA.

International Energy Agency (2001). *Electricity Information*. Paris: OECD/IEA.

Jansen, J. C., Beurskens, L. W. M. and van Tilburg, X. (2006). *Application of Portfolio Analysis to the Dutch Generating Mix: Reference Case and Two Renewables Cases: Year 2030 – SE and GE Scenario*. Petten: ECN (February).

Kendall, M. (1953). The analysis of economic time series. *Journal of the Royal Statistical Society*, 96, 11–25.

Kleindorfer, P. R. and Li, L. (2005). Multi-period, VaR-constrained portfolio optimization in electric power. *Energy Journal*, 26(1), 1–26.

Krey, B. and Zweifel, P. (2005). *An Efficient Energy Portfolio for Switzerland*. Working Paper. Socioeconomic Institute, University of Zurich (March).

Kwan, C. C. Y. (2001). Portfolio analysis using spreadsheet tools. *Journal of Applied Finance*, 11(1), 70–81.

Lehner, B., Czisch, G. and Vassolo, S. (undated). *EuroWasser Section 8: Europe's Hydropower Potential Today and in the Future*. Center for Environmental Systems Research, University of Kassel, Germany.

Pacificorp (2003). *Integrated Resource Plan*. http://www.pacificorp.com/File/File25682.pdf

Paolella, M. and Taschini, L. (2006). *An Econometric Analysis of Emission Trading Allowances*. FINRISK Working Paper 341 (October).

Papapetroul, E. (2001). Oil price shocks, stock market, economic activity and employment in Greece. *Energy Economics*, 23(5), 511–532.

Pethick, D., Calder, R. and Clancy, C. (2002). When the wind doesn't blow. *Energy Power Risk Management*, 6(12).

Sadorsky, P. (1999). Oil price shocks and stock market activity. *Energy Economics*, 21, 449–469.

Sauter, R. and Awerbuch, S. (2002). *Oil Price Volatility and Macroeconomic Activity, a Survey and Literature Review*. Working Paper. IEA–REU (August).

Seitz, N. (1990). *Capital Budgeting and Long-Term Financing Decisions*. Chicago, IL: Dryden Press.

Seitz, N. and Ellison, M. (1995). *Capital Budgeting and Long-Term Financing Decisions*. Chicago, IL: Dryden Press.

Shackle, G. L. S. (1972). *Epistemics and Economics*. Cambridge: Cambridge University Press.

Sharpe, W. (1970). *Portfolio Theory and Capital Markets*. New York: McGraw-Hill.

Sieminski, A. (2007). Oil & natural gas prices, and heating degree days. *Deutsche Bank, Commodities Research*, 5.1.2007.

Sinden, G. (2005). Renewable Resources Characteristics. *Wind power, marine renewables and integration*. http://www.electricitypolicy.org.uk/events/seminars/sinden.pdf

Springer, U. (2003). Can the risks of Kyoto mechanisms be reduced through portfolio diversification: evidence from the Swedish AIJ Program. *Environmental and Resource Economics*, 25(4), 501–513.

Springer, U. and Laurikka, H. (2002). *Quantifying Risks and Risk Correlations of Investments in Climate Change Mitigation*. IWOe Discussion Paper No. 101. University of St Gallen. www.iwoe.unisg.ch/org/iwo/web.nsf

Technology choices and nuclear investment in liberalised electricity markets. http://zzz.cerna.ensmp.fr/Documents/Enseignement/CoursEnergie/FRoques.pdf

Van Wijk, A. J. M. and Coelingh, J. P. (1993). *Wind Potential in the OECD Countries*. University of Utrecht, the Netherlands.

Venetsanos, K., Angelopoulou, P. and Tsoutsos, T. (2002). Renewable energy sources project appraisal under uncertainty: the case of wind energy exploitation within a changing energy market environment. *Energy Policy*, 30, 293–307.

Appendix

Table 5.A1   Fuel cost inputs and economic cost of carbon dioxide (CO_2)

Gas	€4.8/Mbtu
Oil	€41/bbl
Coal	€44/tonne
CO_2	€35/tonne
Uranium	€6/MWh
Biomass	€5.15/GJ

Table 5.A2   Operating and maintenance (O&M) correlation coefficients

Technology	Coal	Gas	Nuclear	Oil	Hydro	Wind	Geothermal	Solar	Biomass
Coal	1.00	0.25	0.00	−0.18	0.03	−0.22	0.14	−0.39	0.18
Gas	0.25	1.00	0.24	0.09	−0.04	0.00	−0.18	0.05	0.32
Nuclear	0.00	0.24	1.00	−0.17	−0.41	−0.07	0.12	0.35	0.65
Oil	−0.18	0.09	−0.17	1.00	−0.27	−0.58	−0.06	−0.04	0.01
Hydro	0.03	−0.04	−0.41	−0.27	1.00	0.29	−0.08	0.30	−0.18
Wind	−0.22	0.00	−0.07	−0.58	0.29	1.00	−0.28	0.05	−0.18
Geothermal	0.14	−0.18	0.12	−0.06	−0.08	−0.28	1.00	−0.48	−0.70
Solar	−0.39	0.05	0.35	−0.04	0.30	0.05	−0.48	1.00	0.25
Biomass	0.18	0.32	0.65	0.01	−0.18	−0.18	−0.70	0.25	1.00

CHAPTER 6

Portfolio Analysis of the Future Dutch Generating Mix

Jaap C. Jansen and Luuk Beurskens*

Abstract

This chapter presents results of an application of Markowitz Portfolio Theory (MPT) to the future portfolio of electricity generating technologies in the Netherlands in the year 2030. Projections of two base-case generating mixes and general scenario assumptions have been taken from two specific scenarios designed by the Dutch Central Planning Bureau, CPB. The two scenarios are 'Strong Europe' (SE) and 'Global Economy' (GE). For each scenario, a base-case reference policy variant as well as two alternative renewables-oriented policy variants are introduced. This chapter focuses on the electricity cost–risk dimension of the Dutch portfolio of generating technologies and the potential for additional deployment of renewable generating technologies to enhance the efficiency of base-case generating mixes in year 2030. The major results of this study are as follows. (1) In both scenarios, the base-case generating mix is not very efficient. Graphical analysis suggests that diversification may yield up to 20% risk reduction at no extra cost. (2) Promotion of renewable energy can greatly decrease the portfolio cost risk. Defining mixes without renewables results in significantly riskier mixes in the absence of concomitant significant changes in portfolio costs. (3) Because of its relatively low cost risk and high potential, large-scale implementation of offshore wind can reduce cost risk of the Dutch generating portfolio. In an SE world large-scale implementation of offshore wind is projected to have a downward effect on Dutch electricity prices in 2030.

Acknowledgements

The authors gratefully acknowledge financial support for this study from the Dutch Ministry of Economic Affairs (EZ). With the usual disclaimer, the authors, who during the period 2002–2006 collaborated intensively with Shimon in a

*Energy research Centre of the Netherlands (ECN) Petten, the Netherlands

range of portfolio-related research activities, would like to put on record the great intellectual inspiration bestowed on them by Shimon Awerbuch.

6.1 Introduction

The Dutch Ministry of Economic Affairs (EZ) has been assessing the socio-economic impact of the Dutch renewables stimulation targets and policies over the period up to the year 2020. To that effect, the Dutch Central Planning Office (CPB), in association with the Energy research Centre of the Netherlands (ECN), has performed a social cost–benefit analysis of possible large-scale implementation of offshore wind in the Dutch continental shelf (Verrips et al., 2005). This study arrived at generally negative conclusions on the merits of early implementation of large-scale offshore wind in the Dutch continental shelf.

This chapter considers the merits of large-scale implementation of offshore wind and other renewables-based generation technologies over a longer period, up to the target year 2030. The CPB scenario assumptions from the report mentioned above are used, without any adjustments but for a longer time-horizon. As an appraisal approach, portfolio analysis, i.e. Markowitz Portfolio Theory (MPT), is applied instead of conventional cost–benefit analysis. Application of portfolio analysis, as against cost–benefit analysis, allows the portfolio price risk dimension to be integrated in a quantitative fashion into the appraisal of distinct generating mixes.

The MPT approach is applied to future Dutch generating mixes for the year 2030, evaluating risk against two CPB scenarios, i.e. Strong Europe (SE) and Global Economy (GE). For each scenario, three policy variants are evaluated:

- the base or 'zero' variant, which does not include wind power, but includes small shares of other renewables;
- an alternative variant articulating offshore wind power;
- another alternative broad-based renewables variant.

6.2 Theoretical framework

Financial portfolio analysis, based on MPT (Markowitz, 1952), builds on the premise that a portfolio of well-chosen assets has reduced risk characteristics when no perfect mutual correlations between the return on each of pair of assets exist. In a similar line of argument, portfolio (cost) risk may be reduced in a portfolio of well-chosen generating technology options as a result of less than perfect correlations between their cost characteristics.

Earlier studies, such as Awerbuch (2000), Awerbuch and Berger (2003) and Berger (2003), suggest that introducing renewables in the generating portfolio may significantly affect overall holding-period return (HPR) risk. This chapter proposes some adjustments in the theoretical framework, among others the introduction of the concept of (portfolio) 'price risk', or rather 'cost risk' replacing the expected returns concept applied in the studies referred to above. In a newly developed model the present authors have developed a *risk–cost-efficient frontier*. This type of efficient frontier shows a graph of risk (expressed in €/MWh) and cost of electricity (COE, expressed in €/MWh) for all efficient portfolios. A portfolio is efficient when a marginal increment in the output of any generation technology does not reduce expected portfolio cost without, at the same time increasing expected cost risk (or the other way around). Underlying efficient portfolios are energy based (i.e. based on shares of constituent electricity generating technologies in terms of electricity generation, e.g. in GWh or TWh, instead of MW capacity). The concept of 'portfolio risk' will be further explained later.

The transformation from a risk–return to a risk–cost-efficient frontier is proposed for the following main reasons:

- 'Return' has quite a different prevailing (financial or physical 'profit') connotation from just the reciprocal value of cost per unit of energy.
- From a mathematical perspective, the reciprocal of portfolio cost is not the same as portfolio return, if the latter is properly defined. (Jansen et al., 2005: Annex F)
- The conversion from portfolio cost to a parameter defined as its reciprocal (dubbed 'portfolio return') makes the link to portfolio risk problematic: the latter cannot be expressed in the same dimensions as the reciprocal parameter of portfolio COE (hereafter expressed in MWh).

Efficient frontiers resulting from applying MPT to portfolios of financial assets depict in a forward-looking way a set of points, each of which corresponds to a particular efficient portfolio. Such an efficient frontier representation brings out two dimensions of underlying efficient portfolios: the projected *portfolio return* in percentage terms per period (*y*-axis coordinate) and *portfolio risk* (*x*-axis coordinate), i.e. the projected standard deviation of portfolio return, both expressed in the same dimension (% per period). Underlying efficient portfolios are composed of a certain efficient, linear combination of individual financial assets from a certain asset universe, with their respective shares in the projected portfolio value as weights. The essential feature of an efficient portfolio is that its (projected) portfolio return cannot be improved without, at the same time, higher portfolio risk exposure. Note that the aforementioned risk concept, as brought out by efficient frontiers of financial portfolios, is quite transparent.

From a societal point of view the crucial question is considered: Which portfolios can yield the lowest expected energy costs at given, acceptable levels of expected risk? To answer this question, ways in which to construct an efficient frontier are sought, showing for the set of efficient portfolios the relationship between the expected portfolio COE (stated briefly: *portfolio cost*) and the expected portfolio COE risk, i.e. *portfolio risk*. Values of portfolio risk should have a transparent interpretation to enable the projection of confidence intervals of portfolio cost. To achieve this, the following three-stage procedure was pursued.

1. For each cost component considered making up the COE of a certain electricity generating technology, determine the expected value and the upper limit value of the two-sided 95% confidence interval. Estimation of such interval is to be based on expert judgment.

2. For each generating technology considered, determine the expected value and the upper limit value of the two-sided 95% confidence interval based on results of step 1.
3. Determine the efficient frontier, based on results of step 2.

Although this procedure does not use the notion of a holding period return, in its elaboration it is fully compatible with MPT applications to portfolios of financial assets.

The portfolio risk indicator emerging from this exercise can be interpreted in a transparent way: it is simply the (expected) standard deviation of portfolio cost. For a specific portfolio cost value it is approximately half the difference between the upper bound value of the projected portfolio cost interval and projected portfolio cost. Moreover, upper bounds of portfolio cost intervals can enable users, e.g. policy makers, to define their risk aversion preferences. For example, if a user wishes to accept, say, 90 €/MWh as a maximum COE with an overshoot risk of 2.5% (on average one case in the right-hand tail rejection area out of 40 cases), the portfolio with the lowest (expected) portfolio cost meeting this condition can be determined.

Hence, by including portfolio COE risk, the MPT approach as applied in this chapter enables policy makers to integrate the *trias energetica* (competitive energy prices, energy supply security, mitigation of adverse environmental impacts) in a quantitative framework. The proposed approach enables policy makers to monitor electricity cost–risk developments using an energy supply security norm as a yardstick, i.e. a preset upper bound to the real COE.

To improve flexibility and overcome obstacles found in earlier spreadsheet-based models, ECN developed a new generic optimization model for determining efficient frontiers. The new model uses the AIMMS[1] dedicated mathematical modelling framework.

For the analysis of cost and risk for a portfolio of electricity generating options, the graphical presentation as shown in Figure 6.1 is used, combined with a table containing some key indicators for cost, risk and composition.

The dotted elliptical area indicates the range of feasible portfolios and the solid line indicates the cost-efficient frontier, comprising all Pareto-efficient[2] combinations of risk and return. Note that the elliptical feasible area is formed under constraints on the different generating options. In an unconstrained world, the feasible area would resemble the well-known boomerang shape also found in financial applications.

Mix Q typically is the global minimum-cost portfolio and mix P is the global minimum-risk portfolio. Mix A represents a target mix for a certain target year. Generating mix A is clearly not efficient, since rearranging could:

- reduce portfolio risk at the same portfolio cost (moving from A to point N), or
- reduce portfolio cost at the same risk (moving from point A to point S), or

---

[1]AIMMS is a dedicated optimization modelling framework developed by Paragon Decision Technology Inc. (http://www.aimms.com).
[2]Pareto efficiency in this context indicates that no improvement in return can be attained without increasing risk and vice versa.

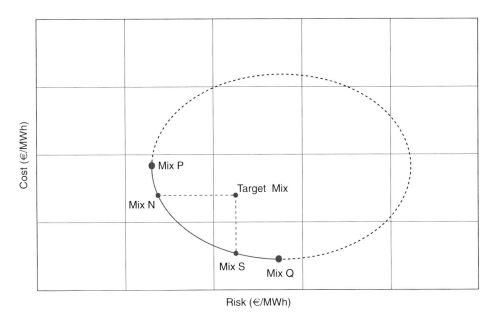

FIGURE 6.1   Example: cost-efficient frontier.

- reduce both (all combinations between point A exclusive and arc NS inclusive, excluding those on lines AN and AS).

An example of characteristic points A, N, S, P and Q is presented in Table 6.1. This table denotes an illustrative example and is not based on real data. Assuming that the costs are distributed independently random, for each portfolio – characterized by its expected portfolio cost and portfolio risk – its maximum portfolio cost within a set probability can be calculated. This figure is presented as 'upper bound at 2.5%' and may be interpreted as the maximum cost that will occur with 97.5% certainty. Examples are given by the figures in the third row.

Policy makers may wish to set norms for maximum portfolio cost in certain milestone years. These norms can be taken as the point of departure for monitoring the evolution of the actual electricity mix and actual technology costs. Based on updated technology costs (cost projections), the maximum portfolio cost in milestone years can be estimated (projected). 'Market failure' (e.g. the predilection of incumbent generators for CCGT i.e. combined cycle gas turbine technology, with attendant high fuel cost risk) may render a country exposed to a supply security risk, considered unduly high by its policy makers. At least for the power sector, portfolio analysis can be used as a tool to monitor the level of energy supply security. Should the estimated portfolio cost in a milestone year exceed the preset norm, this may trigger policies by the public sector to bring about new (replacement or expansive) investments in generating capacity, with – from a socioeconomic cost perspective – low-risk technologies. In a liberalized market, adjustment of market framework conditions can bring this about.

Table 6.1   Example: aggregated results mix A

	Mix P	Mix N–A	Mix A	Mix S–A	Mix Q
Portfolio cost (€/MWh)	28.0	22.0	22.0	13.5	12.5
Portfolio risk σ (€/MWh)	4.0	4.5	10.5	10.5	13.5
Upper bound at 2.5% (€/MWh)	36.0	31.0	43.0	34.5	39.5
Gas CHP (%)	25	30	35	25	25
Coal (%)	25	25	40	30	25
Nuclear (%)	5	5	5	5	5
Renewable wind (%)	20	25	10	30	25
Renewable biomass (%)	25	15	10	10	20

CHP: combined heat and power.

6.3 The Dutch generating mix in 2030

CPB developed long-term scenarios for Europe and uses these scenarios for analysis of energy markets and climate policy (Bollen et al., 2004). Two of these scenarios, GE and SE, have been used by CPB as a basis for a social cost–benefit analysis of large-scale implementation of offshore wind in the Dutch continental shelf (Verrips et al., 2005). Aligning with the latter CPB study, this chapter also uses the long-term CPB scenarios 'Strong Europe (SE)' and 'Global Economy (GE)' as a starting point for long-term portfolio analysis. This section describes a number of input assumptions and presents two alternative policy variants. Furthermore, for scenarios SE and GE, the efficient frontier and risk characteristics are analyzed.

In this analysis of future costs and risks there is a clear distinction between how the world may look like *without* major policy changes and *after* specified changes of policy packages. The first aspect is translated into *scenarios*, which are plausible, consistent descriptions of the future. Scenarios may be regarded as external to the model. As mentioned, this study builds on scenarios constructed by the CPB. The policy aspect is less external, since it defines different approaches or strategies for dealing with external changes. Different policy strategies, including 'business-as-usual', are translated into *policy variants*.

This chapter uses CPB scenarios SE and GE. For each scenario, three variants are considered:

- Reference (0): a reference variant assuming continuation of renewables stimulation policy as currently implemented or whose implentation has already been officially announced (SE0 and GE0). This variant is also referred to as the base case.

- Wind (p1): an intensification of renewables stimulation policy, with the emphasis put on offshore wind energy stimulation (SEp1 and GEp1).
- Biomass (p2): an intensification of renewables stimulation policy, with the emphasis put on a broad variety of relatively cost-effective renewable technologies (SEp2 and GEp2).

In addition to identifying scenarios and policy variants, the model will need some prior information setting the initial situation and restricting possible outcomes. This prior information is translated into a set of *input assumptions*. All input data used in this study have been obtained from, and are consistent with, the data used in CPB's cost–benefit analysis for offshore wind (Verrips et al., 2005). Constraints imposed on the model relate, *inter alia*, to the assumed technical potentials of the distinct renewable generating technologies, because of, for example, resource or authorization (notably, wind power) constraints.

Most technology cost assumptions are similar for both SE and GE. Only onshore wind and offshore wind have distinct cost assumptions. Cost-reducing technical progress for these technologies is assumed to occur at a faster rate (as captured by a lower progress ratio) under SE than under the GE scenario. However, since SE and GE have quite divergent assumptions on carbon dioxide (CO_2) price developments, the resulting total generating costs differ for many, notably fossil-fuel-based, technologies. Furthermore, total electricity demand is assumed to be higher under GE than under SE. Other assumptions are listed below. The feasible range of generating capacities, called energy bounds (see Appendix A), are largely identical in energy terms, except for existing nuclear and coal. The bounds do, however, differ relatively, owing to the higher energy demand in the GE scenario.

6.3.1 The Strong Europe (SE) scenario

Strong international cooperation and important public institutions are key characteristics of the Strong Europe (SE) scenario. In this scenario, European integration proceeds successfully – politically, economically and geographically. Welfare distribution is valued over economic growth and cooperation will result in a stringent climate policy. Up to 2020 a CO_2 price of 11 €$_{2003}$/tonne is assumed, thereafter increasing to 55 €$_{2003}$/tonne in 2030. For gas a price of 4.7 €$_{2003}$/GJ is assumed in 2030. Until 2030, primary energy demand is assumed to increase at a (very) modest rate and CO_2-related emissions would decrease in absolute terms.

6.3.1.1 The SE0 base case

One of the key graphical results of portfolio analysis is construction of the efficient frontier (EF), a graph on which each point represents an efficient portfolio. Portfolio efficiency in this context means that no portfolio with lower costs (in terms of €/MWh) can be obtained without increasing risk.

For the SE0 variant the efficient frontier is depicted in Figure 6.2. Details characterizing special points in this figure are presented in Table 6.2.

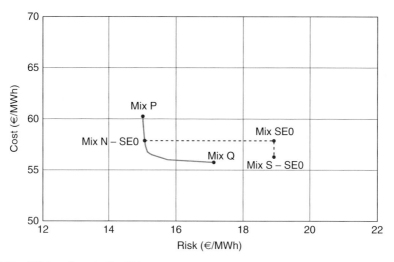

FIGURE 6.2   Efficient frontier for SE0.

Table 6.2   Aggregated results SE0

	Mix P	Mix N–SE0	Mix SE0	Mix S–SE0	Mix Q
Portfolio cost (€/MWh)	60.2	57.9	57.9	56.3	55.7
Portfolio risk σ (€/MWh)	15.0	15.1	18.9	18.9	17.1
Upper bound at 2.5% (€/MWh)	89.7	87.4	95.7	93.4	89.3
Gas CC (%)	18.4	18.4	38.6	41.0	34.4
Gas CHP (%)	37.2	37.2	37.2	38.1	38.1
Coal (%)	12.1	12.7	21.7	11.5	1.5
Nuclear (%)	0.0	0.0	0.0	0.0	0.0
Renewable wind (%)	20.0	20.0	0.0	4.2	20.0
Renewable biomass (%)	10.5	10.5	1.5	4.2	4.9
Renewable other (%)	1.7	1.1	1.0	1.1	1.1

CC: combined cycle; CHP: combined heat and power.

Let us consider each characteristic point:

- Table 6.2 column Mix SE0. The *target mix* set for the SE0 is characterized by (expected) portfolio cost of 57.9 €/MWh and portfolio risk of 18.9 €/MWh. The odds are 1 to 40 (=2.5%) that the target mix will end up in a portfolio electricity cost level exceeding 95.7 €/MWh (two sigma from the mean). As already explained in Section 6.2, the latter type of information may assist policy makers to define levels of cost risk that they consider acceptable. Renewables are poorly represented in the target mix: wind 0%, biomass 2% and other renewables 0%.

- *Point S–SE0* is on the efficient frontier vertically below the target mix. The mix S–SE0 has the same risk as the target mix but its expected electricity costs are lower (56.3 €/MWh). As the target mix is rather risky, point S is situated on the inefficient part of the efficient frontier (not shown in Figure 6.1). Somewhat counter-intuitively, more renewables-based electricity is represented in portfolio S–SE0. Coal (which is costly in the SE0 scenario owing to the CO_2 price) is substituted by gas technologies, wind power and biomass options.
- *Point N–SE0* is on the efficient frontier horizontally to the left of the target mix. The mix N–SE0 has the same cost as the target mix but its expected risk level is much lower (15.1 €/MWh versus 18.9 €/MWh). Renewables are well represented in this low-risk portfolio: wind 20% (representing the total onshore and offshore potential), biomass 10% (also the full potential) and other renewables 1%.
- *Point Q* is the lowest point of the efficient frontier. This point stands for the lowest expected cost portfolio (55.7 €/MWh). Note that its expected risk is appreciably lower than that of the target mix (17.1 €/MWh versus 18.9 €/MWh). As renewables tend to be less cost risky than fossil-fuel-based electricity, while under SE their costs are assumed to come down importantly by 2030, renewables are represented rather well in mix Q: wind 20% (full potential), biomass 4% and other 1%.
- *Point P* is the highest point of the efficient part of the efficient frontier. This point stands for the lowest expected risk portfolio (15.0 €/MWh), but its expected cost is higher than that associated with the target mix (60.2 €/MWh versus 57.9 €/MWh). However, the upper bound at the 2.5% percentile in mix P (89.7 €/MWh) is lower than for the target mix (95.7 €/MWh). As renewables tend to be less cost risky than fossil-fuel-based electricity, renewables are represented quite well in portfolio P: wind 20%, biomass 10% and other 2% (the total renewable potential).

The relatively high expected carbon cost under the SE scenario (55 €/tCO_2 in target year 2030) has a strong impact on costs: even along the efficient frontier no portfolios can be found in the base-case variant with lower expected electricity cost than 55.7 €/MWh. Furthermore, the shape of the efficient frontier is rather hollow, so that over a wide range from bottom right (point Q) to left, large (expected) risk reductions can be obtained at relatively small cost sacrifices (hence slightly higher expected costs), up to a point where the efficient frontier bends steeply upward. The explanation of this shape may relate to almost 'free lunches' that can be obtained initially by moving from Q to the left along the efficient frontier, notably by substitution of gas with coal and biomass co-firing.

6.3.1.2 Variants SEp1 and SEp2

A striking feature under SE is that the target mixes for variants p1 (renewables with offshore wind focus) and p2 (broad-based renewables) are not only much

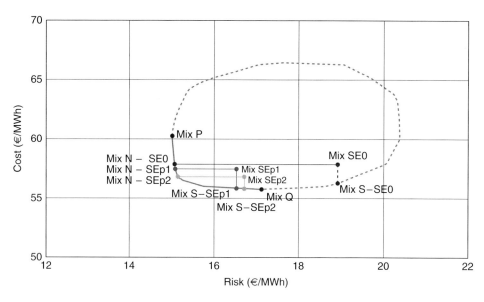

FIGURE 6.3   Efficient frontier for SE0, SEp1 and SEp2.

less risky than for the base-case policy variant, but also characterized by slightly lower expected electricity cost. The carbon factor under SE appears to have a rather high impact, rendering the economics of renewables-based technology vis-à-vis fossil-fuel-based ones much better for renewable electricity generators (RES-E generators). Furthermore, the (expected) portfolio cost–risk differences between target mixes p1 and p2 are rather small: SEp1 has slightly higher costs on the one hand, but slightly lower risk on the other (Figure 6.3; Tables 6.3 and 6.4).

6.3.2 The Global Economy (GE) scenario

The Global Economy (GE) scenario is characterized by strong international coop-eration and an important role for individual responsibility. Economic growth is valued over government interference beyond providing a limited amount of public service. Integration is limited to the economic sphere and cooperation in non-trade issues, such as effective climate policy, fails. Up to 2020 a CO_2 price of 11 €/tonne is assumed, and from 2021 the carbon market is assumed to collapse under the GE scenario with a 0 €/tonne price for CO_2 emission allowances. For gas a price of 4.7 $€_{2003}$/GJ is assumed in 2030. Until 2030, primary energy demand will increase at a steady 2.3%, as will emissions.

6.3.2.1 The GE0 base case

The shape of the efficient frontier under the GE scenario is less concave than under SE, and the frontier is situated lower, particularly with respect to cost. The carbon factor (expected carbon cost in target year 2030 of 0 $€/tCO_2$) is a major underlying factor accounting for the latter feature. As the economics

Table 6.3 Aggregated results SEp1

	Mix P	Mix N–SEp1	Mix SEp1	Mix S–SEp1	Mix Q
Portfolio cost (€/MWh)	60.2	57.5	57.5	55.8	55.7
Portfolio risk σ (€/MWh)	15.0	15.1	16.5	16.5	17.1
Upper bound at 2.5% (€/MWh)	89.7	87.0	89.9	88.2	89.3
Gas CC (%)	18.4	18.4	24.2	28.7	34.4
Gas CHP (%)	37.2	37.2	37.4	38.1	38.1
Coal (%)	12.1	12.7	21.7	7.2	1.5
Nuclear (%)	0.0	0.0	0.0	0.0	0.0
Renewable wind (%)	20.0	20.0	14.0	20.0	20.0
Renewable biomass (%)	10.5	10.5	1.5	4.9	4.9
Renewable other (%)	1.7	1.1	1.1	1.1	1.1

CC: combined cycle; CHP: combined heat and power.

Table 6.4 Aggregated results SEp2

	Mix P	Mix N–SEp2	Mix SEp2	Mix S–SEp2	Mix Q
Portfolio cost (€/MWh)	60.2	56.8	56.8	55.8	55.7
Portfolio risk σ (€/MWh)	15.0	15.2	16.7	16.7	17.1
Upper bound at 2.5% (€/MWh)	89.7	86.5	89.6	88.6	89.3
Gas CC (%)	18.4	18.4	24.2	30.5	34.4
Gas CHP (%)	37.2	37.2	38.1	38.1	38.1
Coal (%)	12.1	13.5	21.7	5.4	1.5
Nuclear (%)	0.0	0.0	0.0	0.0	0.0
Renewable wind (%)	20.0	20.1	13.4	20.1	20.0
Renewable biomass (%)	10.5	9.7	1.5	4.9	4.9
Renewable other (%)	1.7	1.1	1.1	1.1	1.1

CC: combined cycle; CHP: combined heat and power.

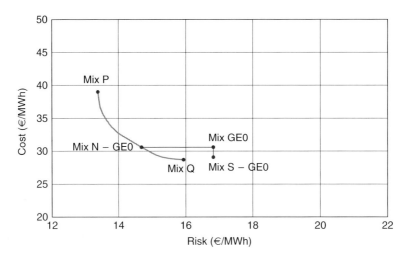

FIGURE 6.4   Efficient frontier for GE0.

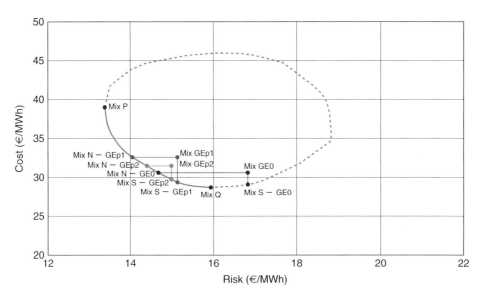

FIGURE 6.5   Efficient frontier and feasible mixes GEp1 and GEp2.

of renewables are much less favorable under GE (again on account of the assumed negligible carbon costs, but for wind also because of assumed slower technological progress), penetration of RES-E is projected to be much slower. The information contained on the special points in Figures 6.4 and 6.5 bears this out.

Only mix N on the efficient frontier, horizontally left from the base-case target mix GE0, and even more so mix P (the least risky portfolio feasible under scenario GE) have an appreciable uptake of RES-E. Under GE RES-E technology tends to be much more expensive than fossil-fuel technology if at typically much lower risk. Hence the Sharpe ratio (cost change per unit of risk change: the slope of the efficient frontier) is initially much less favorable, when moving along the efficient frontier to the left departing from Q. However, in GE the constraints to RES-E deployment imposed upon the model are reached at a much later phase when moving upward along the efficient frontier from Q (bottom right) to P (top left). Hence, on the least risky (upper left) part RES-E is better placed to accommodate risk aversion by moving leftward under GE than under SE.

Compare, for example, mix N under GE (in Table 6.5) with mix N under SE (in Table 6.2) and check the corresponding RES-E shares. The shares of wind (20%) and biomass (10%) in N under SE appear to have increased to their (model-imposed) upper limits, while under GE (wind 7%, biomass 4%) this is clearly not the case. This emphasizes the fact that under GE renewables can accommodate risk reduction at low risk levels better than under SE, where they are already stretched to the limit at low risk levels.

6.3.2.2 Variants GEp1 and GEp2

A remarkable difference between the location of target mixes under the GE variants p1 (wind) and p2 (biomass) and those of their SE counterparts is that under GE their associated expected electricity cost is somewhat higher than the corresponding zero (base-case) target mix. This can be gleaned from Figure 6.4 as well

Table 6.5   Aggregated results GE0

	Mix P	Mix N–GE0	Mix GE0	Mix S–GE0	Mix Q
Portfolio cost (€/MWh)	39.0	30.6	30.6	29.1	28.7
Portfolio risk σ (€/MWh)	13.4	14.7	16.8	16.8	15.9
Upper bound at 2.5% (€/MWh)	65.2	59.4	63.6	62.1	59.9
Gas CC (%)	11.4	11.4	25.1	26.4	15.3
Gas CHP (%)	31.1	31.1	32.8	32.8	32.8
Coal (%)	29.4	45.1	40.0	37.6	48.3
Nuclear (%)	1.1	1.1	1.1	1.1	1.1
Renewable wind (%)	16.8	6.6	0.0	1.3	1.7
Renewable biomass (%)	8.8	3.7	0.1	0.0	0.0
Renewable other (%)	1.4	0.9	0.9	0.9	0.9

CC: combined cycle; CHP: combined heat and power.

Table 6.6   Aggregated results GEp1

	Mix P	Mix N–GEp1	Mix GEp1	Mix S–GEp1	Mix Q
Portfolio cost (€/MWh)	39.0	32.6	32.6	29.4	28.7
Portfolio risk σ (€/MWh)	13.4	14.0	15.1	15.1	15.9
Upper bound at 2.5% (€/MWh)	65.2	60.1	62.2	59.0	59.9
Gas CC (%)	11.4	11.4	18.5	11.4	15.3
Gas CHP (%)	31.1	31.1	32.2	31.1	32.8
Coal (%)	29.4	38.3	35.4	49.8	48.3
Nuclear (%)	1.1	1.1	1.1	1.1	1.1
Renewable wind (%)	16.8	13.4	11.7	2.0	1.7
Renewable biomass (%)	8.8	3.7	0.1	3.7	0.0
Renewable other (%)	1.4	0.9	0.9	0.9	0.9

CC: combined cycle; CHP: combined heat and power.

as from Tables 6.6 and 6.7. Evidently, the costs of deliberate market forcing of RES-E are much higher under GE, where there is no help from the carbon factor.

6.4 Policy implications

In the previous section the reference policy variant and two 'renewables promotion' policy variants were analyzed for MPT efficiency, using the Strong Europe (SE) and Global Economy (GE) scenarios. In line with the assumptions underlying the scenarios, both COE and associated risk are generally lower in GE than in SE, owing to learning rates in technological development and the content of future climate policy. Differences in scenarios are clearly reflected in the shape and position of the feasible areas.

Results of portfolio analysis indicate the following:

- In both scenarios, the base variant is not very efficient and graphical analysis suggests that diversification may yield up to 20% risk reduction at no extra cost.
- Stimulation of renewable energy, as described in policy variants p1 and p2, can greatly improve the cost risk. Even in the GE scenario – the one that is rather unfavorable to a takeoff of renewables-based technology – this can be achieved

Table 6.7 Aggregated results GEp2

	Mix P	Mix N–GEp2	Mix GEp2	Mix S–GEp2	Mix Q
Portfolio cost (€/MWh)	39.0	31.5	31.5	29.7	28.7
Portfolio risk σ (€/MWh)	13.4	14.4	15.0	15.0	15.9
Upper bound at 2.5% (€/MWh)	65.2	59.7	60.8	59.1	59.9
Gas CC (%)	11.4	11.4	16.3	11.4	15.3
Gas CHP (%)	31.1	31.1	32.3	31.1	32.8
Coal (%)	29.4	42.1	37.6	48.3	48.3
Nuclear (%)	1.1	1.1	1.1	1.1	1.1
Renewable wind (%)	16.8	9.6	11.2	3.4	1.7
Renewable biomass (%)	8.8	3.7	0.7	3.7	0.0
Renewable other (%)	1.4	0.9	0.9	0.9	0.9

CC: combined cycle; CHP: combined heat and power.

at little additional cost. For the SE scenario, portfolio cost in the renewables policy variants is lower than that in the zero variant.

- Defining mixes without intensification of renewables stimulation (i.e. the zero variant target mixes) would result in riskier mixes (about 10% risk reduction is possible compared with the alternative policy variants 1 and 2), while portfolio costs would not be materially affected (about 6% cost increase for GEp1, 3% cost increase for GEp2, small cost reduction of 1–2% for SEp1 and SEp2).
- Further optimization beyond the variants evaluated is possible. However, the largest increase has already been realized with the relatively straightforward policy options p1 or p2.
- All in all, the results indicate that intensification of renewables stimulation policy can be justified from a socioeconomic perspective. In the SE scenario, the choice between p1 and p2 depends on risk aversion preferences: p1 is indicated to be slightly less risky but also slightly costlier. In the GE scenario the results presented above indicate that policy variant p2 would be socioeconomically slightly more favorable than p1.

The effects of variation of a number of input parameters on the cost and risk of the generating mixes have been investigated in sensitivity analyses. Owing to uncertainty surrounding cost and risk, the results of this study should be treated with caution. To put these in due perspective, sensitivity analysis is a quite valuable tool. To keep this chapter to a reasonable length, details of the sensitivity analyses are not given, but the main results are presented in the following paragraphs.[3]

The price of carbon (CO_2) is of key importance to the additional cost at which the security of supply in the power sector can be improved by moving toward an increasing share for renewables-based options. A higher carbon price dramatically improves the market position of renewables. An increase in the price of carbon tilts and shifts the efficient frontier upwards.

Owing to the large share of natural gas in the SE0 generating mix, expected portfolio cost and risk increase considerably. Under the assumption of 'high' gas prices (high compared with the CPB SE and GE scenarios), the risk mitigating potential for renewables-based generating options is highly amplified. Hence, the sensitivity of renewables-based generation technologies for the gas price is quite high.

Since biomass is only considered in co-firing and the share is limited, variations in the price have little effect on either costs or risk. With an increasing biomass price, the mix shifts towards a larger share of coal.

Finally, the sensitivity analysis shows that offshore wind – because of its relatively low risk and high potential – can significantly reduce portfolio risk. Under the SE scenario assumptions, tightening the technical offshore wind constraints results in higher coal shares. Also, from 1 GW up to 6 GW every discrete relaxation of the offshore wind constraint by 1 GW increments has the same marginal risk reduction potential.

The results of sensitivity analyses that have been shown in this section indicate that the characteristic of renewables-based technology to reduce portfolio risk is rather robust. This holds not only for broad-based renewables stimulation strategy but also for strategies with a certain focus on offshore wind. Further, the economics of renewables-based generating technologies are quite sensitive to the evolution of the gas price. In this respect, recall that both the GE and the SE scenario assume a rather moderate gas price evolution.

A general observation is that the large distances of target mixes from their corresponding efficient frontier under the distinct scenario variants and the uncertainties underlying the technology cost and potential assumptions suggest that it is difficult for policy makers to impose the right framework conditions on the market that lead to socially optimal portfolios. Nevertheless, under scenarios of rising real-term fossil fuel prices and increasingly binding carbon constraints, it would seem appropriate to reduce long-term (electricity) cost risk and long-term cost rises by renewables R&D and market stimulation.

6.5 Conclusions

Technology costs have been chosen in accordance with the cost–benefit analysis study for offshore wind (Verrips et al., 2005). Input data have been composed with utmost attention and care, but the true future costs remain highly dependent

---

[3] For sensitivity analyses see Jansen et al. (2006).

on external factors. Scenario parameters such as reference mixes, CO_2 price and gas price assumptions have been chosen in line with the above-mentioned study and could be the subject of discussion.

Risk estimates were derived following a predefined methodology, and projections of long-term cost and risk for generating options specifically and portfolios at large remain difficult, even under the most up-to-date approaches. Furthermore, fuel correlations and technology parameter correlations are indicative and based on expert judgements.

Of all predefined target portfolios for the year 2030, none is efficient in the sense deployed in this study: for each portfolio, reductions in either cost or risk, or both, are possible. In most cases, risk reductions and cost reductions can be obtained by increasing the share of renewable generating options (notably wind power and biomass). These opportunities can be quantified as a 20% risk reduction and a 4% cost reduction (Tables 6.8 and 6.9). Defining mixes without renewables results in riskier mixes (about 10% risk reduction is possible) (Tables 6.10 and 6.11).

The outcome is very sensitive to CO_2 price assumptions. In the SE scenario, with prices of 55€/tonne, the renewables options become much more competitive than in the GE scenario, with zero carbon costs. The relative importance of gas-fuelled power plants (58% in GE0 and 76% in SE0) poses a quite serious cost risk for the Dutch electricity sector. Renewables can considerably reduce cost risk of

Table 6.8 Potential diversification effect GE0

	GE0	Minimum	Reduction (%)
Portfolio risk	16.8	13.4 (mix P)	20
Portfolio cost	30.6	28.7 (mix Q)	6

Table 6.9 Potential diversification effect SE0

	SE0	Minimum	Reduction (%)
Portfolio risk	18.9	15.0 (mix P)	21
Portfolio cost	57.9	55.7 (mix Q)	4

Table 6.10 Potential diversification effect GEp1/GEp2

	Mix GE0	Mix GEp1	Reduction (%)	Mix GEp2	Reduction (%)
Portfolio risk	16.8	15.1	10	15.0	11
Portfolio cost	30.6	32.6	−6	31.5	−3

Table 6.11   Potential diversification effect SEp1/SEp2

	Mix SE0	Mix SEp1	Reduction (%)	Mix SEp2	Reduction (%)
Portfolio risk	18.9	16.5	13	16.7	12
Portfolio cost	57.9	57.5	1	56.8	2

the generating portfolio. The impact on risk and cost is strongly dependent on the scenario assumptions (notably the CO_2 price, the gas price and, to a lesser extent, the coal price) and the cost assumptions of renewables.

The analysis approach set out in this report is based on the methodology explained in Berger (2003) and Awerbuch and Berger (2003), and pioneered by Shimon Awerbuch in the 1990s. Several methodological refinements have been proposed. These have been implemented in this study, and some also in other ongoing or recently concluded research projects. The following contributions have been presented in this report:

- the introduction of an advanced notion of the efficient frontier based on cost;
- the use of energy-based instead of generating capacity-based portfolios;
- the expression of risk in terms of costs instead of a percentage rate;
- consistent determination of risk associated with generating costs for distinct technologies.

This chapter has documented some major improvements in one-period analysis of generating technology portfolios through the application of MPT. Focal research issues to enhance the reliability further and widen the scope of applications for the MPT approach in the domain of electricity and energy mix portfolios include:

- improving the use of historical cost information to derive projections of future risk values, such as incorporating generalized autoregressive conditional heteroscedastic (GARCH) techniques (e.g. Humphreys and McClain, 1998);
- improving the methodology to derive the projected correlation matrix, showing the assumed interrelationships between portfolio cost components;
- improving the allowance made for the cost impacts of penetration of intermittent renewable resources, which warrants, *inter alia*, a segmentation of the power market (into peak, intermediate and base load categories) and renewable resources (e.g. average wind speed categories, average insolation categories), and specification of contributions to ancillary power provision services;
- expanding the cost component on pollutant emissions with inclusion of the cost of non-GHG polluting emissions such as NO_x and SO_2. In the cost projections presented in this paper only the costs of CO_2 are considered;
- conversion from one-period analysis to multiperiod analysis, permitting not only the identification of efficient portfolios in a certain target year but also the determination of optimal trajectories for rebalancing portfolios from the base year to the target year. This would warrant specification of generation plant vintage years. Some leads are presented in Steinbach (2001)and Kleindorfer and Li (2005).

References

Awerbuch, S. (2000). Getting it right: the real cost impacts of a renewables portfolio standard. *Public Utilities Fortnightly*, 15 February.

Awerbuch, S. and Berger, M. (2003). *EU Energy Diversity and Security: Applying Portfolio Theory to Electricity Planning and Policy-Making*. Paris: International Energy Agency (February).

Awerbuch, S., Jansen, J. C. and Beurskens, L. (2004). *Building Capacity for Portfolio-Based Energy Planning in Developing Countries*. Final Report. London–Paris: Renewable Energy & Energy Efficiency Partnership (REEEP) and UNEP (August).

Awerbuch, S., Stirling, A. C., Jansen, J. C. and Beurskens, L. (2006). Portfolio and diversity analysis of energy technologies using full-spectrum risk measures. In Leggio, K. B., Bodde, D. L. and Taylor, M. L. (Eds), *Managing Enterprise Risk: What the Electric Industry Experience Implies for Contemporary Business*. Oxford: Elsevier.

Berger, M. (2003). *Portfolio Analysis of EU Electricity Generating Mixes and its Implications for Renewables*. Ph.D. Dissertation, Technische Universität Wien, Vienna.

Bollen, J., Manders, T. and Mulder, M. (2004). *Four Futures for Energy Markets and Climate Change*. The Hague: CPB.

Humphreys, H. B. and McClain, K. T. (1998). Reducing the impacts of energy price volatility through dynamic portfolio selection. *Energy Journal*, 19(3), 107–131.

Jansen, J. C., Beurskens, L. W. M. and van Tilburg, X. (2006). *Application of Portfolio Analysis to the Dutch Generating Mix. Reference Case and Two Renewables Cases: Year 2030 – SE and GE Scenario*. ECN-C-05-100. Energy research Centre of the Netherlands (February).

Kleindorfer, P. R. and Li, L. (2005). Multi-period VaR-constrained portfolio optimization with applications to the electric power sector. *Energy Journal*, 26(1), 1–26.

Markowitz, H. (1952). Portfolio selection. *Journal of Finance*, 7(1), 77–91.

Steinbach, M. C. (2001). Markowitz revisited: mean-variance models in financial portfolio analysis. *Society for Industrial and Applied Mathematics Review*, 43(1), 31–85.

Verrips, A., de Vries, H. J., Seebregts, A. J. and Lijesen, M. (2005). *Windenergie op de Noordzee – Een Maatschappelijke Kosten Baten Analyse*. The Hague: CPB (September).

Appendix A

Input assumptions

This appendix presents a concise overview of the assumptions used in this chapter. Tables 6.A1 to 6.A5

Table 6.A1 Technology specific upper and lower bounds of electricity generation (TWh, 2030)

	SE		GE	
	Lower bound	Upper bound	Lower bound	Upper bound
Gas CC	35.3	96.1	11.4	46.8
Gas CHP	71.3	72.9	31.1	32.8
Coal	0.0	83.3	0.0	55.7
Nuclear	0.0	0.0	1.1	1.1
Renewable wind	0.0	38.4	0.0	16.8
Renewable biomass	0.0	20.1	0.0	8.8
Renewable other	0.0	3.4	0.0	1.5

CC: combined cycle; CHP: combined heat and power.

Table 6.A2 Estimated fuel costs (€/GJ, 2030)

	Mean	High
Gas	4.70	10.00
Coal	1.70	3.00
Uranium	2.22	3.00
Biomass (co-firing)	5.00	7.00
Biogas (co-firing)	0.00	2.00
Biomass small	4.00	6.00

Table 6.A3 Correlations fuel costs, expert opinions

	Gas	Coal	Uranium	Biomass	Renewable
Gas	1.0	0.7	0.2	0.4	0.0
Coal	0.7	1.0	0.4	0.4	0.0
Uranium	0.2	0.4	1.0	0.1	0.0
Biomass	0.4	0.4	0.1	1.0	0.0
Renewable	0.0	0.0	0.0	0.0	1.0

Table 6.A4   Correlations non-fuel costs, expert opinions

	Investment	Variable O&M	Fixed O&M	CO$_2$
Investment	0.5	0.0	0.0	0.0
Variable O&M	0.0	0.5	0.0	0.0
Fixed O&M	0.0	0.0	0.5	0.0
CO$_2$	0.0	0.0	0.0	1.0

O&M: operation and maintenance.

Table 6.A5   CO2 costs/emission estimates

	CO$_2$ emission (kg/GJ)	Mean (€/t)	High (€/t)
Gas	56.1		
Coal	94.7		
CO$_2$ price SE		55.0	85.0
CO$_2$ price GE		0.0	30.0

Appendix B

Technology characteristics

Figure 6.B1 and Figure 6.B2

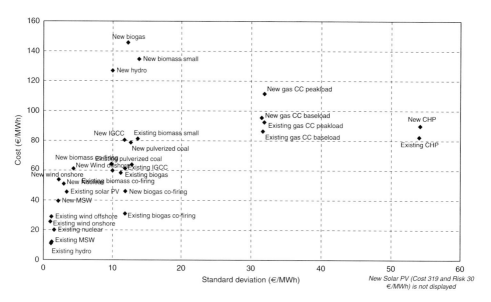

FIGURE 6.B1   Technology characteristics SE (high gas). New solar PV (cost 319 and risk 30 €/ MWh) is not displayed. CC: combined cycle; CHP: combined heat and power; MSW: municipal solid waste; PV: photovoltaics.

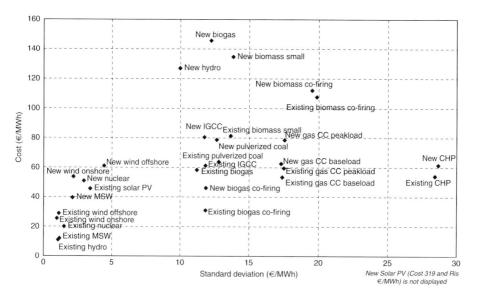

FIGURE 6.B2   Technology characteristics SE (high biomass). New solar PV (cost 319 and risk 30 €/MWh) is not displayed. CC: combined cycle; CHP: combined heat and power; IGCC: integrated gasification combined cycle; MSW: municipal solid waste; PV: photovoltaics.

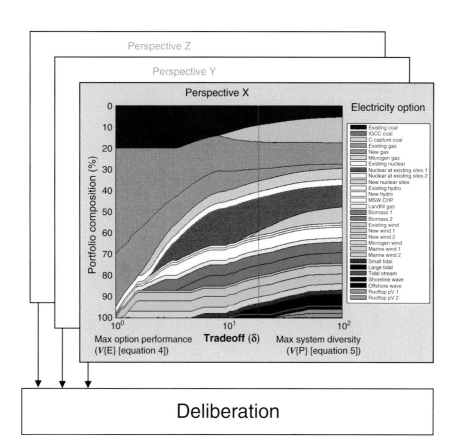

PLATE 1 Illustrative performance–diversity tradeoffs for UK electricity portfolios. IGCC: integrated gasification combined cycle; MSW: municipal solid waste; CHP: combined heat and power; PV: photovoltaic.

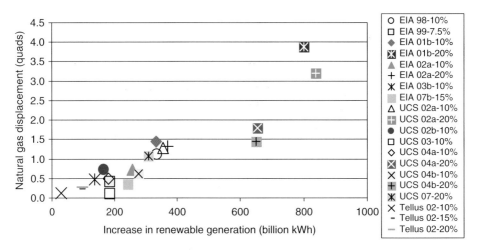

PLATE 2 Forecast natural gas displacement in 2020.

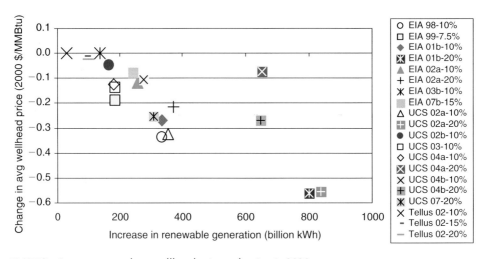

PLATE 3 Forecast natural gas wellhead price reduction in 2020.

CHAPTER 7

The Role of Wind Generation in Enhancing Scotland's Energy Diversity and Security

A Mean-Variance Portfolio Optimization of Scotland's Generating Mix

Shimon Awerbuch[*]**, in collaboration with Jaap C. Jansen and Luuk Beurskens**[**]

Abstract

The UK Energy White Paper sets targets for decarbonization and the deployment of wind and other renewable electricity generating alternatives. The Scottish Executive is committed to increasing renewable energy shares in order to progress on the White Paper objectives. Fossil fuel independence, reliance on domestic sources and enhanced energy security are additional motivating factors for these objectives. Much to its credit, the Scottish Executive is pushing forward with the adoption of wind and other renewables in spite of the widespread belief that these technologies cost more and that increasing their share of the generating mix must therefore increase overall generating costs. The Executive's efforts are especially notable since risk and other externalities, as subsequently discussed, tend to drive market participants to overinvest in fossil technologies relative to wind. The idea that a more costly technology in the mix must raise overall generating cost may seem obvious and compelling. Nonetheless, it is flawed. Energy planning represents an investment decision problem. Investors commonly apply portfolio theory to manage risk and maximize portfolio performance under a variety of unpredictable economic outcomes. This report describes essential portfolio theory ideas and discusses their application to Scotland's electricity generating mix. The report illustrates how wind and other renewables can benefit the Scottish generating mix. Efficient generating portfolios include greater shares of wind, which enhance energy security and also reduce overall

[*]SPRU Energy Research Group, University of Sussex, Brighton, UK
[**]ECN – Energy research Centre of the Netherlands, Petten, the Netherlands

Analytical Methods for Energy Diversity and Security © 2008 Elsevier Ltd.
978-0-08-056887-4

generating cost. The optimal results indicate that compared with National Grid projected mixes, there exist generating mixes with larger wind shares at equal or lower expected cost and risk.

7.1 Least-cost versus portfolio-based approaches in generation planning

Wind and other renewables provide clean generating alternatives, and hence offer an effective climate change mitigation mechanism. Yet policy makers are concerned because of the widespread perception that increasing the deployment of wind will raise the overall cost of generating electricity.

Electricity policy and capacity planning are largely conceived using *least-cost* principles, under which policy makers evaluate generating alternatives using their *stand-alone* costs. These approaches consistently bias in favor of risky fossil alternatives, while understating the true value of wind and similar fixed-cost, low-risk, passive, capital-intensive technologies.

Today's dynamic and highly uncertain future requires better techniques that reflect market risk. Financial investors know that a diversified asset portfolio provides the best means of hedging risk. Given today's uncertainty about future technology cost and performance, it makes sense also to shift electricity policy and planning from its current emphasis of evaluating alternative technologies, to evaluating alternative generating *portfolios* and strategies. Mean-variance portfolio (MVP) theory, an established part of modern finance theory, is highly suited to the problem of planning and evaluating Scotland's electricity generating portfolio.

MVP evaluates generating alternatives not on the basis of their *stand-alone* cost, but on the basis of their portfolio cost, i.e. their contribution to overall portfolio generating cost relative to their contribution to overall portfolio risk. At any given time, some alternatives in the portfolio may have higher costs while others have lower costs, yet over time, the astute combination of resources serves to minimize overall expected generation cost relative to the risk. This report describes an MVP-based analysis that examines the effect of increasing the share of wind generation in Scotland. The results suggest that Scotland's electricity generating mix will benefit from additional wind shares, even under an assumption that it costs more than other alternatives on a stand-alone basis.

Although counterintuitive, the idea that adding more costly wind can actually reduce portfolio-generating cost is consistent with basic finance theory and derives from the fact that wind generating costs do not correlate, or co-move with fossil prices. Wind generating costs are essentially fixed or *riskless* over time and are independent of fossil fuel fluctuations Adding wind therefore helps to *diversify* the generating mix and enhance its cost–risk performance. The operating costs of a generating mix containing 30% wind will fluctuate a lot less year-to-year than one with no wind. This idea, which most investors intuitively understand, is widely interpreted as 'Don't put all your eggs in one basket.'

7.1.1 Portfolio-based planning for electricity generation

Portfolio optimization focuses on generating costs and their risk. Future fossil fuel and other outlays are random statistical variables that move unpredictably over time. No one knows for sure what the price of gas will be next year, just like nobody knows what the stock markets will do. Estimating the generating cost of a particular portfolio presents the same problems as estimating the expected return to a financial portfolio. It involves estimating cost from the perspective of its market risk.

Portfolio optimization locates generating mixes with the lowest expected cost at every level of risk. Risk is measured in the standard finance fashion as the year-to-year variability[1] of technology generating costs. The projected year 2010 generating mix developed by the National Grid Transco (NGC) serves as a benchmark or target starting point. The optimized results indicate that it is possible to improve on the cost–risk properties of the NGC 2010 mix, i.e. there exist other mixes with larger wind shares that exhibit equal or lower expected cost and risk levels.

7.1.1.1 Current and projected Scotland capacity

Most of Scotland's capacity growth is projected to be in the form of onshore wind (Figure 7.1), which provides Scotland with an opportunity to diversify its mix away from fossil generation. The NGC projected 2010 capacity mix includes 4.1 GW of new onshore wind, which combined with existing wind capacity, totals 5.4 GW of onshore wind. Though significant, it exploits less than half of Scotland's sizeable 11.5 GW wind resource, more than the country's total 2004 generating capacity of 10.5 GW.

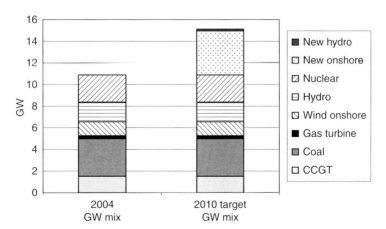

FIGURE 7.1   Current and projected generating capacity for Scotland. CCGT: combined cycle gas turbine.

[1] Measured as the statistical standard deviation.

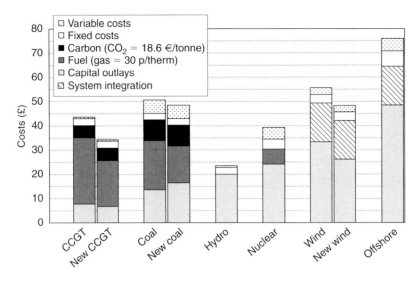

FIGURE 7.2   Base-case stand-alone technology costs. CCGT: combined cycle gas turbine.

7.1.1.2 Technology generating cost

Figure 7.2 shows generating costs for existing capacity as well as new entrants. New entrant costs are based on figures provided by Credit Suisse–First Boston (CSFB, 2005).[2] Natural gas price has been revised from CSFB's 27.0 p/therm to 30.3 p/therm, which is still well below recent levels. The cost of carbon dioxide (CO_2) has also been revised from CSFB's €15/tonne to €18.6/tonne, again, below recent price levels. Embedded generation costs are based on these adjusted CSFB costs, but are further adjusted to reflect fuel efficiencies and construction cost differences between old- and new-generation technologies. CSFB does not include system integration costs for wind, which are estimated at £16/MWh based on a widely cited NGC study (Dale et al., 2004).

7.2 Portfolio optimization of Scotland's generating mix

Portfolio optimization compares the risk–return properties of the projected NGC target mix to a set of optimal portfolios that minimize cost and risk. The results indicate that wind diversifies the mix without raising cost, in spite of the fact that its assumed stand-alone cost (Figure 7.2) exceeds that of other new entrants'.

Portfolio optimization does not advocate for particular generating mixes, but rather displays the risk–cost tradeoffs among various mixes. The results of the

---

[2] Offshore wind costs provided by Airtricity. All generating costs expressed in terms of £/MWh, which can be divided by 1000 to obtain the equivalent p/kWh cost.

analysis presented here are illustrative in the sense that they do not represent a specific capacity expansion plan and generally ignore any requirement to optimize technologies to the load–duration curve. They are meant to illustrate the so-called *portfolio effect*: as long as the mix can be reshuffled over time, adding wind and other fixed-cost technologies has the effect of diversifying the generating mix and reducing its expected cost.

In deregulated environments, investment decisions are make by individual power producers who evaluate only their own direct costs and risks and do not reflect the effects that their technologies may have on overall portfolio performance. Wind investors, for example, cannot capture the risk-mitigation benefits they produce for the generating portfolio, which leads to underinvestment in wind relative to levels that may be more optimal from a customer or societal perspective. Some investors prefer the risk–reward menu offered by fuel-intensive technologies, such as combined cycle gas turbine (CCGT), which have very low capital costs. Given sufficient market power, these investors may be able to externalize fuel risks onto customers. In the presence of market power, these investors do not bear the full risk effects they impose onto the generating mix, which may lead to overinvestment in gas relative to what is more optimal from a total portfolio perspective.

The charts accompanying the subsequent discussion show the portfolio generating cost and risk for a number of scenarios. An infinite number of possible mixes exists on each chart, although only a small set of typical mixes is located and shown.

7.2.1 The base case

Expressing technology costs without their market risk means little. Figure 7.3 shows the base-case cost and risk of each existing and new-entrant technology along with the estimated cost–risk of the NGC 2010 target mix.[3] The chart depicts what is a highly cautious set of costs for onshore and offshore wind. The assumed offshore wind cost, £76/MWh, is nearly 50% higher than similar costs used in the Netherlands and elsewhere.

7.2.1.1 Base-case portfolio optimization results

Figure 7.4 shows the base-case cost and risk for the NGC target as well as the optimized results. The NGC generating mix has an overall cost of £45.10 with a risk of 4%. While there exists an infinite number of portfolios with superior cost–risk properties, four typical mixes are focused on, located on the *efficient frontier*, i.e. the location of all optimized portfolios. The NGC mix lies above the efficient frontier. Any generating portfolio that lies below or to the left of the target represents an improvement in cost and risk. For example, mix N is superior to the target mix. It has the same cost, but lower risk. Mix S is also more desirable, since it

---

[3] In Figure 7.3, 'Nuclear' represents exiting plant, for which capital investment risk is zero so that total risk reflects primarily fuel and other operating costs. Decommissioning risks and the cost–risk of long-term waste disposal are ignored.

reduces the target's cost by about 6% (£42.6/£45.1; Figure 7.4, table insert) without increasing risk.[4]

In the NGC mix, 23% of total electricity produced comes from onshore wind. Mix N, by comparison, contains 31% onshore wind plus 2% offshore wind, in

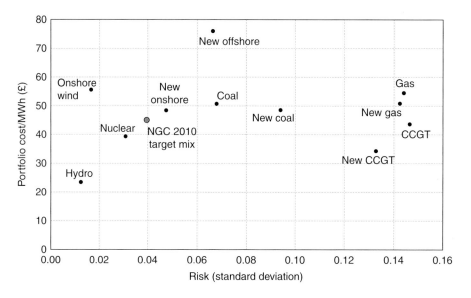

FIGURE 7.3   Base-case technology cost and risk. CCGT: combined cycle gas turbine.

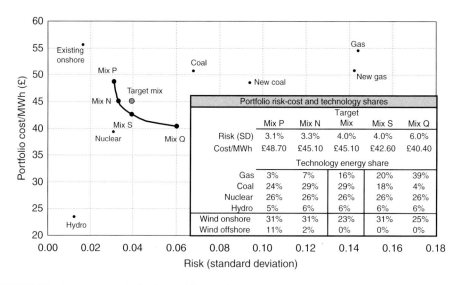

FIGURE 7.4   Base-case optimized portfolio results.

[4] Nuclear is constrained to its 2004 level, which equates to 26% of total generation in 2010.

spite of the fact that wind costs more than new gas. Mix P, which costs about 0.5 p/kWh more than the target, contains 11% offshore wind. These typical mixes would no doubt contain larger wind shares were it not for an arbitrary constraint imposed on the results: onshore wind was limited to two-thirds of Scotland's total available resource. At this level of penetration onshore wind represents 31% of the total mix.

7.2.2 Case II: accelerated (minimum 10%) offshore wind deployment

In the base case, mix S (and Q) contains no offshore wind and mix N contains only a minimal amount. Portfolio theory suggests that other mixes that do contain offshore wind are likely to exist at the same risk level. These mixes are located by searching for optimized solutions that include a given minimum offshore wind share. This minimum wind share is arbitrarily set to 10%. The results (Figure 7.5) indicate that increasing offshore wind to at least 10% of the mix leaves the costs of mix P and mix N unchanged, although the risk of mix N rises to 4%, i.e. mix N shifts to the right from its base-case risk of 3.3%, so it is virtually co-located with the NGC target mix.

Accelerating offshore wind in this manner also raises the cost of mix S by about 0.25 p/kWh, from £42.6/MWh in the base case to £45.1/MWh in Figure 7.5. Mix S is therefore now also virtually co-located with mix N and with the NGC target. In general, therefore, accelerating offshore wind to 10% of the mix produces only minimal cost–risk effects on the overall generating system. While the cost–risk of mixes N and S is almost identical to the NGC target, their technology

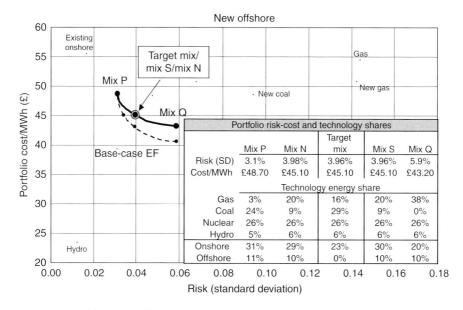

FIGURE 7.5 Portfolio cost and risk: accelerated offshore wind case.

makeup differs. Mixes N and S comprise essentially identical technology shares, which include 10% offshore wind along with about 30% onshore wind. The total wind share of these optimized mixes is therefore 40%, as compared to 23% for the NGC target (which contains no offshore wind). The results of this case therefore indicate that offshore wind can be increased to at least 10%, without raising cost–risk relative to the NGC target. The cost of this move relative to the base case is a minimal 0.25 p/kWh.

7.2.3 Case III: higher 'current outlook' natural gas prices

This case balances highly cautious wind assumptions with higher assumed natural gas prices, which are now taken as 40 p/therm, the approximate current cost of a four-year forward. Since CO_2 cost is correlated with natural gas prices, its value is also raised to €24/tonne (Figure 7.6). System charges remain unchanged at £16/MWh of wind output.

Under the assumptions of this case, generating costs for NGC target mix rise to £47.5 from £45.1 in the base case.[5] Similar cost increases are observed for the four typical mixes, which now generally contain smaller gas generation shares (Table 7.1). As for the previous cases, the optimized mixes N and S show a superior cost–risk relative to the target mix. Optimized onshore wind shares remain unchanged from the base case,[6] although offshore wind increases in mixes P and N.

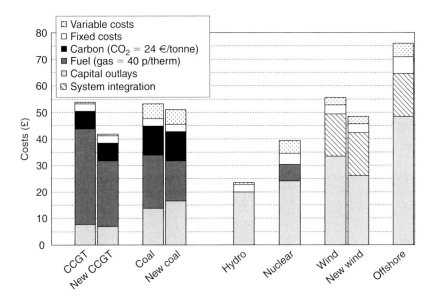

FIGURE 7.6   Technology costs for 'outlook gas' case.

[5] Target mix costs did not change for the accelerated offshore wind case.
[6] Onshore wind is already at its maximum limit, the exception being mix Q, where onshore wind remains at 25%.

7.2.3.1 The basis for a 'no regrets' Scottish wind policy

Figure 7.7 compares the range of optimal solutions for Case III ('current outlook' natural gas prices) with those of Case II, accelerated offshore wind. The efficient frontier (the location of optimal mixes) for these two cases is very similar. This suggests that if gas prices remain at current levels, it makes considerable economic sense for the Scottish Executive to pursue policy options that accelerate offshore wind deployment.

The message of the analysis therefore is that with current market expectations, a mix such as N, which contains 5% offshore wind (Table 7.1), represents a 'no regrets' policy: it costs no more than the target mix yet it lowers risk. If

Table 7.1 Case III: Outlook gas portfolio details

	Mix P	Mix N	Target mix	Mix S	Mix Q
Risk (SD)	3.2%	3.3%	4.1%	4.1%	5.1%
Cost/MWh	£ 49.80	£ 47.50	£ 47.50	£ 44.50	£ 43.50
			Technology energy share		
Gas	2%	4%	16%	20%	25%
Coal	23%	29%	29%	18%	18%
Nuclear	26%	26%	26%	26%	26%
Hydro	5%	6%	6%	6%	6%
Wind onshore	31%	31%	23%	31%	25%
Wind offshore	13%	5%	0%v	0%	0%

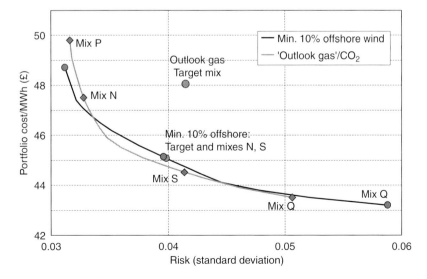

FIGURE 7.7 Optimal solutions: cases II and III.

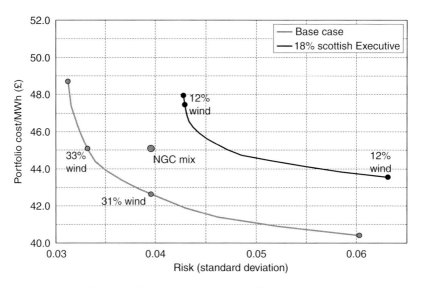

FIGURE 7.8   Cost–risk for Scottish Executive 2010 renewables target.

expected high gas prices are sustained, a Scotland mix consisting of about 31% onshore wind and 10% offshore wind will generally outperform the NGC target mix and will perform at least as well as other mixes with less wind.

7.2.3.2 The cost of not fully exploiting Scotland's onshore wind potential

The Scottish Executive has set a 2010 renewables target of 18% of electricity generated (Scottish Executive, 2001). Hydro shares represent about 6% of generation, so that the Executive's target effectively implies a 12% goal for wind. Lower wind shares generally increase cost–risk. Indeed, relative to the base case, limiting wind to the Executive's 12% target raises cost and risk significantly (Figure 7.8).

While the Scottish Executive has set *targets*, not limits, the message of the analysis is clear: failing to exploit Scotland's wind resources significantly raises cost and risk.

7.3 Conclusions: implications for Scotland's capacity planning

Today's dynamic and uncertain energy environment requires portfolio-based techniques that reflect market risk and de-emphasize stand-alone generating costs. MVP theory is well tested and ideally suited to evaluating national electricity strategies. It helps to identify solutions that enhance energy diversity and security and are therefore more robust than arbitrarily mixing technology alternatives.

Portfolio analysis reflects the cost interrelationship (covariances) among generating alternatives, which is crucial for correctly evaluating generating portfolios. The analysis described here does not represent or advocate for a particular capacity expansion plan. Rather, its purpose is to demonstrate that increasing the share of wind in Scotland generally lowers overall generating costs, even if it is believed that wind costs more than gas. The results suggest that the NGC 2010 mix, although it reflects relatively significant wind shares, may not go far enough. Larger wind shares appear to insulate better the generating mix from systematic risk of gas (and coal) price movements, which have historically been quite correlated.

This report presents a series of optimized portfolio results for several scenarios, using a highly cautious set of cost estimates for wind. It compares these optimized portfolios to the NGC-2010 generating mix, which projects an onshore wind share of 23% and offshore wind of 0%. The optimized results presented here strongly suggest that without increasing cost or risk, onshore wind can be increased to at least 31% of electricity generation: half again as much as the NGC target and nearly 75% more than envisioned by the Scottish Executive's 2010 targets. Even with a highly cautious cost of £76/MWh (compared to gas at 30p/therm) offshore wind shares can rise to at least to as much as 10% (2GW of capacity) without increasing cost.

Moreover, if natural gas prices remain in the range of 40p/therm, as futures prices indicate, a mix of about 31% onshore and 10% offshore wind provides the basis for a no-regrets wind policy for Scotland. The evidence suggests that this mix will outperform the NGC-2010 mix and will perform at least as well as other portfolios with less wind.

Against this backdrop, the Scottish Executive 18% 2010 renewables targets may not be sufficiently aggressive. Indeed, the analysis indicates that reducing wind shares from their optimized levels (31% onshore, 5–10% offshore) significantly increases the cost and the risk of the Scottish mix.

In deregulated environments, investment decisions are made by individual power producers who evaluate only their own direct costs and risks, but do not reflect the effects that their technologies may have on overall generating portfolio performance. For example, wind investors cannot capture the risk-mitigation benefits they produce for the overall portfolio, which leads to underinvestment in wind relative to societally optimal levels. Some investors, however, may prefer the risk menu offered by fuel-intensive gas CC turbines, which have low initial costs. Given sufficient market power, these investors may be able to externalize fuel risks onto customers so that they do not bear the full risk effects they impose onto the generating mix. This would lead to overinvestment in gas relative to what is more optimal from a portfolio or societal perspective.

Given the high degree of uncertainty about future energy prices, the relative value of generating technologies must be determined not by evaluating alternative resources, but by evaluating alternative resource portfolios. Energy analysts and policy makers face a future that is technologically, institutionally and politically complex and uncertain. In this environment, MVP techniques help to establish renewables targets and portfolio standards that make economic and policy sense. They also provide the analytical basis that policy makers need to devise efficient generating mixes that maximize security and sustainability. MVP analysis shows that contrary to widespread belief, attaining these objectives need not increase cost. In the case of Scotland, increasing the share of wind, even if it is believed to cost more on a stand-alone basis, reduces portfolio cost–risk and enhances energy security.

References

Credit Suisse–First Boston (2005). *Electricity Handbook*, Vol. 10 (20 July).

Dale, L., Milborrow, D., Slark, R. and Strbac, G. (2004). Total cost estimates for large-scale wind scenarios in UK, Energy Policy, 32, 1949–1956.

Scottish Executive (2001) *Scotland's Renewable Energy Potential Beyond 2010: A Consultation Paper*. www.scotland.gov.uk/library5/environment/renewenergy2010.pdf

Further reading

Awerbuch, S. (2000). Getting it right: the real cost impacts of a renewables portfolio standard. *Public Utilities Fortnightly* (15 February).

Awerbuch, S. (2005). Portfolio-based electricity generation planning: policy implications for renewables and energy security. *Mitigation and Adaptation Strategies for Global Change* (in press).

Awerbuch, S. and Berger, M. (2003). *Energy Security and Diversity in the EU: A Mean-Variance Portfolio Approach*. IEA Report No. EET/2003/03. Paris: International Energy Agency (February). http://library.iea.org/dbtw-wpd/textbase/papers/2003/port.pdf

Berger, M. (2003). *Portfolio Analysis of EU Electricity Generating Mixes and Its Implications for Renewables*. Ph.D. Dissertation, Technischen Universität Wien, Vienna (March).

Bolinger, M., Wiser, R. and Golove, W. (2004). Accounting for fuel price risk when comparing renewable to gas-fired generation: the role of forward natural gas prices. *Energy Policy*. Vol. 34, 706–720.

Brealey, R. A. and Myers, S. C. (any edition) *Principles of Corporate Finance*. New York: McGraw-Hill.

Fabozzi, F., Gupta, F. and Markowitz, H. (2002). The legacy of modern portfolio theory. *Journal of Investing*, 11(Fall), 7–22.

Seitz, N. and Ellison, M. (1995). *Capital Budgeting and Long-Term Financing Decisions*. Chicago, IL: Dryden Press.

CHAPTER 8

Generation Portfolio Analysis for a Carbon Constrained and Uncertain Future

Ronan Doherty*, Hugh Outhred** and Mark O'Malley***

Abstract

Many modern electricity systems are faced with the challenge of reducing greenhouse gas emissions and dealing with increasing and more volatile fuel prices. Adequately dealing with these issues requires the evolution of suitable generation portfolios. However, doubts remain over whether the liberalized marketplace will deliver such portfolios. Analysis is undertaken to try to determine how the generation portfolio on the all-Ireland system may evolve by 2020. Resulting portfolios are examined with respect to the impact of carbon costs on the portfolio development and, in particular, wind energy. An assessment is made of the exposure of the portfolios to fuel price volatility and how portfolios may be diversified to avoid this. The analysis tries to gain insight into future generation portfolios with the aim of informing how policy instruments may be tailored to address these issues.

Key words: Energy resources, environmental factors, fuel diversity, generation planning, power system economics.

Acknowledgements

This work has been conducted in the Electricity Research Centre, University College Dublin, Ireland, which is supported by Electricity Supply Board (ESB), ESB National Grid, Commission for Energy Regulation, Cylon, Airtricity and Enterprise Ireland. The authors gratefully acknowledge Garth Bryans, Morgan Bazilian of SEI, and John FitzGerald and Niamh McCarthy of the ESRI for their help during this work.

*Electricity Research Centre, University College Dublin, Ireland
**School of Electrical Engineering and Telecommunications, University of New South Wales, Sydney, Australia
***Electricity Research Centre, University College Dublin, Ireland

Analytical Methods for Energy Diversity and Security © 2008 Elsevier Ltd.
978-0-08-056887-4 All rights reserved.

8.1 Introduction

Modern electricity systems are faced with many challenges, such as pressure to reduce greenhouse gas emissions and increasing fuel prices and fuel price volatility. In the past such issues could have been considered in central generation resource planning in an attempt to best meet the future needs of consumers, the economy and society (Khatib, 2003). With the recent onset of market liberalization in many systems, there has been a corresponding de-emphasis on central planning and it remains unclear whether market forces will deliver suitable generation portfolios to deal with such issues (Bergin et al., 2005).

Wind generation is seen in many countries as an important means to reduce greenhouse gas emissions (Neuhoff, 2005). However, the characteristics of wind generation differ from those of conventional generation and doubts remain over whether the features of wind generation are reflected properly in electricity markets that were designed to suit conventional generation.

As the proportion of gas-fired generation increases in many systems, concerns grow about the overreliance on gas, which can exhibit volatile price patterns. This is of even more concern in systems that import a large proportion of their fossil fuel needs, and it has been suggested that in many liberalized markets there is little incentive for investors or utilities to diversify generation resources (FitzGerald, 2002; Roques et al., 2005).

These issues pose a serious challenge to policy makers in many systems, including the all-Ireland system. Responding to these issues in the current environment may be more challenging than it has been in the past, as it must, in general, be done in parallel with an electricity market. A combination of market design features, direct or indirect subsidies, or even new state-owned generation may be necessary to address these issues fully. However, the correct regulatory interventions or policies cannot be initiated without analysis and insight into how a system evolves into the future.

This chapter attempts to gain insight into possible generation portfolios on the all-Ireland system in 2020, with an aim of informing how policy may be tailored to address the issues mentioned above. The all-Ireland system has an installed capacity of approximately 8000 MW and is currently in the process of introducing a new electricity market (All-Ireland Project, 2005). The system is currently struggling with its emissions targets (Energy Policy Statistical Support Unit of Sustainable Energy Ireland, 2005) and has an increasing dependence on gas-fired generation. Ireland also has one of the best wind resources in the world. This chapter uses a methodology developed in Doherty (2005) and Doherty et al. (2006) to determine least-cost-generation portfolios for various scenarios in 2020. Issues of load duration, plant utilization and system capacity are dealt with in the analysis and the unique characteristics of wind generation are also accounted for. The resulting portfolios can be viewed as what may result from a liberalized market if no intervention takes place and are assessed with respect to the issues highlighted above.

Section 8.2 gives a brief outline of the generation options and the least-cost portfolio optimization algorithm. Analysis in Section 8.3 aims to gain insight into the impact of carbon costs and the role of wind generation, and the resultant effects on the generation portfolios. Section 8.4 focuses on analysing portfolios with insufficient generation diversity and exposure to fuel price volatility and investigates possible responses. Conclusions are given in Section 8.5.

8.2 Generation options and portfolio optimization

Least-cost portfolio analysis was undertaken for the 2020 all-Ireland system. It was assumed that, of the current generation capacity, only 800 MW of interconnection and 509 MW of hydro generation will remain in 2020. It was assumed that there is a maximum of 3800 MW of usable wind generation resource, enough to serve 22% of electricity demand. Further details of the least-cost portfolio analysis can be found in Doherty (2005) and Doherty et al. (2006).

8.2.1 Generator inputs

An extensive list of generator data was gathered for this work. Unit sizes, characteristics, efficiencies and costs were gathered from several sources (Commission for Energy Regulation, 2004; ESB National Grid, 2004; Royal Academy of Engineering, 2004). Table 8.1 shows the generators, costs efficiencies and characteristics assumed achievable in the all-Ireland system by 2020.

8.2.2 Fuel prices

Two fuel price scenarios are used in this work, a low fuel price scenario which is based on 2005 fuel prices and a high fuel price scenario based on projected 2020 fuel prices (Table 8.2). The fuel price scenarios were compiled from several sources (European Commission, 2003; Commission for Energy Regulation, 2004; ESB National Grid, 2004; Royal Academy of Engineering, 2004). The most notable feature in the high fuel price scenario, compared to the low, is the relatively high price of gas compared to the other fuels.

8.2.3 Generation adequacy

It is essential that each power system has enough capacity to serve the load to the extent defined by a system reliability criterion. Intermittent, non-dispatchable sources of generation, such as wind generation, make a different contribution to the generation adequacy of a system compared with conventional dispatchable generation. Capacity credit studies (Doherty, 2005; Doherty et al., 2006) were undertaken for each generation type to ensure that each portfolio has sufficient

Table 8.1 Generation costs and characteristics

Plant type	Notional size of installation (MW)	Plant life (years)	Build time (years)	Average efficiency (%)	Capital cost (€/MW)	O&M (€/MW p.a.)	CO_2 emissions (t CO_2/MWh)	Availability (%)	Resource limited (MW)
Coal PF	1,000 (3 × 333 MW)	30	4	37	1,479,200	34,800	0.92	84	–
Coal IGCC	800 (2 × 400 MW)	25	5	48	1,761,321	69,000	0.71	84	–
Peat FB	150	25	4	37	1,223,807	55,200	1.15	87	1,000
OCGT	110	20	1	43	518,411	36,000	0.47	92	–
CCGT	390	20	2	56	537,500	50,000	0.36	88	–
Wind 1 (onshore)	30 (15 × 2 MW)	20	2	–	981,475	34,800	0.00	–	1,200
Wind 2 (mix)	30 (15 × 2 MW)	20	2	–	1,028,775	54,250	0.00	–	2,600
Biomass and biogas 1	10	20	2	–	2,418,750	80,000	0.00	78	70
Biomass and biogas 2	10	20	2	–	3,386,250	90,000	0.00	78	50
Biomass and biogas 3	10	20	2	–	4,353,750	90,000	0.00	78	500

O&M: operation and maintenance; CO_2: carbon dioxide; PF: pulverized fuel; IGCC: integrated gasification combined cycle; FB: fluidized bed; CCGT: combined cycle gas turbine; OCGT: open cycle gas turbine.

Table 8.2 Fuel price scenarios

Fuel	Low (€/GJ)	High (€/GJ)
Gas	4.31	5.54
Coal	1.59	1.65
Peat	2.64	3.12

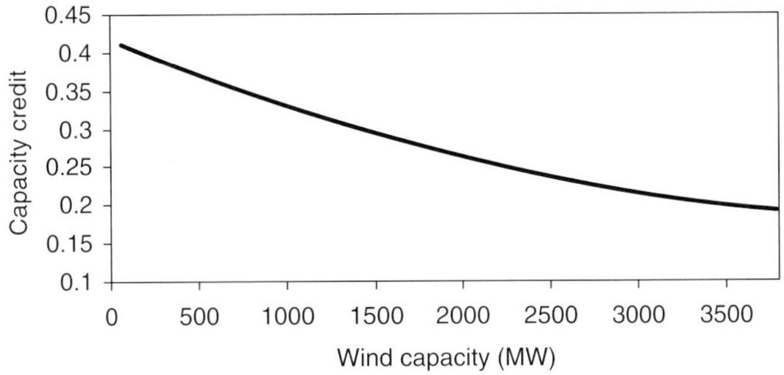

FIGURE 8.1 Capacity credit of wind generation.

capacity to meet a loss of load expectation (LOLE) of eight hours per year (ESB National Grid, 2003). Figure 8.1 shows the capacity credit of wind generation as a function of the installed wind capacity.

8.2.4 Least-cost portfolio optimization

A linear programming portfolio optimization algorithm (Doherty, 2005; Doherty et al., 2006) was used to find the mix of generation technologies that, for a given set of inputs, results in the load being met at least cost. Issues of plant utilization, load duration and generation adequacy are included in the analysis. However, temporal system aspects and unit startup factors are not included. The algorithm optimizes the installed capacity of each type of generation in the portfolio and optimizes how they are utilized with respect to the load duration curve. The resulting portfolios are presented and analyzed in the following sections.

8.3 Carbon costs and the role of wind generation

8.3.1 Least-cost generation portfolio results

Least-cost portfolios for the all-Ireland in 2020 system were examined for various fuel price and carbon cost scenarios to gain insight into desirable generation portfolios. Tables 8.3 and 8.4 show the installed capacities of the least-cost portfolios.

Table 8.3 Portfolios with increasing carbon tax for low fuel prices

Plant type	Installed capacity (MW)					
	0 €/t CO$_2$	10 €/t CO$_2$	20 €/t CO$_2$	30 €/t CO$_2$	40 €/t CO$_2$	50 €/t CO$_2$
Coal PF	7,060	0	0	0	0	0
Coal IGCC	0	0	0	0	0	0
Peat FB	0	0	0	0	0	0
OCGT	1,732	2,278	2,572	2,156	2,469	2,405
CCGT	0	6,289	5,826	6,241	5,805	5,782
Wind 1 and 2	0	600	1,200	1,200	1,800	2,400
Biomass 1, 2 and 3	0	0	0	0	0	0
Interconnection	800	800	800	800	800	800
Hydro	509	509	509	509	509	509
Total	10,101	10,476	10,907	10,906	11,383	11,896

CO$_2$: carbon dioxide; PF: pulverized fuel; IGCC: integrated gasification combined cycle; FB: fluidized bed; OCGT: open cycle gas turbine; CCGT: combined cycle gas turbine.

Table 8.4 Portfolios with increasing carbon tax for high fuel prices

Plant type	Installed capacity (MW)					
	0 €/t CO$_2$	10 €/t CO$_2$	20 €/t CO$_2$	30 €/t CO$_2$	40 €/t CO$_2$	50 €/t CO$_2$
Coal PF	7,060	6,560	5,229	0	0	0
Coal IGCC	0	0	0	0	0	0
Peat FB	0	0	0	0	0	0
OCGT	1,732	1,838	2,469	2,381	2,374	2,374
CCGT	0	0	576	5,765	5,700	5,630
Wind 1 and 2	0	1,200	1,800	2,800	3,800	3,800
Biomass 1, 2 and 3	0	0	0	0	0	70
Interconnection	800	800	800	800	800	800
Hydro	509	509	509	509	509	509
Total	10,101	10,907	11,383	12,255	13,183	13,183

CO$_2$: carbon dioxide; PF: pulverized fuel; IGCC: integrated gasification combined cycle; FB: fluidized bed; OCGT: open cycle gas turbine; CCGT: combined cycle gas turbine.

The cost of carbon in the all-Ireland system may be based on the European traded cost of carbon and may also include an additional factor reflecting the penalties incurred by not meeting international emissions targets. The cost of carbon is included here in the form of a carbon tax. If regulatory bodies can ensure that the cost of carbon is properly reflected in the marketplace, it is reasonable to assume, given the inputs, that these are the sorts of generation portfolio that the industry will be heading toward in the year 2020.

It can be seen that a combined cycle gas turbine (CCGT)-based system is found to be least cost once the carbon tax is $10 \euro /$ tonne of carbon dioxide (tCO_2) and above for the low fuel price scenario and $30 \euro /tCO_2$ and above for the high duel price scenario. This is consistent with the industry in Ireland at present, where most proposed generation projects are for the development of new CCGTs. It can also be seen that the optimal penetration of wind power increases as expected with increasing carbon tax. The increased gas price in the high fuel price scenario accelerates the role of wind as a means of reducing carbon emissions in a least-cost manner.

8.3.2 Emissions

Under the Kyoto Protocol, the Republic of Ireland must limit its increase in greenhouse gas emissions to 13% above 1990 levels in the period 2008–2012. In 2003, greenhouse gas emissions in Ireland were 25% above 1990 levels (Energy Policy Statistical Support Unit of Sustainable Energy Ireland, 2005). It can be seen from Tables 8.3 and 8.4 that properly reflecting the cost of carbon in the marketplace is important to give a signal as to the appropriateness of coal-based generation. This factor has a large impact on emissions. Given the long lifespan of generation plant, inappropriate coal plant operating during periods of high carbon cost may cause significant and unnecessary cost to the system and wider economy. Figure 8.2 shows the effect of the various carbon taxes on the emissions from the generation portfolios. This is expressed as a percentage of the scenarios with no carbon tax.

Even in a gas-based system, the presence of a carbon tax plays an important role with respect to emissions by signalling the appropriate penetration of wind

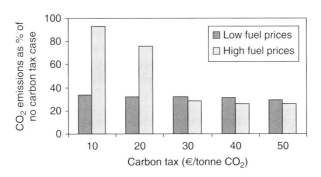

FIGURE 8.2   Emission from portfolios for various carbon taxes and fuel price scenarios.

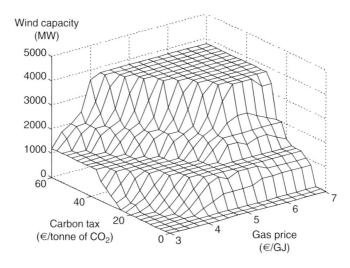

FIGURE 8.3   Wind capacity in least-cost portfolio for various carbon tax and gas price.

generation. For the all-Ireland system in 2020 it was found that 3800 MW of wind generation could result in a reduction in CO_2 of 21% from a purely gas-based system.

8.3.3 Role of wind generation in portfolios

An examination of the role of wind generation in least-cost portfolios was undertaken for a large range of gas price and carbon tax scenarios. These two variables have a large impact on the generation portfolios and have a significant uncertainty associated with them. It can be seen from Figure 8.3 that wind generation plays a significant role in least-cost portfolios for a large range of scenarios. It can also be seen that for a considerable number of scenarios the optimal wind capacity was found to be the maximum amount assumed available, 3800 MW.

The Republic of Ireland currently aims to serve 13.2% of electricity from renewable sources by 2010 (European Union, 2001). The analysis here assumes that there are no regulatory or market factors obstructing the development of wind generation and that the cost of carbon is fully reflected in that marketplace. This approach would be correct macroeconomic practise but is currently not the case in the all-Ireland system in 2005. The results show that under these conditions there would be a significant development of renewable energy for many of the future scenarios. This approach would help toward meeting renewable energy targets, perhaps without the need for additional subsidy.

8.3.4 Effect of increasing wind capacity

Analysis was carried out to examine the effect of increasing wind capacity on the mix of remaining generation in the least-cost portfolios. Figure 8.4 shows the installed capacities of the generation in the least-cost portfolios with increasing

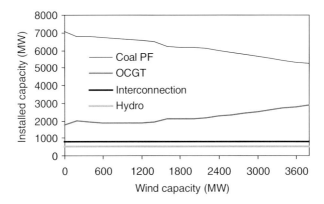

FIGURE 8.4 Installed capacities of generation in least-cost portfolios against wind capacity for the low fuel price, no carbon tax scenario. PF: pulverized fuel; OGCT: open cycle gas turbine.

wind generation for the low fuel price scenario with no carbon tax. It can be seen that the increasing wind capacity causes a decrease in the amount of base loaded plant and an increase in the amount of peaking capacity required in least-cost portfolios. This is due to the change that wind generation causes in the net-load duration curve. Similar trends were found for portfolios that had CCGTs as the base loaded plant. This behavior is in contrast to the impact that wind capa-city has on existing portfolios, where it normally displaces the units with higher incremental costs. To insure efficient generation portfolios as wind capacity increases it becomes more important that the correct signals for reserve, midload plant and peaking plant are provided in the marketplace in the long term.

8.4 Uncertainty and portfolio diversification

8.4.1 Background

The all-Ireland system relies heavily on imported fuel for electricity production. In 2003, in the Republic of Ireland, 88% of electricity produced was from imported fuels (IEA, 2003). This means that Ireland may be exposed to price hikes and even possible shortages in supply as a result of economic and political changes in other countries. Assessing the detrimental effects that such events may have on the economy of the island is a difficult task and an important factor may be the level of exposure of competing economies to similar events.

The results given in Section 8.3 and analysis of the industry in Ireland at present suggest that it is very likely that gas-fuelled plant will become the dominant means of electricity production in the future. Analysis carried out by FitzGerald (2002) suggested that in 2001, in the Republic of Ireland, gas consumption accounted for 0.38% of gross national product (GNP). By 2010 this figure may rise significantly. If gas were supplying 80% of electricity needs at the high fuel price scenario given in Section 8.2, then gas consumption may account for up to 1.01% of GNP. A significant gas price shock at this stage would result in significant cost to the economy and a loss of competitiveness with respect to economies with lower gas price exposure. It is generally accepted that diversification of fuel resources will serve to reduce the exposure to such risk.

8.4.2 Diversity

Correctly quantifying by how much to diversify generation resources and determining what to diversify with is a challenging problem faced by policy makers. There is little agreement on the best approach on which to base generation diversification, with the large degree of unquantifiable uncertainty about the future proving conceptually challenging. However, two approaches have emerged as being possibly suitable.

Awerbuch and Berger (2003) adopt the approach of mean variance portfolio theory to create generation portfolios that can be analyzed on a risk-return basis. This approach requires probabilistic quantification of the uncertainty of various factors. The analysis includes fuel price risk, and the authors derive a cross-correlation matrix for the price of electricity generated from the various fuel types.

Sterling (1994) argues that mean-variance portfolio (MVP) theory is not appropriate for dealing with exposure to fuel price fluctuations, as they have no pattern. The author states that diversification is a response to ignorance rather than quantifiable risk and suggests that diversity should be quantified using the Shannon–Wiener index. The author seeks diversity as a goal in itself, rather than as a means of reducing something specific.

Each approach has its strengths and weaknesses (Roques et al., 2005). In this section both approaches are examined in the context of the all-Ireland system in 2020.

8.4.2.1 Mean-variance portfolio theory

This approach requires the mean and standard deviation of the cost of electricity produced by the various fuel types. These were derived from historic fuel prices and it was found that electricity produced from gas and coal plant had a standard deviation of 8€/MWh and 4.2€/MWh, respectively, over the period considered. The correlation coefficient was found to be 0.3. It is assumed that the standard deviation is zero for the cost of electricity produced from peat, biomass and wind generation. These values and assumptions are in line with the literature (FitzGerald, 2002; Awerbuch and Berger, 2003; Bergin et al., 2005).

8.4.2.2 The Shannon–Wiener index

The Shannon–Wiener index is mainly used in ecology as a measure of diversity. Sterling (1994) suggests that the index is also suitable for examining diversity in generation portfolios, as it does not require any 'pretence to knowledge' over the future in terms of probabilistic measures. The Shannon–Wiener index H is defined as in Equation (1), where p_n is the proportion of generation represented by the generation type n:

$$H = \sum_{i=N}^{N} -p_n \ln(p_n) \tag{1}$$

With just one generation type the index has the value of zero. With two equal generation elements it has a value of 0.69; this rises to 1.1 with three equal elements and rises above 2 with seven equal elements.

8.4.3 All-Ireland portfolio illustration

Both approaches to assessing the diversity or exposure to risk of the generation portfolio were applied to the all-Ireland system in 2020 for the high fuel price scenario and the 30€/tCO_2 carbon tax. The least-cost portfolio optimization was run for a wide range of portfolio options in order to search the space. Figures 8.5 and 8.6 show the results of the analysis. They plot the cost of electricity of the generation from a particular portfolio versus the standard deviation of the electricity cost and the Shannon–Wiener index respectively. The x-axis of Figure 8.6 has been reversed for ease of comparison. Table 8.5 shows the makeup of the portfolios shown in Figures 8.5 and 8.6. Also shown is the standard deviation of the electricity cost and the Shannon–Wiener index for the current all-Ireland generation portfolio and the efficient frontiers. An efficient frontier is the frontier at which the price cannot be reduced any further without accepting an increase in the volatility or a decrease in diversity. In reality, desirable portfolios should be at or near the efficient frontiers.

What can be noticed from the application of both techniques to the all-Ireland system is that although the methods are conceptually different at a high level when they are applied to a scenario with limited options they result in similar outcomes. The main difference between the two techniques it that the MVP technique is less favorable toward gas-based generation, as the inputs suggest that gas-based generation is more likely to be problematic.

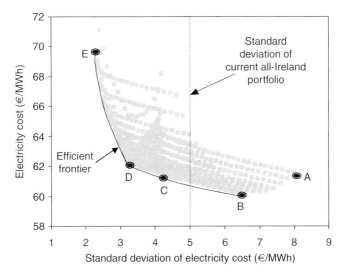

FIGURE 8.5   Mean-variance portfolio analysis.

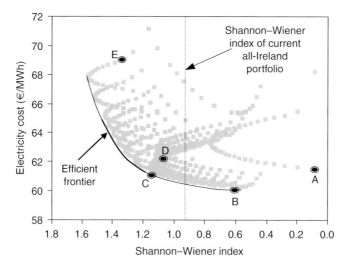

FIGURE 8.6   Shannon–Wiener index portfolio analysis.

Table 8.5   Diversity analysis: significant portfolio

Plant type	Installed capacity for portfolio (MW)				
	A	B	C	D	E
Coal PF	0	0	0	0	0
Coal IGCC	0	0	2,307	4,180	3,506
Peat FB	0	0	0	0	1,000
OCGT	2,013	2,381	2,374	2,374	2,374
CCGT	6,778	5,765	3,394	1,451	721
Wind 1 and 2	0	2,800	3,800	3,800	3,800
Biomass 1, 2 and 3	0	0	0	70	474
Interconnection	800	800	800	800	800
Hydro	509	509	509	509	509
Total	10,100	12,255	13,184	13,184	13,184

PF: pulverized fuel; IGCC: integrated gasification combined cycle; FB: fluidized bed; OCGT: open cycle gas turbine; CCGT: combined cycle gas turbine.

In general, a cost-reflective market should result in electricity being served at least cost. For this scenario this would correspond to portfolio B and would result in a relatively low level of fuel diversity and possible high levels of exposure to gas price spikes. If regulatory bodies do not ensure that the cost of carbon is reflected in the marketplace then this may result in less wind generation, causing less diversity and heavier reliance on gas, i.e. portfolio A. In both cases the generation portfolio is heavily reliant on generation from gas and considerably less diverse than the current all-Ireland generation portfolio.

It can be seen that portfolios that are more diverse than the current portfolio are achievable and may not necessarily come at a significant increase in cost relative to the least-cost portfolio. Portfolios C and D include some integrated gasification combined cycle (IGCC) coal plant which decreases the reliance on gas, and which only slightly increases the cost of electricity. (IGCC plant was found to be more economic than pulverized fuel plant owing to the high carbon tax.) Portfolio C, which lies on the efficiency frontier in both sets of analyses, looks appropriate for this scenario. In this portfolio gas plant and coal plant each serve about 32% of the energy demand, while wind serves 22% and interconnection and hydro serve about 14%.

8.4.4 Insuring diversity in generation portfolios

Analysis in FitzGerald (2002) and Roques et al. (2005) suggests that, in systems where gas generation is generally setting the market price, utilities are unlikely to invest in non-least-cost technologies for the sake of diversity. Given this and the current observable trends in the industry, it is difficult to envisage how the future all-Ireland generation portfolios will maintain adequate diversity in the liberalized marketplace unless there is some sort of targeted intervention. However, the form that this intervention should take remains unclear. Direct state subsidies to specific industries have occurred in the past but may no longer be appropriate and may undermine the establishment of a fully liberalized marketplace. One approach may be to develop a market instrument that would provide an incentive to diversify, possibly based on one of the metrics used in the analysis here. Assessing how much diversity is appropriate or assessing the economic value of increased diversity is a challenging problem. Such an instrument would also have to be consistent and systematic to provide the correct long-term signals for investment. The new market structure in the all-Ireland system includes a capacity payment mechanism (All-Ireland Project, 2005), which could be thought of as catering for elements of 'public good' or elements that may be insufficiently dealt with in the market. An option may be to deal with the portfolio diversity issue within this mechanism by somehow weighting payments with respect to diversity.

8.5 Conclusions

This chapter presented high-level analysis of how the generation portfolio on the all-Ireland system may evolve by 2020. Portfolios were assessed in relation to the issues surrounding the reduction of CO_2 emissions and the exposure of fuel price shocks.

Analysis suggested that fully reflecting the cost of carbon in the marketplace was important in terms of signalling for the appropriate generation capacity in the long term. It is suggested that the integration of wind energy is the main route to increasing renewable energy and decreasing emissions in the all-Ireland system. Reflecting the cost of carbon in the marketplace is important for wind generation but it is essential that other issues are also addressed. Markets, which have been designed to suit conventional generation, may place unnecessary obstacles to the development of wind generation and it is important that these obstacles are removed. The present results showed that wind generation also causes a change in the makeup of the remaining generation in the least-cost portfolio. With wind generation it becomes more important that the correct long-term signals for reserve, midload plant and peaking plant are provided in the marketplace.

Two methods of assessing generation diversity were applied to the all-Ireland system in 2020. Despite being conceptually different approaches, the outcomes, in a system with limited options, were quite similar. It appears that direct intervention will be necessary in the all-Ireland system to insure that the system does not become overdependent on gas-fired generation. It may be desirable to design a market instrument to provide an incentive for diversity. However, valuing diversity and designing the appropriate mechanism will be a challenging task.

To some extent various market instruments, such as carbon taxes and diversity inducements, may provide participants with conflicting incentives (Bergin et al., 2005). This however, reflects the competing priorities of policy makers facing multiple objectives. If there is just one clear, well-designed instrument reflecting each objective then the marketplace should deliver the least-cost solution to the specified objectives.

References

All-Ireland Project (2005). *Single Electricity Market (SEM) High Level Design Decision.* http://www.allislandproject.org

Awerbuch, S. and Berger, M. (2003). *Applying Portfolio Theory to EU Electricity Planning and Policy-Making.* http://www.iea.org (February)

Bergin, A., FitzGerald, J., Keeney, M., O'Malley, E. and Scott, S. (2005). *Aspects of Irish Energy Policy.* http://www.esri.ie

Commission for Energy Regulation (2004). *Best New Entrant Price 2005.* http://www.cer.ie

Doherty, R. (2005). *New Methods for Planning and Operating Modern Electricity Systems with Significant Wind Generation.* Ph.D. Thesis, University College Dublin.

Doherty, R., Outhred, H. and O'Malley, M. (2006). Establishing the role that wind generation may have in future generation portfolios. *IEEE Transactions on Power Systems*, 21, 1415–1422.

Energy Policy Statistical Support Unit of Sustainable Energy Ireland (2005). *Energy in Ireland 1990–2003: Trends, Issues and Indicators.* www.sei.ie

ESB National Grid (2003). *Generation Adequacy Report 2004–2010.* http://www.eirgrid.com

ESB National Grid (2004). *Impact of Wind Power Generation in Ireland on the Operation of Conventional Plant and the Economic Implications.* http://www.eirgrid.com

European Commission Directorate-General for Energy and Transport (2003). *European Energy and Transport – Trends to 2030.* http://europa.eu.int

European Union (2001). *Directive 2001/77/EC of the European Parliament and the Council of 27 September 2001 on the Promotion of Electricity Produced from Renewable Energy Sources in the Internal Electricity Market.* http://europa.eu.int/

FitzGerald, J. (2002). *The Macro-Economic Implications of Gas Dependence.* http://www.esri.ie

International Energy Agency (2003). *Energy Policies of IEA Countries – Ireland 2003 Review.* http://www.iea.org

Khatib, H. (2003). *Economic Evaluation of Projects in the Electricity Supply Industry.* No. 44. IEE Power and Energy Series. Paris: IEA.

Neuhoff, K. (2005). Large-scale deployment of renewables for electricity generation. *Oxford Review of Economic Policy,* 21, 88–110.

Roques, F., Newbery, D., Nuttall, W., Connors, S. and de Neufville, R. (2005). The diversification value of nuclear power as part of a utility technology mix when gas and carbon prices are uncertain. In *Proceedings of the IAEE European Energy Conference,* Bergen (August).

Royal Academy of Engineering (2004). *The Costs of Generating Electricity.* www.raeng.org.uk

Sterling, A. (1994). Diversity and ignorance in electricity supply investment. *Energy Policy,* 22, 195–216.

CHAPTER 9

The Economics of Renewable Resource Credits

Christiaan Hogendorn[1] and **Paul Kleindorfer**[2]

Acknowledgements

The authors acknowledge helpful comments by Shimon Awerbuch and Jonathan Lesser on an earlier draft of this chapter, as well as comments by an anonymous reviewer. This chapter is dedicated to the memory of Shimon Awerbuch, whose influence in the development of market-based approaches to renewable energy development was profound.

9.1 Introduction

Liberalization of electricity markets has provided consumers with a choice of power suppliers. In turn, this means that power suppliers can differentiate their products to appeal to consumer tastes. While electricity is in some ways a commodity product, one major source of potential differentiation is the method of generating the power. 'Green' or renewable sources of power, e.g. wind, solar, biomass and hydro, cause less pollution and produce less toxic waste than non-green alternatives. Some renewables also clearly generate less carbon dioxide (CO_2) emissions per kWh of energy delivered, and they have the potential therefore to assist in meeting national or regional targets to reduce CO_2 emissions. As analyzed in the pioneering work of Awerbuch and Berger (2002), renewables also contribute to the diversity and local sourcing of energy and therefore improve the security of domestic and regional energy supply. Renewable energy sources have been given great impetus by Renewable Portfolio Standards in many US states

[1] Associate Professor of Economics, Wesleyan University, Middletown, CT, USA
[2] Anheuser-Busch Professor of Management Science (Emeritus), The Wharton School, University of Pennsylvania, Philadelphia, PA, USA

Analytical Methods for Energy Diversity and Security © 2008 Elsevier Ltd.
978-0-08-056887-4 All rights reserved.

and in regional agreements such as the European Union (EU) Directive 2001/77/ EC, which set a renewable energy target to increase the portion of electricity generated from renewable sources in the EU to 22% by 2010.

As a result of both consumer preferences and national targets for renewable energy production, there has been a growth in interest in renewable energy credits or 'tradable green certificates' (TGCs) and other approaches to promote the development and deployment of renewable generation technologies. These have taken several forms, including the nature of the technologies covered by these credits (e.g. specifically for biofuels, solar or wind or covering a broader class of renewable energy sources) and in terms of the reporting or payment obligations attached to these credits.[1] There are many differences in national programs to promote renewable energy, by either discouraging non-renewables or encouraging renewables. However, specific programs can take many forms, including the following:

- *The generic renewable portfolio standard (RPS) approach applied to generators*: Constraints are imposed on generators to include in their generation mix a certain percentage of specific types of renewables [measured in terms of total megawatt hours (MWh) sold into the grid], where satisfying this constraint can be accomplished by owning/leasing the generation source, by contracting for it through bilateral or tolling contracts, or by TGCs.
- *The generic RPS approach applied to suppliers/distributors*: Constraints are imposed on suppliers/distributors to include in their supply portfolios a certain percentage of specific types of renewables (measured in terms of total MWh sold to consumers in their territory), where satisfying this constraint can be accomplished by bilateral agreements with generators (who are then required to provide proof of the particular RPS characteristics they advertise), or by TGCs.
- *The generic RPS approach applied to consumers*: Constraints are imposed on customers to include in their consumptions portfolios a certain percentage of specific types of renewables (measured in terms of total MWh bought by these consumers), where satisfying this constraint can be accomplished by showing that they have purchased energy from suppliers who satisfy the appropriate RPS characteristic (who are then required to provide proof of the particular RPS characteristics they advertise), or by TGCs.
- *The generic carbon tax approach*: For each tonne of CO_2 emitted, a payment is made to a carbon fund or a carbon emissions permit is obtained from a tradable emissions permit market, where such markets may be subject to banking, grandfathering and airshed totals, as well as other administrative rules. The logic here derives from the SO_2 and NO_x markets, used to control acid rain and precursors of atmospheric ozone.
- *The generic Renewable Energy Feed-in Tariff (REFIT) approach*: Payments are made to renewable generators according to a certain agreed schedule, set in advance

[1] See Lauber (2004), Holt and Bird (2005) and Ford et al. (2007) for details of some of the different approaches to renewable energy credits implemented in the USA and Europe.

in consultation with stakeholders and enforced by regulation (with audits and sanctions). This schedule can either be a single payment per MWh actually delivered to the grid, or it could be a multipart payment consisting of a fixed payment per MW/year as well as a payment per MWh delivered to the grid.[2] Both the capacity charge (if any) and the energy charge for REFIT programs are often implemented administratively (i.e. by setting a fixed subsidy per MW or MWh for a specific period). The guaranteed charges can either be paid for through government subsidies or mandated by regulatory fiat or statute, and paid ultimately by consumers.[3]

- *Mixed or hybrid approaches*: The above approaches can be combined in various ways. For example, taxes on fossil fuel plants can be used to provide funds for REFIT programs. RPS-type programs can also be coupled with carbon taxes or with REFIT programs. Programs can also be mandatory (regulatory fiat) or voluntary (consumer driven). The RPS/TGC approach was a voluntary scheme in its original incarnation in California, but 18 states in the USA and some EU countries have imposed mandatory targets and associated TGC arrangements, notably Belgium, Italy, Sweden and the UK.

The point developed in this chapter is that implementing any of these programs that involves either assuring RPS standards or providing subsidies to renewable energy sources per MWh produced can be accomplished efficiently by TGCs. It is also argued that if the government wishes to provide subsidies (as opposed to mandating renewables use through one of the RPS approaches), it can do so through participation in the TGC market. The use of TGCs to implement payments per MWh of energy generated by renewables does not preclude the government providing additional per MW-year subsidies to encourage entry of specific types of renewables (where the MW subsidy could itself be set by a separate market-based mechanism such as auctions; Lesser and Su, 2007).

A further approach, which continues in use in several jurisdictions, is the use of direct subsidies for renewables implemented through grants or projects. While this might have been a reasonable approach in the very early days of renewable generation, it is out of step with current maturing markets for renewable generation and supporting equipment. The use of grants and projects to implement renewable subsidies encourages rent-seeking and bureaucratic intervention rather than integration of renewables within the overall energy market. TGCs and most REFIT programs, on the other hand, allow the government to provide subsidies to renewable energy sources in a manner that can be integrated with normal market operations. Under TGCs, for example, the government can purchase a quantity of TGCs on the open market, perhaps guaranteeing a minimum price for these for a particular period to provide needed price stability for potential investors (Agnolucci, 2007).

---

[2] On the functioning of two-part tariffs for TGC markets, see Lesser and Su (2007).

[3] A well-known example of the regulatory fiat approach has been the German law, which sets floors for wholesale renewable energy prices as a percentage of the final retail tariff.

The idea of TGCs was apparently first proposed by the Enron Corporation, under the title of 'green tags' (Enron, 1997; Holt and Bird, 2005). Green certificate programs (i.e. TGCs) simply separate the market for the environmental characteristics of electricity from the electricity itself and allow trading in both the environmental good and the electricity good. It is based on the idea that the consumer's desire can be separated into two components: a desire for general-purpose electricity and a desire that more electricity be generated from green sources. Since an electron generated by a green generator is identical to an electron from any other type of generator, the consumer presumably does not care who actually receives the green electron. They do care, however, that the green electron was generated and displaced a non-green electron.

Each megawatt hour of energy generated by a green generator earns one TGC. The tag is not attached to the electricity in any way; it simply guarantees that 1 MWh of green electricity was generated and injected into the transmission grid (and hence used by someone whose identity may be unknown). The TGC can then be sold by the green generator to a power marketing company. The power marketer, in turn, advertises (or verifies to its regulator) that its power is $X\%$ green, based on how many TGCs it has purchased. It may not be that the electrons its customers receive are actually from the generators that sold the tags, but those customers are assured that their electricity payments were used for the generation of green electricity.

The first TGC exchange that the authors were aware of was launched in the USA in the Automated Power Exchange (APX), one of California's competitive power exchanges. Buyers and sellers who use the APX market to trade generic electric power may also purchase 'green tickets' that represent 1 MWh of green power. More recently, several other TGC marketers have entered in various US states; the Department of Energy maintains a list of these marketers, with some 25 states now either allowing or mandating the use of TGCs for purposes of verifying compliance with portfolio standards or to advertise to customers the green content of power marketing offers.[4] The situation in the EU was noted above, and remains an exemplary standard in the global arena for attempts to implement the Kyoto Protocol and beyond. The situation in Asia and Oceana has also developed quickly and market-based approaches to renewable energy supply, including TGCs, have been implemented in Australia, Japan and South Korea, with other countries moving to join this trend.[5]

This chapter examines the economics of the TGC market, including the integration of TGC markets with the generic approaches described above to renewable resource credits. In particular, the use of the TGC mechanism to implement government subsidies for green power producers is analyzed. Amundsen and Mortensen (2001) modelled the Danish TGC scheme and showed how the price

[4] The list is available online at the DOE Green Power Network website, www.eren.doe.gov/greenpower/
[5] See, for example, http://www.ec-asean-greenippnetwork.net/dsp_page.cfm?view=page &select = 231 for a survey of TGC implementation worldwide.

of TGCs is affected by various parameters and by a simultaneous market for CO_2 emission permits (see also Morthorst, 2003, for a similar market-level approach). The basic model presented here is similar to theirs at the aggregate market level. However, this approach does not begin at the market level, but rather is grounded in traditional demand theory and in the preferences of citizens/customers who will be affected in the long run by climate change and other impacts that provide the rationale for the shift to renewable energy in the first place.

The plan of the chapter is as follows. An ideal scheme is analyzed in which there are only two types of generation, renewable (or 'green') and non-renewable (or 'non-green'), and in which each kWh of green power generated at a particular plant is verifiable without error. In the next section, a simple model of TGCs is presented and the effects of TGCs are analyzed at the market equilibrium, when the government plays no role other than authenticating green power production. Section 9.3 analyzes and compares the impacts of various government policies to promote renewable energy, including Pigouvian taxes on non-green power, subsidies of green power and government purchase of TGCs. In Section 9.4, some of the details of practical policies for implementing TGCs and some open research questions are discussed.

9.2 Tradable green certificates in the electricity market

This section examines the effects of a stylized TGC system in a simple market setting. Those consumers who value green power sufficiently choose to purchase TGCs, and their revealed preference for TGCs is conveyed through these purchases. The TGC market is assumed to operate separately from the power market. Green generators receive the market-clearing price of a TGC for every kWh of energy they produce. The interaction of the two markets is described in detail below.

9.2.1 The consumer's problem

Consumers may purchase one of two goods, non-green power and green power, where non-green power consumption is denoted by x and green power consumption by y. Non-green power produces pollution: if the total amount of non-green power produced is X, then the amount of pollution is $G(X)$. Green power is assumed not to produce pollution.

A typical consumer's utility function is developed according to two principles:

- Energy produced from green sources is identical to that produced from non-green sources. Thus, x and y are perfect substitutes in terms of their ability to provide energy to perform household and industrial tasks.
- Consumers fall into different types indexed by θ, where high-θ consumers are more concerned about the pollution caused by non-green power.

A rational consumer must value both types of electricity the same in terms of their ability to do work, and any value placed on green power must ultimately derive from concerns about the pollution caused by non-green power. Consumer θ's utility function is represented as:[6]

$$U(x + y, \theta) + V(y, G(X), \theta) + M \qquad (9.1)$$

The functions U and V represent willingness to pay for electricity consumption and for environmental attributes of electric power, respectively. The final term M is just the Hicksian aggregate utility for all other goods in the economy with M taken as the numeraire good. The first argument of U, $x + y$, is the total amount of energy consumed. As noted above, the two types of electricity are perfect substitutes, so they enter the utility function additively. The function V, reflecting the consumer's valuation of the environmental attributes of green power, depends on total green power consumed and on the total amount of pollution $G(X)$, where X is the aggregate non-green power consumed, and G is a non-negative and monotonically increasing function.[7]

Why does green power consumption, y, enter the utility function as a separate argument? One possibility is that the level of green power consumption is apparent to others, in which case *prestige benefits* might accrue to the buyer (Harbaugh, 1998). Alternatively, consumers may receive utility in the form of a *warm glow* for having done their part to help reduce $G(X)$.[8] Finally, it may be that some type of *social norm* develops to encourage purchases of green power (Bernheim, 1994; Kreps, 1997). It is possible, of course, that none of these 'social' effects is present and that, because of free-rider problems, the only rationale for green energy purchase is government-mandated purchases by consumers (as in Sweden, for example). Even in the absence of such social effects, it seems reasonable to assume that there are real perceived costs associated with increased use of non-renewable energy, providing the basic public choice rationale for the move to renewables. The various possibilities are considered below.

Assume that $U(x + y, \theta)$ and $V(y, G(X), \theta)$ conform to the following conventional assumptions about willingness to pay:

$$U_1 > 0, U_{11} < 0, \; V_1 \geq 0, \; V_{11} \leq 0, \; V_2 > 0 \qquad (9.2)$$

---

[6] Here, separability is assumed between the willingness to pay for electric power per se and for the environmental attributes of electric power. In particular, the same symbol y is used to represent both the consumption good 'green energy' and the environmental attributes derived from y units of this good. Several other more general utility functions were analyzed, including further interaction terms across electric and environmental attributes. Not much is gained, and considerable complexity is added, from this more general treatment. Therefore, the separable case (9.1) is used throughout.

[7] The issue of the scope of damage of emissions is discussed in Section 9.4. The easiest assumption is that the physical effects of pollution are the result of aggregate pollution, although consumers may be affected non-uniformly by aggregate pollution.

[8] Palfrey and Prisbrey (1997) give results suggesting that warm glow is positive for most people.

The assumption that V_1 is non-negative implies that green power has a non-negative social effect over and above providing energy to do work. A consumer's type θ influences the strength of his or her preference for green power. The utility function (9.1) can be represented by many common functional forms. For example, the utility functions

$$U(x + y, \theta) = A(x + y)^\alpha;$$
$$V(y, G(X), \theta) = B\theta((y - \tau)^+)^\beta G(X)^\gamma, \; y > 0, \; V(0, \cdot,) = 0 \qquad (9.3)$$

where, for any real number z, $z^+ = \max(z, 0)$, $0 < \alpha$, $\beta < 1$; $\gamma > 0$; and $A > 0$, $B > 0$ would satisfy the assumed properties for U and V. (This example also includes a threshold level of consumption of green power τ. The absence or presence of a threshold level does not affect any of the theoretical results presented.)

Now let the price of x be p, and let the price of y be $p + g$, where g is the price of the TGC associated with each unit of y. Given the assumed utility function (9.1), the consumer's problem is

$$\text{Max } U(x + y, \theta) + V(y, G(X), \theta) - px - (p + g)y \qquad (9.4)$$

In the electricity market, total non-green energy X is large relative to each consumer's consumption x, and the pollution level $G(X)$ is therefore relatively insensitive to changes in x brought about by price changes. Therefore, it is assumed that consumers do not take account of the effect of x on G in their consumption decisions, and that a change in total pollution G brought about by changes in x does not significantly change the marginal utility of consuming green power (i.e. that $V_{12} \approx 0$). Taking first-order conditions:

$$\partial U / \partial x = U_1 \leq p, \text{ with equality if } x > 0 \qquad (9.5)$$

$$\partial U / \partial y + \partial V / \partial y = U_1 + V_1 \leq p + g, \text{ with equality if} \qquad (9.6)$$
$$y > 0$$

Taking p and g as given, as well as $G(X)$, it is straightforward to solve the conditions above. The solution is obtained by first solving Equation (9.6) for y and then using this in (9.5), assumed to hold as an equality, for x. One obtains

$$y(g, G(X), \theta) = [V_1^{-1}(g, G(X), \theta)]^+ \qquad (9.7)$$

$$x(p, g, G(X), \theta) = [U_1^{-1}(p, \theta) - y(g, G(X), \theta)]^+ \qquad (9.8)$$

where $z^+ = \text{Max}[z, 0]$ and where the inverse functions in (9.7) and (9.8) are defined by the inequalities

$$U_1[U_1^{-1}(p, \theta), \theta] \leq p; \; V_1[V_1^{-1}(g, G(X), \theta), G(X), \theta] \leq g \qquad (9.9)$$

In case some mandated percentage of renewables were required to be purchased by each consumer (as in the current case of Sweden), then a similar problem to (9.4) above would be obtained, subject to the additional constraint that $\xi x \leq (1 - \xi)y$, where ξ is the required lower bound on the fraction of renewable energy consumed. The key, as in any economic analysis of efficiency, is to trace the consequences of policies such as these mandated minimum proportional purchases back to the underlying preferences and choices of the affected economic agents. The following analysis relies on the direct formulation (9.1 − 9.9). However, more interventionist or 'paternalistic' policies could be appropriate to guide consumer choice, if lack of information, myopia or other behavioral biases undermine efficient energy choice by consumers (Kunreuther et al., 1998; Thaler and Sunstein, 2003).

For some utility functions, $V_1(0,G(X),\theta) = \infty$ for every θ. In this case, some green power will be purchased by every consumer. In general, however, one would expect a possibly significant proportion of consumers to have $y(g,G(X),\theta) = 0$, especially if there were a non-zero connection charge or information cost to obtain green power. The reader can check directly from (9.7) and (9.8), and the assumptions about U and V, that the necessary and sufficient First order conditions (FOCs) (9.6) hold. In particular, it follows from (9.8) and $U_1^{-1}(p,\theta) > 0$ that if $y(g,G(X),\theta) = 0$, then $x(p,g,G(X),\theta) > 0$, so that some power is consumed by every consumer. Also, the comparative statics yield the expected results[9] for x and y, namely:

(i)   Total power, $x(p,g,G(X),\theta) + y(g,G(X),\theta)$, is monotonic decreasing in p, and unaffected by g.
(ii)   Non-green power $x(p,g,G(X),\theta)$ is monotonic non-increasing (non-decreasing) in p (in g).
(iii)   Green power $y(g,G(X),\theta)$ is monotonic non-increasing in g, unaffected by p.

The weak monotonicity results reflect the fact that some consumers may not purchase a particular form of power unless p or g is sufficiently low. Note that implementing TGCs effectively reduces g from positive infinity to some finite amount, so property (ii) indicates that a TGC system will increase the amount of green power consumed.

9.2.2 Market demand

The market demand functions $X(p,g)$ and $Y(p,g)$ depend on the distribution of the consumer types, where the number of consumers of type θ is denoted as $dF(\theta)$, so that:

$$X(p,g) = \int x(p, g, G(X(p, g)), \theta)dF(\theta) \tag{9.10}$$

$$Y(p,g) = \int y(g, G(X(p, g)), \theta)dF(\theta) \tag{9.11}$$

A lemma is now presented that indicates that the market demand functions X and Y preserve the properties of the individual demand functions x and y. The

[9]See the Appendix for all proofs of propositions and some technical details, including those related to comparative statics.

proof of this lemma is simplified by the assumption that $V_{12} = 0$. This assumption is more than is needed to aggregate demand, but it allows a better comparison of government policies in Section 9.3.

Lemma 1: If $V_{12} = 0$, all the comparative static properties of the individual demand functions x and y are preserved in the market demand functions X and Y. In particular, one can write $Y(g)$ instead of $Y(p,g)$.

9.2.3 Supply and market equilibrium

Generation of power takes place in a perfectly competitive market in which there are two types of firm (or equivalently two separable types of generation). Those that produce non-green power have aggregate cost function $C_N(Q)$, while those that produce green power have $C_G(Q)$. Let the industry-level marginal costs be denoted by $c_N(Q)$ and $c_G(Q)$. It is assumed that $c_N < c_G$ at the equilibrium quantity; if it were not, then no one would ever buy non-green power. Clearly, profit-maximizing firms will adjust output until marginal cost and price are equated, so that at the industry level (see also Amundsen and Mortensen, 2001):

$$c_N(Q_N) = p; c_G(Q_G) = p + g \qquad (9.12)$$

The market equilibrium quantities $X^* = X(p,g)$ and $Y^* = Y(g)$ are given by the simultaneous solution of the consumer and firm problems as:

$$c_N(X(p,g)) = p; c_G(Y(g)) = p + g \qquad (9.13)$$

Assuming that marginal costs are increasing, in accord with merit-order dispatch, the usual comparative statics are obtained. Market equilibrium quantities will increase as technological progress decreases cost or as exogenous conditions increase demand.

9.2.4 Effects of tradable green certificates

The market equilibrium conditions make clear some important effects of implementing the TGC proposal. First, the price of a TGC is always the difference in cost between green and non-green power. This means the TGC market is efficient in the sense that consumers equate their social effects of buying green power with the extra cost of that power.

Second, if there are enough high-θ consumers, a large amount of green power could be produced. If there were any learning curve or scale effects in green power, this might decrease its cost. It is shown below in the discussion of producer subsidies (which operate the same as a cost decrease) that the price of TGCs would fall, which would encourage more consumption of green power. Since G enters as a pure externality in consumption, it would seem that there is a case to be made for government intervention, a topic discussed in the next section.

9.3 Government intervention in the green power market

From a social point of view, the TGC market could increase welfare in two ways. First, it would allow people to achieve their social objectives of decreasing pollution by contributing to the production of green power. Second, and presumably more important, it could result in a substitution away from non-green power, thus reducing the pollution externality $G(X)$. The following sections consider interventions by a government agency whose primary goal is reducing the externality.

9.3.1 The goal of environmental policy

The full social planner's problem takes account of the pollution externality *and* allocates green power in such a way as to maximize the 'social effect' for each consumer. If the planner allocates $x(\theta)$ and $y(\theta)$ to consumers of type θ, then the welfare function is:

$$
W = \int [U(x(\theta) + y(\theta), \theta) + V(y(\theta), G(X), \theta)]dF(\theta) - C_N(X) - C_G(Y)
$$
$$
X = \int x(\theta)dF(\theta) \qquad Y = \int y(\theta)dF(\theta)
$$
(9.14)

The conditions for social optimum are[10]

$$
\frac{\partial U}{\partial x} + \frac{\partial V}{\partial G}\frac{dG}{dX} = c_N \qquad \forall\,\theta
$$
$$
\frac{\partial U}{\partial y} + \frac{\partial V}{\partial y} = c_G \qquad \forall\,\theta
$$
(9.15)

Since the market equilibrium sets $\partial U/\partial y + \partial V/\partial y = p + g = c_G$, the conditions for the social optimum (9.15) make it clear that the TGC market equilibrium given by (9.13) is efficient; the planner does not want to increase or decrease the amount of green power produced. The problem comes in the non-green market, where the free-market price p does not reflect the pollution externality $(\partial V/\partial G)(dG/dX) < 0$. An environmental policy maker thus has an incentive to intervene in the market in such a way as to reduce the production and consumption of non-green power.

Clearly, the first-best solution is a Pigouvian tax on non-green power equal to the amount of the pollution externality. Despite its seeming simplicity, such a scheme has several well-recognized problems. First, the information requirements required to set the tax are enormous. Second, the externality is not really caused by non-green electricity alone, but actually by all sources of pollution (e.g.

---

[10] Here, only the case in which every consumer has both $x(\theta)$, $y(\theta) > 0$ is considered. An appropriate inequality system would be obtained if for some consumers $x(\theta) = 0$ or $y(\theta) = 0$ were optimal.

CO_2 by carbon-based fuels used in all sectors of the economy). Thus a comprehensive (e.g. carbon) tax would be required to achieve the true social optimum. Third, and most important, the scale of this intervention is enormous, causing major changes in the economy. Thus, unsurprisingly, we have seen only piecemeal sectoral approaches to global pollution regulations.

9.3.2 Government intervention with a fixed budget

If Pigouvian taxation of non-green power is not an option, an alternative environmental policy might involve extra spending in the green power sector. The budget, B, available for such spending might come from general taxation, or it might come from a less-than-Pigouvian tax on the non-green power industry. If the budget comes from general government revenues, there would be some cost associated with the deadweight loss from taxation.

In the following sections, several policy options are considered for using the budget B to reduce the externality $G(X)$. The first options are the traditional tools of a producer or a consumer subsidy. These could be implemented through various types of REFIT programs. As argued in Section 9.1, implementing such subsidies through grants or projects is excluded from consideration here, as such an approach encourages rent-seeking and bureaucratic intervention rather than providing incentives, based on actual renewable energy produced. As will be shown, REFIT policies implemented using TGCs allow a more efficient market-based approach for the government to provide subsidies to renewable energy sources in a manner that avoids these problems. Under a REFIT/TGC approach, the government can purchase TGCs on the open market. The efficiency of two different approaches, per-unit subsidy and direct-buy, to implementing the government purchases, is described and evaluated. Each of these policies could be financed from general revenues or from taxes on the non-green power sector.

It is important to note that in all of the following policy analysis, it is assumed that the consumers' utility functions are unchanged, i.e. given by (9.1). Crucially, the 'social effect' of buying green power persists even as the government makes interventions in the market. This ignores the possibility of the 'crowding out' of individual altruism by government subsidy (Frey and Oberholzer-Gee, 1997). Analysis of this and other behavioral 'anomalies' would be interesting avenues for research, but are ignored here.

9.3.3 Traditional producer and consumer subsidies financed from general revenue

Suppose the government offers a per-unit subsidy of h to the producers of green power. The new green power supply function becomes $c_G(Q_G) - h$. Accordingly, the market equilibrium conditions become

$$c_N(X(p,g)) = p; c_G(Y(g)) - h = p + g \qquad (9.16)$$

Since the government cannot spend more than its renewable energy budget B (and it is assumed that the budget is small enough that it does spend all of it), then there is an additional condition that

$$B = hY(g) \qquad (9.17)$$

Proposition 1: An increase in the subsidy h increases equilibrium Y, decreases equilibrium X, and decreases both p and g in equilibrium.

The effects of a per-unit consumer subsidy for any green power purchased would be similar. A per-unit consumer subsidy would shift the demand curves for X and Y from $X(p,g)$ to $X(p,g - k)$ and from $Y(g)$ to $Y(g - k)$. The resulting market equilibrium would be determined from (9.13) by

$$c_N(X(p,g - k)) = p; c_G(Y(g - k)) = p + g \qquad (9.18)$$

The effect of 'k' plays essentially the same role as the '$- h$' in (9.16), with the same effect on the output of green and non-green power. If the government has budget B available for this subsidy, then there is the additional constraint that

$$B = kY(g - k) \qquad (9.19)$$

Proposition 2: An increase in the subsidy k increases equilibrium Y and decreases equilibrium X.

If the government always spends its entire budget B on subsidy, define the equilibrium X and Y as a function of B as $X^S(B)$ and $Y^S(B)$. We then have the following corollary:

Corollary 1: For either type of subsidy, an increase in the government budget B increases equilibrium Y and decreases equilibrium X.

9.3.4 Government direct purchase of tradable green certificates, financed from general revenue

Having set up the TGC market, the government has an additional option for encouraging green power. It could simply enter the market and buy TGCs without actually consuming the associated electricity. (Additionally, one can imagine non-profit organizations buying TGCs in a similar manner, as the American Lung Association has done in the market for sulfur emissions rights.)

If the government purchases a TGC without consuming the associated electricity, then it is as if this residual electricity were 'converted' from green to non-green, at least in the eyes of a consumer. The government's buying activity can be viewed as an increase in green power demand and a simultaneous increase in the supply of non-green power. However, this increased supply of non-green power does not, in actual fact, add to the external pollution effect $G(X)$. The formulation of this problem is that government chooses to buy some number of TGCs ψ. Consumers' value functions are then given by

$$U(x + y, \theta) + V(y, G(X - \psi), \theta) + M \qquad (9.20)$$

where the term $G(X - \psi)$ makes it clear that the ψ units appear as non-green power to the consumer but do not actually cause any pollution externality.

The solution to the consumer's problems produces demand functions for non-green and green power as before, X and Y. However, the actual production of green power includes the units purchased by the government, i.e. green power generators produce $Y + \psi$ units. And since the government does not actually consume the ψ units of electricity, they are added to the supply of non-green power. Thus, non-green generators need only produce $X - \psi$ units to satisfy demand.

The solution to the consumer's problem is the same as before. Market equilibrium is determined by the quantities $X(\psi)$ and $Y(\psi)$ that solve the equilibrium conditions

$$c_N(X(p,g) - \psi) = p\, c_G(Y(g) + \psi) = p + g \qquad (9.21)$$

where ψ is the number of TGCs purchased by the government. In a budget-constrained environment, the price of TGCs is $g(\psi)$, the solution to (9.21), so the number of TGCs ψ that would be purchased by the government with a total budget B would be determined by

$$B = g(\psi) \cdot \psi \qquad (9.22)$$

Proposition 3: An increase in the government direct buy ψ decreases (crowds out) private consumption of green power Y but increases the total amount of green power produced, $Y + \psi$. An increase in ψ increases X, the apparent amount of non-green power sold, but actual production of non-green power, $X - \psi$, will fall.

Corollary 2: An increase in the government budget decreases (crowds out) private Y. This effect is less than one for one, so an increase in the government budget increases total green power $Y + \psi$.

9.3.5 Comparison of policies

How do the subsidy and direct-buy policy instruments compare with one another? Two issues are present. First is the effect of alternative policies in reducing $G(X)$, for a given budget [i.e. for a given B, which yields smaller non-green power, $X^S(B)$ or $X^D(B)$?]. More generally, accounting for changes in demand for green power under the two regimes as well as for non-green power, which of the two regimes yields greater overall welfare [as measured by (9.7)]?

It is not possible to say in general whether $X^S(B)$ is larger or smaller than $X^D(B)$. Nor is it possible to say anything in general about the comparative welfare effects at equilibrium of these two policies. The effectiveness of the policies depends on the elasticities of demand and supply, where the elasticity of demand, in turn, depends on the distribution of consumer types as well as the form of the utility function.

For the case where the supply curves are horizontal and consumer types are well behaved, the comparison can be made more definitely.

Proposition 4: Suppose the c_N' and c_G' are constant. Let $\varepsilon_G = -\dfrac{\partial Y}{\partial g}\dfrac{g}{Y}$ be the elasticity of demand for green power with respect to the price of TGCs. Then a budget B spent on a direct buy of TGCs will result in more green power (and therefore less non-green power) than the same budget spent on a subsidy if and only if $\varepsilon_G < 1$.

This proposition suggests that direct government intervention in the TGC market and the supply side (as in the REFIT programs focused on suppliers) would be particularly effective in regions where consumers are relatively unresponsive to price changes in making their green power decisions. One such market could be comprised of green-power buyers who are mostly committed environmentalists and therefore not much influenced by price considerations. However, a market might also have this feature if most people were not interested in TGCs (or perhaps were unaware of the impact of their energy purchases on $G(X)$ and therefore generally unconcerned about green power). Thus, a TGC market could be an effective tool for government intervention even in markets where consumers do not participate extensively. Of course, determining the optimal 'budget' and associated subsidies for the case of an uninformed public in whose name the government is to act remains a deep puzzle for public economics.[11]

9.4 Institutional considerations and discussion

There are several institutional considerations that will be important in the ultimate design of TGC markets. These are discussed below under several headings: alternative forms of the tradable green certificate market; engendering and maintaining public trust in the institution; and efficiency issues in the public–private partnership likely to emerge in regulating environmental aspects of electrical power.

9.4.1 Alternative forms of the tradable green certificate market

As noted in Section 9.1, various forms of credit and constraints to promote renewables have emerged as policy instruments, including REFIT-type credits and RPS-type constraints. The basic issue addressed here is that any efficient approach to these credits or constraints will require authentication of the actual production and delivery of energy. It is argued that the TGC approach can accomplish this authentication and provide simultaneously the basic instrument for targeting per-unit energy subsidies to renewable energy sources. The traditional approach is to

---

[11]See Sen (1995) and Thaler and Sunstein (2003) for an introduction to this topic. In the context of renewable energy, see the discussion of this and related dynamic problems of supporting the right tempo for technological change in Newell et al. (2006).

determine the per-unit subsidy for energy produced by administrative fiat, e.g. as a direct payment to the renewable generators per kWh produced or through some pricing formula relative to other energy sources (e.g. at the level of 90% of final retail price, as in the German case). Even if subsidies are determined in part by administrative or political policies, using TGCs to implement credits would allow the price for renewables to be integrated with the overall power market. Issues such as transmission, location of facilities, reliability and timing of power injections would then all be dealt with, as they should be, through the normal rules of the market, not by administrative fiat.

A number of hybrid solutions, including many that follow the REFIT model, can be implemented through the TGC approach. The rationale for doing this is two-fold. First and foremost, it is to enlist the discipline of the market in an ongoing fashion to provide proper motivation for investor choices of renewable technology. After all, these need to fit as part of an overall portfolio of generation technology choices, including for the foreseeable future a significant percentage of non-green power. Second, if renewables are to be valued as part of market-level portfolio management, as foreseen by Awerbuch and Berger (2002) and others in the electric power area (e.g. Kleindorfer and Li, 2005), then their price must be determined in a transparent manner and as free as possible from short-term political whims. Note that, as per the discussion following Equation (9.15), the green power market is economically efficient, and the market failure that we are trying to correct through TGCs comes from the externality in the non-green power market. Thus, any type of subsidy to green power is a second-best solution to the real problem. However, in the absence of a Pigouvian tax on non-green power, government intervention in the green power market can be a reasonable corrective for the externality associated with the pollution caused by non-green power. As this externality is not likely to disappear, non-green power will continue to have lower private costs than its full social costs, so a continuing intervention makes sense if the transactions costs of implementing such an intervention do not outweigh the benefits. If the intervention is implemented through a well-functioning TGC market, it should be relatively free of 'grantsmanship' and other transactions costs compared with other types of government policies.[12]

9.4.2 Maintaining public trust

Another issue that has been raised, and is critically important for environmental disclosure, is verifiability and understandability of the institutions designed to facilitate environmental disclosure. There remain significant concerns about public understanding and acceptance of the TGC mechanism, possibly rooted in a general mistrust of large energy firms in the wake of the Enron bankruptcy and the general disregard of former monopolistic firms of consumer rights. In the authors' view, this matter needs to be taken seriously, but it only will be if there are proper means of authenticating energy actually produced by one or another

---

[12] See also Finon and Perez (2007) for a detailed argument on this issue.

renewable generation facility. TGCs, if properly monitored, can help to reinforce both the measurement of renewable energy produced and determining who is consuming this energy. From the point of both economic and social accountability, TGCs can play an important and positive role.

9.4.3 Efficiency issues

This chapter has pointed on several occasions to the key efficiency aspect of TGCs, and that is its confluence with market discipline and conduct. It is also compatible with various regulatory interventions and tax/subsidy schemes, without the typical grantsmanship that underlies these approaches to date in a world of project-based subsidies and continues to be associated with any purely administrative approach to the problem of determining appropriate levels of renewable energy credits. As pointed out, several designs are feasible within the context of TGC itself, and this discussion has not provided any ranking of these alternative designs. The primary reason is that it is not likely that there is a dominant design, but rather that the particular approach to the implementation of TGCs will depend on the relative importance of several factors that could affect the desirable structure of the TGC market. Some of the principal factors include deadweight losses, externalities and the perceived need to promote price stability for renewables as an infant industry.

9.4.3.1 Deadweight losses

Both the subsidy and the direct purchase schemes may be financed out of general tax revenue, with the generation of this tax revenue creating a deadweight loss $D(B)$, a function of the level of the subsidy budget for renewables. As this deadweight loss increases, these schemes become less attractive relative to the simultaneous tax on non-green power and subsidy to green power. Furthermore, even among the schemes that depend on general tax revenue, which scheme is optimal will depend on the demand and cost parameters. As deadweight loss increases, it becomes more costly to implement a government intervention, and therefore picking the optimal intervention becomes even more crucial if specific welfare improvements are to be achieved.

9.4.3.2 Magnitude of the externality

The economic motivation for any of the government interventions discussed above is the reduction of the pollution externality $G(X)$. If the magnitude of this externality increases for given X, the optimal level of government intervention will clearly increase, i.e. for tax revenue-financed interventions, the optimal budget, B, will increase. The authors do not conjecture that the ranking of the different schemes is sensitive to the size of the externality, but the magnitude of the difference between the best and worst schemes could very well increase. In any case, the major implication of non-green power externalities and their correction through government intervention is that it will be critical to continue to assess the cost of these externalities and the willingness to pay at a societal level to control them. The TGC mechanism provides a transparent approach that reflects the total cost of government intervention in a manner that could facilitate the necessary political and public discussion to address this matter in an ongoing, open manner.

9.4.3.3 Price stabilization and investment recovery

A key issue raised by both renewable energy investors and certain environmental groups is the need for assurance for long-term capital recovery of investments in renewables. Indeed, as noted above, this is one of the primary reasons for the support of the renewable energy investors for REFIT-type programs. In the context of TGCs, the government could guarantee minimum or limit maximum prices for the TGCs, and do so over some extended period, to assure increased stability in the renewables sector and to encourage investment. However, the history of such interventions (e.g. minimum price controls in agriculture and price caps in electric power markets) is dotted with many examples of economic inefficiency, both in the short run and in terms of long-run technology choices (e.g. Newell et al., 2006). If subsidies are to be provided by the government for renewables, then using TGCs would at least assure a level playing field through the market (in terms of kWh of energy actually produced and used).

9.4.3.4 Infant industries

It is possible that some green power generation techniques, such as wave and tidal power, are in such early stages of the innovation process that they would not emerge from private entrepreneurship even in the presence of RPS and/or REFIT policies supported by TGCs. As such, they may qualify as 'infant industries' needing additional government subsidies to be successful. The presence of the TGC market would still be a great benefit to avoiding 'grantsmanship' in this context, however. Because a TGC market provides information on the currently prevailing cost differences between green and non-green power across a range of technologies, it provides a baseline competitive cost level for any new green power source. Arguments for additional subsidies could then be made, based exclusively on non-environmental infant industry arguments such as imperfections in capital markets, lack of coordination in complementary investments, or other positive externalities such as technology spillovers.

TGCs are, in the end, a direct result of the increasing interest by people everywhere in environmental disclosure. This has driven both energy consumers and environmentalists to be concerned with a credible measurement of the environmental attributes of the power supplied and consumed. Several approaches have been analyzed here, under the general heading of tradable green certificates, for achieving the requisite balance between the efficiency of market-based approaches and public trust, and their mutual compatibility with the usual instruments of tax/subsidy by the government in addressing externalities.

A number of issues remains open at this point. This discussion has focused on market-based energy payments. One could also examine supplementary capacity payments for renewables, just as such capacity payments have been introduced for generation capacity in general. Hopefully, such capacity payments would themselves be market based, e.g. based on auctions of desired renewable capacity quotas as suggested by Lesser and Su (2007), rather than set administratively. In any case, there is a natural integration of TGC markets and market-based capacity payments, just as there is for normal power markets and capacity payments. A second issue is the treatment of peak-load effects. In principle, there is no problem making the TGC prices time dependent or seasonal, and the normal clearing mechanism for power markets underlying the above analysis would apply straightaway. The integration of TGCs with long-term contracting, portfolio selection involving renewables and derivative instruments based on TGCs are interesting further topics to be explored. There is obviously much left to be said on this subject.

References

Agnolucci, P. (2007). The effect of financial constraints, technological progress and long-term contracts on tradable green certificates. *Energy Policy*, 35(6), 3347–3359.

Amundsen, E. S. and Mortensen, J. B. (2001). The Danish green certificate system: some simple analytical results. *Energy Economics*, 23, 489–509.

Awerbuch, S. and Berger, M. (2002). *Energy Diversification and Security in the EU: Mean-Variance Portfolio Analysis of the Electricity Generating Mix and its Implications for Renewables*. Paris: International Energy Agency.

Bernheim, B. D. (1994). A theory of conformity. *Journal of Political Economy*, 102 (October), 841–877.

Enron, (1997). *Green Tags: A Resource Disclosure System for Electric Power Consumers*. Houston, TX: Enron Corporation (April).

Finon, D. and Perez, Y. (2007). The social efficiency of instruments of promotion of renewable energies: a transaction-cost perspective. *Ecological Economics*, 62(1), 77–92.

Ford, A., Vogstad, K. and Flynn, H. (2007). Simulating price patterns for tradable green certificates to promote electricity generation from wind. *Energy Policy*, 35(1), 91–111.

Frey, B. S. and Oberholzer-Gee, F. (1997). The cost of price incentives; an empirical analysis of motivation crowding-out. *American Economic Review*, 87 (September), 746–755.

Harbaugh, W. T. (1998). The prestige motive for making charitable transfers. *American Economic Review*, 88(2), 277–282.

Holt, E. and Bird, L. (2005). *Emerging Markets for Renewable Energy Certificates*. Golden, CO: National Renewable Energy Laboratory.

Kleindorfer, P. R. and Li, L. (2005). Multi-period, VaR-constrained portfolio optimization in electric power. *The Energy Journal*, 26(1), 1–26.

Kreps, D. M. (1997). Intrinsic motivation and extrinsic incentives. *American Economic Review*, 87 (May), 359–364.

Kunreuther, H., Onculer, A. and Slovic, P. (1998). Time insensitivity for protective measures. *Journal of Risk and Uncertainty*, 16, 279–299.

Lauber, V. (2004). REFIT and RPS: options for a harmonised community framework. *Energy Policy*, 32(12), 1405–1414.

Lesser, J. A. and Su, X. (2007). *Design of an Economically Efficient Feed-In Tariff Structure for Renewable Energy Development*. Working Paper, Bates White LLC, Washington, DC.

Morthorst, P. E. (2003). A green certificate market combined with a liberalized power market. *Energy Policy*, 31, 1393–1402.

Newell, R. G., Jaffe, A. B. and Stavins, R. N. (2006). The effects of economic and policy incentives on carbon mitigation technologies. *Energy Economics*, 28(5), 563–578.

Palfrey, T. R. and Prisbrey, J. E. (1997). Anomalous behavior in public goods experiments: how much and why?. *American Economic Review*, 87 (December), 829–846.

Sen, A. (1995). Environmental evaluation and social choice: contingent valuation and the market analogy. *Japanese Economic Review*, 46(1), 23–37.

Thaler, R. H. and Sunstein, C. R. (2003). Libertarian paternalism. *American Economic Review (Papers and Proceedings)* 93(2), 175–179.

Appendix

Comparative statics results

The first order conditions for a consumer who buys positive quantities of both non-green and green power are

$$U_1(x + y, \theta) = p \qquad\qquad U_1(x + y, \theta) + V_1(y, G(X), \theta) = p + g \qquad\text{(9.A1)}$$

Differentiating (9.A1) with respect to p (treating $G(X)$ as fixed) gives

$$U_{11}\left(\frac{\partial x}{\partial p} + \frac{\partial y}{\partial p}\right) = 1 \qquad U_{11}\left(\frac{\partial x}{\partial p} + \frac{\partial y}{\partial p}\right) + V_{11}\frac{\partial y}{\partial p} = 1 \qquad\text{(9.A2)}$$

Solving both equations in (9.A2) simultaneously implies that

$$V_{11}\frac{\partial y}{\partial p} = 0 \Rightarrow \frac{\partial y}{\partial p} = 0 \qquad\text{(9.A3)}$$

The left equation in (9.A2) combined with (9.A3) and the assumption that $U_{11} < 0$ implies

$$\frac{\partial x}{\partial p} = \frac{1}{U_{11}} < 0 \qquad\text{(9.A4)}$$

Differentiating (9.A1) with respect to g gives

$$U_{11}\left(\frac{\partial x}{\partial g} + \frac{\partial y}{\partial g}\right) = 0 \qquad U_{11}\left(\frac{\partial x}{\partial g} + \frac{\partial y}{\partial g}\right) + V_{11}\frac{\partial y}{\partial g} = 1 \qquad\text{(9.A5)}$$

Solving the two equations in (9.A4) simultaneously gives

$$\frac{\partial y}{\partial g} = \frac{1}{V_{11}} < 0 \qquad\text{(9.A6)}$$

Then (9.A6) and (9.A5) together imply that

$$\frac{\partial x}{\partial g} = \frac{1}{V_{11}} > 0 \qquad\text{(9.A7)}$$

Note that (9.A6) and (9.A7) together mean that

$$\frac{\partial x}{\partial g} + \frac{\partial y}{\partial g} = 0 \qquad\text{(9.A8)}$$

Finally, if we solve (9.A1) simultaneously and differentiate with respect to G, the pollution level, we get

$$V_{11}\frac{\partial y}{\partial G} + V_{12} = 0 \qquad\text{(9.A9)}$$

Since we have assumed that V_{12} is close to zero, then y and x are approximately invariant with G.

Lemma 1: If $V_{12} = 0$, all the comparative static properties of the individual demand functions x and y are preserved in the market demand functions X and Y.

Proof: Differentiating (9.10) with respect to p and recalling that x is assumed to change very little with G,

$$\frac{\partial X}{\partial p} = \int \left(\frac{\partial x}{\partial p} + \frac{\partial x}{\partial G} \frac{\partial G}{\partial X} \frac{\partial X}{\partial p} \right) df(\theta) = \int \frac{\partial x}{\partial p} dF(\theta) \qquad (9.A10)$$

Since $\partial x / \partial p \le 0$, (9.A10) implies $\partial X / \partial p \le 0$. By similar reasoning with respect to g,

$$\frac{\partial X}{\partial g} = \int \frac{\partial x}{\partial g} dF(\theta) \qquad (9.A11)$$

The results for $Y(p,g)$ are analogous.

Proposition 1: An increase in the subsidy h increases equilibrium Y and decreases equilibrium X.

Proof: The derivatives of the equilibrium conditions in (9.16) with respect to h are

$$c_N' \left(\frac{\partial X}{\partial p} \frac{dp}{dh} + \frac{\partial X}{\partial g} \frac{dg}{dh} \right) = \frac{dp}{dh} \qquad (9.A12)$$

$$c_G' \left(\frac{\partial Y}{\partial g} \frac{dg}{dh} \right) - 1 = \frac{dp}{dh} + \frac{dg}{dh} \qquad (9.A13)$$

Solving (9.A14) for dg/dh gives

$$\frac{dg}{dh} = \frac{dp}{dh} A \qquad (9.A14)$$

where

$$A = \left(1 - c_N' \frac{\partial X}{\partial p} \right) \left(c_N' \frac{\partial X}{\partial g} \right)^{-1} \qquad (9.A15)$$

From the comparative statics, $\partial X / \partial p < 0$ and $\partial X / \partial g > 0$. The marginal cost function, c_N, is upward sloping, implying $c_N' > 0$.[13] Together these conditions imply $A > 0$. Substituting (9.A16) into (9.A15) gives

$$\frac{dp}{dh} = - \left(1 + A - c_N' A \frac{\partial Y}{\partial g} \right)^{-1} \qquad (9.A16)$$

---

[13] To simplify the proofs, the limiting case of constant marginal costs is ignored.

The right-hand side of (9.A16) is negative because $\partial Y/\partial g < 0$. Combining this result with (9.A14) implies that

$$\frac{dg}{dh} < 0 \qquad\qquad \frac{dp}{dh} < 0 \tag{9.A17}$$

Returning to (9.A12) and simplifying,

$$c_N'\left(\frac{\partial X}{\partial p}\right) = \frac{dp}{dh} \tag{9.A18}$$

Since c_N' is positive, (9.A18) implies that $dX/dh < 0$. It remains to consider dY/dh. This is given by

$$\frac{dY}{dh} = \frac{\partial Y}{\partial g}\frac{dg}{dh} \tag{9.A19}$$

Both terms on the right-hand side of (9.A19) are negative, so $dY/dh > 0$. ∎

Proposition 2: An increase in the subsidy k increases equilibrium Y and decreases equilibrium X.

Proof: The derivatives of (9.18) with respect to k are

$$c_N'\left(\frac{\partial X}{\partial p}\frac{dp}{dk} + \frac{\partial X}{\partial g}\left(\frac{dg}{dk} - 1\right)\right) = \frac{dp}{dk} \tag{9.A20}$$

$$c_G'\left(\frac{\partial Y}{\partial g}\left(\frac{dg}{dk} - 1\right)\right) = \frac{dp}{dk} + \frac{dg}{dk} \tag{9.A21}$$

Solving (9.A20) for dg/dk gives

$$\frac{dg}{dk} = A\frac{dp}{dk} + 1 \tag{9.A22}$$

where A is defined by (9.A15). Substituting (9.A22) into (9.A21) and rearranging gives

$$\frac{dp}{dk} = -\left(1 + A - c_G'A\frac{\partial Y}{\partial g}\right)^{-1} \tag{9.A23}$$

Since $\partial Y/\partial g < 0$, every term on the right-hand side of (9.A23) is positive hence the whole expression is negative: $dp/dk < 0$. Rewriting (9.A20) gives

$$c_N'\left(\frac{\partial X}{\partial k}\right) = \frac{dp}{dk} \tag{9.A24}$$

and since $dp/dk < 0$, it must be that $dX/dk < 0$. Substituting (9.A22) into dY/dk gives

$$\frac{dY}{dk} = \frac{\partial Y}{\partial g}\frac{dp}{dk}A \tag{9.A25}$$

which is positive: $dY/dk > 0$. ∎

Corollary: An increase in the government budget increases equilibrium Y and decreases equilibrium X.

Proof: Differentiate (9.16) with respect to B:

$$c_G' \left(\frac{dY}{dB} \right) - \frac{dh}{dB} = \left(\frac{dp}{dh} + \frac{dg}{dh} \right) \frac{dh}{dB} \tag{9.A26}$$

In the case of a producer subsidy, differentiate (9.17) with respect to B:

$$\frac{dh}{dB} = \frac{1}{Y(p, g) + h \dfrac{\partial Y}{\partial g} \dfrac{dg}{dh}} \tag{9.A27}$$

All terms are positive by (9.A17), and an increase in h decreases X as shown in Proposition 1.

Substitute (9.A27) and simplify with (9.A14) and (9.A16):

$$\frac{dY}{dB} = \frac{A \dfrac{\partial Y}{\partial g}}{h \dfrac{\partial Y}{\partial g} A - Y \left(1 + A - c_G' A \dfrac{\partial Y}{\partial g} \right)} \tag{9.A28}$$

All terms are negative, so dY/dB is positive. In the case of a consumer subsidy, the proof is very similar. ∎

Proposition 3: An increase in the government direct buy ψ decreases (crowds out) private consumption of green power Y but increases the total amount of green power produced, $Y + \psi$. An increase in ψ increases X, the apparent amount of non-green power sold, but actual production of non-green power, $X - \psi$, will fall.

Proof: The derivatives of (9.21) with respect to ψ are:

$$c_N' \left(\frac{\partial X}{\partial p} \frac{dp}{d\psi} + \frac{\partial X}{\partial g} \frac{dg}{d\psi} - 1 \right) \frac{dp}{d\psi} \tag{9.A29}$$

$$c_G' \left(\frac{\partial Y}{\partial g} \frac{dg}{d\psi} + 1 \right) = \frac{dp}{d\psi} + \frac{dg}{d\psi} \tag{9.A30}$$

This proof first shows that $dp/d\psi < 0$, and then that $dp/d\psi + dg/d\psi > 0$. These facts can then be used to derive all the stated results about $dX/d\psi$ and $dY/d\psi$.

Solving (9.A29) for $dg/d\psi$ gives

$$\frac{dg}{d\psi} = A \frac{dp}{d\psi} + \left(\frac{\partial X}{\partial g} \right)^{-1} \tag{9.A31}$$

where A is defined in (9.A15); it was shown following (9.A15) that $A > 1$. Substitute (9.A31) into (9.A30), and use the comparative static result (9.A8):

$$\frac{dp}{d\psi} = \frac{-1}{\frac{\partial X}{\partial g}\left(1 + A - c_G' A \frac{\partial Y}{\partial g}\right)} \tag{9.A32}$$

All terms in the denominator are positive, so $dp/d\psi < 0$.

Next we turn to the sign of $dp/d\psi + dg/d\psi$. Using (9.A31) gives

$$\frac{dp}{d\psi} + \frac{dg}{d\psi} = (1 + A)\frac{dp}{d\psi} + \left(\frac{\partial X}{\partial g}\right)^{-1} \tag{9.A33}$$

Substitute (9.A32) into (9.A33) and simplify

$$\frac{dp}{d\psi} + \frac{dg}{d\psi} = \frac{c_G' A}{\left(1 + A - c_G' A \frac{\partial Y}{\partial g}\right)} \tag{9.A34}$$

The numerator and denominator are both positive, so $dp/d\psi + dg/d\psi > 0$, and from the previous result that $dp/d\psi < 0$, it is now shown that $dg/d\psi > 0$.

The first claim of the proposition is that the government direct buy crowds out private consumption of Y. The effect of ψ on Y is given by

$$\frac{dY}{d\psi} = \frac{\partial Y}{\partial g}\frac{dg}{d\psi} \tag{9.A35}$$

Since $dg/d\psi > 0$, (9.A35) is negative.

The second claim of the proposition is that the crowding out is less than one for one. The fact that the sum $dp/d\psi + dg/d\psi > 0$ implies that both sides of (9.A30) must be positive. For the left-hand side of (9.A30) to be positive requires that $\frac{dY}{d\psi} + 1 > 0$, which proves the claim.

The third claim of the proposition is that increasing ψ increases the apparent amount of non-green power sold, X. The effect of ψ on X is given by

$$\frac{dX}{d\psi} = \frac{\partial X}{\partial p}\frac{dp}{d\psi} + \frac{\partial X}{\partial g}\frac{dg}{d\psi} \tag{9.A36}$$

Since $dp/d\psi < 0$ and $dg/d\psi > 0$, both terms of (9.A36) are positive, and $dX/d\psi > 0$.

The fourth claim of the proposition is that although X is increased, the actual production of non-green power is reduced. The fact that $dp/d\psi < 0$ implies that both sides of (9.A29) are negative. For the left-hand side of (9.A29) to be negative requires that $\frac{dx}{d\psi} - 1 < 0$, which proves the claim. ∎

Corollary: An increase in the government budget decreases (crowds out) private Y. This effect is less than one for one, so an increase in the government budget increases total green power $Y + \psi$.

Proof: Differentiate (9.21) with respect to B:

$$c'_G \left(\frac{dY}{dB} + \frac{d\psi}{dB} \right) = \left(\frac{dp}{d\psi} + \frac{dg}{d\psi} \right) \frac{d\psi}{dB} \qquad (9.A37)$$

Differentiate (9.22) with respect to B:

$$\frac{d\psi}{dB} = \frac{1}{\dfrac{dg}{d\psi} \psi + g(\psi)} \qquad (9.A38)$$

From Proposition 3, the denominator is positive. Substitute (9.A38) into (9.A37) and simplify with (9.A32) and (9.A34):

$$\frac{dY}{dB} + \frac{d\psi}{dB} = \frac{A}{\psi c'_G A + \psi \left(\dfrac{\partial X}{\partial g} \right)^{-1} + g \left(1 + A - c'_G A \dfrac{\partial Y}{\partial g} \right)} \qquad (9.A39)$$

All terms are positive, so (9.A39) is positive. Thus an increase in B increases ψ which decreases Y by Proposition 3, but it increases $Y + \psi$. ∎

Proposition 4: A budget B spent on a direct buy of TGCs will result in more green power (and therefore less non-green power) than the same budget spent on a subsidy if and only if the demand for green power is inelastic with respect to the price of TGCs.

Proof: A direct buy causes a larger increase in Y if (9.A39) is greater than (9.A28) when both are evaluated where $\psi = h = 0$ initially. That comparison readily simplifies to:

$$1 > -\frac{\partial Y}{\partial g} \frac{g}{Y} \qquad (9.A40)$$

Frontier Applications of the Mean-Variance Optimization Model for Electric Utilities Planning

CHAPTER 10

Efficient and Secure Power for the USA and Switzerland

Boris Krey[*] and Peter Zweifel[**]

Abstract

In this chapter, portfolio theory is applied to power technologies of the USA and Switzerland. A current user view is adopted to determine the efficient frontier of generation technologies in terms of expected return and risk. Since shocks in generation costs per kWh (the inverse of expected returns) are correlated, seemingly unrelated regression estimation (SURE) is applied to filter out the systematic components of the covariance matrix. Since some of the portfolios of particular interest (minimum variance, maximum expected return) call for a high share of one technology, security of supply becomes an issue. Shannon–Wiener and Herfindahl–Hirschman indices are calculated to determine the tradeoff between efficiency and security of supply. Results suggest that risk-averse utilities (and ultimately, consumers) in the USA would have gained from adopting a feasible portfolio containing more coal, gas and oil at a price of a somewhat reduced security of supply. In the case of Switzerland, the realistic portfolio consists of nuclear, storage hydro, run of river and solar, with shares identical to those of the actual portfolio in 2003. Therefore, the current mix of Swiss generating technologies in Switzerland may be deemed efficient.

Key words   Efficiency frontier, Herfindahl–Hirschman index, portfolio theory, power generation, seemingly unrelated regression estimations (SURE), Shannon–Wiener index.

Acknowledgements

This research was supported by CORE, the Federal Energy Research Commission under the supervision of the Swiss Federal Office of Energy. The authors would like to thank Andreas Gut, Matthias Gysler, Lukas Gutzwiller,

[*] Research Associate, Socioeconomic Institute, University of Zurich, Zurich, Switzerland
[**] Professor of Economics, Socioeconomic Institute, University of Zurich, Zurich, Switzerland

Tony Kaiser, Michel Piot and Pascal Previdoli, as well as the participants of the INFRATRAIN Autumn School 2006 in Berlin, the IAEE International Conference 2006 in Potsdam and the annual SSES meeting 2006 in Lugano, for many helpful comments. Remaining errors are our own.

10.1 Introduction

Efficient portfolios of assets maximize expected return for any given level of risk or alternatively minimize expected risk for every level of expected return. This founding concept of finance, developed by Markowitz (1952), can be applied to a portfolio of real assets as well. Power companies, holding a portfolio of power generation technologies, face the task of achieving maximum expected return (defined as kWh/US $[1]) for any given level of risk (defined as the standard deviation of expected return), or put the other way around, the minimum risk for every level of expected return in terms of kWh/US $. In this way, they contribute to the attainment of widely recognized objectives of energy policy, i.e. the provision of electricity in an economical way while minimizing the overall risk of cost variability. This calls for taking into account the correlations between costs and therefore expected returns[2] of different power generation technologies.

For example, fossil-fuel-generated electricity faced dramatic cost fluctuations during the past decade, mainly caused by an oil price surge exceeding 300%[3] since 1999. In contrast, power generated by storage hydro fluctuated by less than 5% in Switzerland, mainly because of a stable price of water use. A portfolio containing both generation technologies therefore reduces risk considerably. Indeed, portfolio mixes in 2003 containing a larger share of gas power (for the USA) and the same share of nuclear (for Switzerland), combined with new-renewable generation technologies such as wind (for the USA) and solar (for Switzerland) serve to increase greatly the expected returns for both countries while keeping risk more or less constant. However, this concentration on mainly two technologies implies a reliance on two primary energy sources, which may jeopardize security of supply.

Apart from containing an international comparison, the present contribution has three novel features. First, while most of the published research adopts the investor's point of view that characterizes financial analysis, this work takes the current user's point of view. For financial investors, the current price of a share

[1] The definition of expected return adopted in this study is similar to Awerbuch (2004), who used the definition kWh/cent.

[2] As outlined in Awerbuch (2006b) and Awerbuch and Berger (2003), generation cost is nothing but the inverse of expected return. Therefore results are unaffected by whether portfolio optimization is based on maximizing expected return or minimizing cost, both ways leading to the same outcome.

[3] Source: WTRG Economics (www.wtrg.com).

is irrelevant for the composition of their portfolio. All that counts is its future increase in value. By way of contrast, a utility must consider the current cost per kilowatt hour of the inputs it intends to use. Second, correlations between unobserved shocks influencing the cost of electricity generation technologies (the inverse of expected return) are taken into account, improving the efficiency of estimates. This clearly differs from previous contributions, where these correlations are not accounted for. Third, the security of supply issue is also addressed. Indeed, an efficient portfolio of electricity generation technologies may call for a high share of one particular technology (and hence energy source) if unit costs are strongly correlated, obviating diversification effects. However, such a solution would be deemed to impart excessive risk to the provision of electricity in the eyes of most policy makers. To reflect this concern, indices of concentration are calculated in order to depict a possible tradeoff between efficiency and security of supply with regard to electricity generation technologies.

This study is structured as follows. Section 10.2 presents a short review of key literature on portfolio theory as applied to power generation technologies and on the measurement of supply security. In Section 10.3, the theory of efficient power generation portfolios from a current user's point of view is laid out. Because common shocks (such as weather) impinge on generation costs (the inverse of expected return), the seemingly unrelated regression estimation method is adopted. In Section 10.4, seemingly unrelated regression estimation (SURE)-based efficient frontiers are constructed for the USA and Switzerland, with emphasis on solutions with special features, i.e. minimum variance (MV), same expected return (SER), same variance (SV) and maximum expected return (MER) portfolios. Shannon–Wiener (SW) and Herfindahl–Hirschman (HH) indices will also be calculated to see whether US and Swiss power generation technologies are sufficiently diversified. Conclusions are offered in Section 10.5.

10.2 Literature review

Portfolio theory and the concept of diversification were introduced by Markowitz (1952). Efficient portfolios maximize expected return for a given amount of risk (which is measured by the variance or standard deviation of the return of the portfolio). Equivalently, they minimize risk for a given expected return. As stated by Fabozzi et al. (2002), portfolio theory continues to be the most important tool for constructing efficient portfolios for financial assets.

More recently, portfolio theory has also been applied to real assets, such as those related to energy generation. According to Jansen et al. (2006), the first application to energy is due to Bar-Lev and Katz (1976), who examined whether US power utility companies are efficient users of fossil fuel. Costs of inputs are 'as burned', including overheads resulting from transportation expenses, heating of oil lines, stock cleaning, and fuel handling facilities for coal, fuel storage, inventory and maintenance. Compared with the efficient frontier, actual operations by electric utilities are characterized by a relatively high rate of expected returns, combined with an excessive amount of risk however. The authors argue that utilities could move toward the efficient frontier by purchasing fuels at a higher but guaranteed (i.e. futures market) price.

Adegbulugbe et al. (1989) examine the long-term optimal structure of energy supply in Nigeria. They use a multiperiod linear programming model of the total energy system to minimize direct fuel costs while achieving certain developmental objectives. Results indicate that gas and petroleum should play an important role in the future Nigerian energy mix, with coal limited to a very small share as long as its costs of production and transportation are as high as they were at the time. Nuclear power and solar energy are not part of the efficient frontier at all.

A major limitation characterizing the contributions of both Bar-Lev and Katz (1976) and Adegbulugbe et al. (1989) is that they fail to account for time-varying covariances in energy prices. In addition, they neglect possible correlations between shocks impinging on primary energy prices. Finally, only the unit costs of fuels enter calculations, causing other private costs (current operation, use of capital) to be disregarded, let alone social costs (health and global warming).

Humphreys and McClain (1998) tackle at least three previous limitations by (1) filtering out the systematic components of the covariance matrix of energy prices over time, (2) using a more comprehensive definition of private cost, and (3) including external costs. As to (1), their estimated variances and covariances are derived from generalized autoregressive conditional heteroscedastic (GARCH) models. By applying GARCH, the authors try to filter out systematic changes in volatility in response to shocks. Their results suggest that a shift away from oil toward natural gas would reduce overall volatility at a given rate of expected return (in terms of reduced cost of power). Focusing on changes rather than levels, Humphreys and McClain adopt the conventional financial portfolio approach, i.e. the investor's point of view. However, producers are mainly interested in the level of prices they have to pay for their inputs, with expected future changes being of secondary importance. Finally, as is true of all other studies, the authors fail to control for unobserved shocks affecting several generation technologies at the same time.

More recently, there has been research singling out electricity. Berger et al. (2003) use Markowitz theory to examine existing and projected generation technology mixes in the European Union. According to their study, renewables that are characterized by high fixed but low variable costs (such as wind) figure prominently in efficient portfolios, owing to their favorable expected returns and diversification effects. A weakness of this study is its database. Important components of cost are proxied by business indicators such as the S&P 500 index. In addition, neither external costs nor common unobserved shocks are taken into account.

Roques et al. (2005, 2006) apply stochastic optimization to determine whether nuclear power may serve as a hedge against uncertain gas and carbon prices. However, high and uncertain capital cost as well as potential construction and licensing delays cause the role of nuclear to be limited. Rather than estimating correlations between unit costs, the authors resort to the use of arbitrary correlation scenarios. This arbitrariness is crucial because the stronger the (positive) correlation between the cost of nuclear power and other technologies, the weaker its diversification effect.

Jansen et al. (2006) again apply Markowitz theory to determine efficient portfolios of power-generating technologies for the Netherlands in the year 2030. Their results suggest that diversification may yield a risk reduction of up to 20% at no extra loss in expected returns.

Portfolio analysis assumes shocks to be stochastic. However, cost hikes may be the result of concerted behavior on the part of suppliers who have market power. The risk of collusion is the higher the smaller the number of suppliers, which in turn varies directly with the number of energy sources. Based on this line of argument, measures of concentration such as the SW and HH indices have been increasingly applied in studies related to power generation technologies. The SW index (a measure of entropy) reflects diversity, while the HH index measures market concentration. Both indices permit evaluation of the security of supply of different power generating technologies thanks to a greater number of suppliers. They therefore complement the mean-variance portfolio approach for policy makers who fear purchases of primary energy to be exposed to collusion or monopoly – a consideration of relevance especially in the markets for natural gas and uranium.

Grubb et al. (2005) explore the relationship between low-carbon objectives and strategic security of supplies in the context of the UK power system by calculating both SW and HH indices. They identify a complementarity between the two objectives in that a reduction of carbon intensity is uniformly associated with greater long-term diversity in UK power generation. However, they neglect stochastic shocks altogether, which could cause a tradeoff between Markowitz efficiency and protection from market power.

Doherty et al. (2005) complement their portfolio analysis with the SW index to assess the fuel portfolio of a power plant in Ireland as of 2020. Not surprisingly, the plant's efficient minimum variance portfolio contains a much more diversified mix of generation technologies than the maximum expected return portfolio, while the SW index favors the minimum variance alternative. However, this study is based on a covariance matrix of returns that has not been purged of extreme shocks and therefore may lack stability.

10.3 Methodology

10.3.1 Real asset portfolio estimation

Owners of a real asset portfolio seek to maximize its expected return at a given risk or alternatively to minimize risk given their expected return. In more formal terms, the expected return of a real asset portfolio $E(R_p)$ consisting of m risky assets is given by

$$E(R_p) = \sum_{i=1}^{m} w_i E(R_i) \qquad (10.1)$$

where w_i is the share of asset i and $E(R_i)$ its expected return. In the present case of five components, the portfolio standard deviation (σ_p) involves the variances and correlation coefficients in the following way:

$$
\sigma_p = \left(
\begin{array}{l}
w_1^2\sigma_1^2 + w_2^2\sigma_2^2 + w_3^2\sigma_3^2 + w_4^2\sigma_4^2 + w_5^2\sigma_5^2 + 2w_1w_2\rho_{12}\sigma_1\sigma_2 \\
+2w_1w_3\rho_{13}\sigma_1\sigma_3 + 2w_1w_4\rho_{14}\sigma_1\sigma_4 + 2w_1w_5\rho_{15}\sigma_1\sigma_5 + 2w_2w_3\rho_{23}\sigma_2\sigma_3 \\
+2w_2w_4\rho_{24}\sigma_2\sigma_4 + 2w_2w_5\rho_{25}\sigma_2\sigma_5 + 2w_3w_4\rho_{34}\sigma_3\sigma_4 + 2w_3w_5\rho_{35}\sigma_3\sigma_5 \\
+2w_4w_5\rho_{45}\sigma_4\sigma_5
\end{array}
\right)^{\frac{1}{2}}
\tag{10.2}
$$

where $\rho_{ij} = \mathrm{cov}_{ij}/(\sigma_i\sigma_j)$, $i,j = 1, \ldots, 5$, are correlation coefficients and σ_i are individual standard deviations. In the case of the USA, the five sources are oil, coal, gas, nuclear and wind. Accordingly, Equations (10.1) and (10.2) become

$$
E(R_{USp}) = w_{Oil}E(R_{Oil}) + w_{Coal}E(R_{Coal}) + w_{Gas}E(R_{Gas}) + w_{Nuclear}E(R_{Nuclear})
$$
$$
+ w_{Wind}E(R_{Wind})
\tag{10.3}
$$

$$
\sigma_p = \left(
\begin{array}{l}
w_{Oil}^2\sigma_{Oil}^2 + w_{Coal}^2\sigma_{Coal}^2 + w_{Gas}^2\sigma_{Gas}^2 + w_{Nuclear}^2\sigma_{Nuclear}^2 + w_{Wind}^2\sigma_{Wind}^2 \\
+2w_{Oil}w_{Coal}\rho_{Oil,Coal}\sigma_{Oil}\sigma_{Coal} + 2w_{Oil}w_{Gas}\rho_{Oil,Gas}\sigma_{Oil}\sigma_{Gas} \\
+2w_{Oil}w_{Nuclear}\rho_{Oil,Nuclear}\sigma_{Oil}\sigma_{Nuclear} + 2w_{Oil}w_{Wind}\rho_{Oil,Wind}\sigma_{Oil}\sigma_{Wind} \\
+2w_{Coal}w_{Gas}\rho_{Coal,Gas}\sigma_{Coal}\sigma_{Gas} + 2w_{Coal}w_{Nuclear}\rho_{Coal,Nuclear}\sigma_{Coal}\sigma_{Nuclear} \\
+2w_{Coal}w_{Wind}\rho_{Coal,Wind}\sigma_{Coal}\sigma_{Wind} + 2w_{Gas}w_{Nuclear}\rho_{Gas,Nuclear}\sigma_{Gas}\sigma_{Nuclear} \\
+2w_{Gas}w_{Wind}\rho_{Gas,Wind}\sigma_{Gas}\sigma_{Wind} + 2w_{Nuclear}w_{Wind}\rho_{Nuclear,Wind}\sigma_{Nuclear}\sigma_{Wind}
\end{array}
\right)^{\frac{1}{2}}
\tag{10.4}
$$

   Portfolio theory determines not a single best mix but an efficient frontier containing an infinite number of solutions. The optimal solution depends on consumer preferences, which reflect risk aversion. In Figure 10.1, let there be only two generation technologies, GT1 and GT2. By assumption, GT1 (e.g. solar generated power) has low expected return (measured as kWh/US $) but low volatility of unit cost. By way of contrast, GT2 has much higher expected return but is more risky (e.g. run of river generated power). If the correlation between the two generation technologies is less than perfect, the efficient frontier runs concave. The lower the correlation coefficient, the stronger this portfolio effect.[4] In Figure 10.1, the efficient frontier formed by GT1 and GT2 with its high expected return but also high volatility runs concave rather than linear, permitting holders of this power portfolio to profit from a diversification effect (Awerbuch and Berger, 2003). Although adding GT3 may not look attractive at first owing to low expected returns, this technology is so little correlated with GT1 and especially GT2 that it causes the efficient frontier to become more concave. One example

---

[4] Awerbuch (2006b) claims portfolio effects to become pronounced when correlation coefficients are below about 0.6.

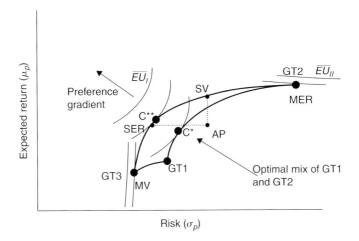

FIGURE 10.1 Efficient portfolios of generation technologies (GT). AP: actual portfolio; MV: minimum variance; SER: same expected return; SV: same variance; MER: maximum expected return.

of this effect can be found in Awerbuch (2006a), who shows that by adding risky wind generation to the existing power mix of Scotland, a substantially reduced portfolio standard deviation can be attained. Indeed, wind generation costs in Scotland do not correlate with fossil prices, causing a marked diversification effect (Awerbuch, 2006a).[5]

In order to determine the optimal portfolio (to be selected among the efficient ones), knowledge of the consumer's preferences would be necessary. Along an indifference curve $\overline{EU}$, expected utility is held constant. The preference gradient of Figure 10.1 indicates a risk-averse decision maker who likes a higher expected return but dislikes volatility. Evidently, the optimum allocation is given by the highest valued indifference curve that is still an element of the efficient frontier. For the frontier composed of GT1 and GT2 only, this optimum is depicted by point C*. If GT3 is indeed available, C** becomes the new optimum, with both higher expected return and less volatility. Clearly, C** dominates C*, demonstrating the future contribution to welfare that can be expected from the availability of additional energy technologies thanks to improved diversification.

In the absence of information about utilities' and societies' degree of risk aversion, some solutions that do not depend on this information are of interest. First, a very risk-averse decision maker is predicted to prefer the minimum variance (MV) portfolio, which coincides with GT3 in Figure 10.1. Another solution of importance is the same expected return (SER) portfolio. It contains generation technology mixes that are reshuffled as to have the same expected return as the

---

[5] Awerbuch (2006b) also refers to Brealey and Myers (1994), who show that adding riskless government bonds yielding as little as 3% to a stock portfolio with a rate of return of 8% still serves to raise the expected return at any level of risk.

actual portfolio (AP)[6] while being on the efficient frontier. An SER thus offers a much lower volatility than the current portfolio. Third, the same variance (SV) portfolio contains the mix of technologies that is as risky as AP, but being located on the efficient frontier generates more expected return. Fourth, an almost risk neutral decision maker (represented by indifference curves $\overline{EU}_{II}$) will opt for the maximum expected return (MER) portfolio (GT2 in the example). These four portfolios along the efficient frontier (MV, SER, SV and MER) permit the choice for utilities and policy makers to be narrowed down.

Finally, the Sharpe ratio (SR), a measure of return to risk, can be used for the same purpose. It is given by

$$SR = ER_p / \sigma_p \qquad (10.5)$$

where $E(R_p)$ is the expected return of the efficient portfolio (Equation 10.1), while σ_p represents the volatility (measured as standard deviation of the expected return) of the efficient portfolio (Equation 10.2). A higher value of the SR is preferred over a lower one.

Some solutions (MV and MER) involve only one technology (GT3 and GT2, respectively; Figure 10.1). However, contrary to financial markets, where investors can allocate their entire wealth to one asset, opting for a single generation technology often is not feasible. For example, a portfolio containing photovoltaic generation only would have to be excluded unless a very long planning horizon is adopted. This consideration calls for imposing constraints, as in Section 10.4 below.

10.3.2 Seemingly unrelated regression estimation[7]

In view of Equation (10.2), portfolio risk σ_p depends on individual standard errors σ_i and the correlations between returns ρ_{ij}. As argued before, it is important to derive estimates of the covariance matrix (i.e. of σ_i and σ_{ij}) that are reasonably time-invariant. In each time series of electricity generation costs considered, this calls for the estimation of predicted values $\hat{R}_{i,t} = R_{i,t} - \hat{u}_{i,t}$ that do not contain a systematic shift. Such values can be computed from the residuals $\hat{u}_{i,t}$ of the following autoregressive process of order j (sometimes complemented by a time variable)

$$R_{i,t} = \alpha_{i0} + \sum_{j=1}^{m} \alpha_{ij} \cdot R_{i,t-j} + u_{i,t} \qquad (10.6)$$

where $R_{i,t}$ is the (return) for technology i in year t, α_{i0} is a constant for technology i, α_{ij} is the coefficient of the return lagged j years, $R_{i,t-j}$ is the dependent variable (rate of return) lagged j years, and $u_{i,t}$ is the error term for technology i in year t.

---

[6] The actual portfolio contains the de facto power mix as of 2003 (see Section 10.4.2).
[7] This section is based on Krey and Zweifel (2006).

If the shocks $u_{i,t}$ causing volatility in $R_{i,t}$ were uncorrelated across technologies, one could estimate the expected return for each electricity-generating technology separately to obtain residuals $\hat{u}_{i,t}$ and hence values for $\hat{R}_{i,t}$. However, as shown by previous research (Krey and Zweifel, 2006), error terms are significantly correlated across energy sources. This constitutes information that can be exploited for improving the efficiency of estimation, typically resulting in sharper estimates of the parameters α_{ij}, of residuals $u_{i,t}$, and hence of the σ_i and σ_{ij} making up the covariance matrix of returns. The pertinent econometric method is called seemingly unrelated regression estimation (SURE). The SURE model consists of m regression equations (m is the number of electricity generation technologies), each of which satisfies the requirements of the standard regression model. The assumption that is specific to SURE is that the covariance matrix $E(\mathbf{uu'})$ is not diagonal, with I the $m \times m$ identity matrix.

$$E(\mathbf{uu'}) = \begin{bmatrix} \sigma_{i,i}I & \sigma_{i,k}I \\ \sigma_{k,j}I & \sigma_{k,k}I \end{bmatrix} \tag{10.7}$$

By way of contrast, traditional ordinary least squares (OLS) estimation would be appropriate if the disturbance terms of technologies i and k were not correlated. However, this does not hold for US and Swiss power technologies (see Section 10.4.3), giving rise to the covariance matrix shown in Equation (10.8) for the case of the USA:

$$\Omega = E(\mathbf{uu'}) = \begin{bmatrix} \sigma_{OilOil}I & \sigma_{OilGas}I & \sigma_{OilNucl}I & \sigma_{OilWind}I & \sigma_{OilCoal}I \\ \sigma_{GasOil}I & \sigma_{GasGas}I & \sigma_{GasNucl}I & \sigma_{GasWind}I & \sigma_{GasCoal}I \\ \sigma_{NuclOil}I & \sigma_{NuclGas}I & \sigma_{NuclNucl}I & \sigma_{NuclWind}I & \sigma_{NuclCoal}I \\ \sigma_{WindOil}I & \sigma_{WindGas}I & \sigma_{WindNucl}I & \sigma_{WindWind}I & \sigma_{WindCoal}I \\ \sigma_{CoalOil}I & \sigma_{CoalGas}I & \sigma_{CoalNucl}I & \sigma_{CoalWind}I & \sigma_{CoalCoal}I \end{bmatrix} \tag{10.8}$$

In sum, SURE allows one to estimate simultaneously the expected returns of all power generation technologies in one regression while taking into account the possible correlation of error terms across equations. This approach is novel and has, to the authors' knowledge, not been applied in previous research concerned with portfolios of real assets.

10.3.3 Shannon–Wiener index

Whereas up to this point, shocks to expected returns were considered to be stochastic, measures of concentration reflect a concern about supplier strategies. The fewer technologies a power system relies upon, the fewer (as a rule) the number of suppliers, and the more the system is exposed to the (non-stochastic) effects of collusion and monopoly. One measure of concentration (or rather diversity) is entropy, also known as the SW index, given by:

$$SW = \sum_{i=1}^{m} - p_i \ln(p_i) \tag{10.9}$$

where p_i ($i = 1, \ldots, m$) is the proportion of generation represented by the ith type of generation technology. A value below 1.00 indicates a system that is highly concentrated and therefore subject to the risk of collusion or monopoly, leading to interrupted supply and/or price hikes.

10.3.4 Herfindahl–Hirschman index

Another measure of concentration and therefore of security of supply is the HH index. This index is calculated according to:

$$HH = \sum_{i=1}^{m} p_i^2 \qquad (10.10)$$

where p_i is the share of the ith technology, usually expressed as a percentage. Therefore, $HH = 10,000$ in the case of a monopoly. Conversely, a value $HH < 1000$ is taken by antitrust authorities as indicating no concentration. In the present context, a value of $HH > 1800$ is interpreted as being problematic in terms of exposure to supply risk (Grubb et al., 2005).

Stirling (1998) prefers the SW index over the HH index, primarily because the mathematical properties of the SW index are more readily derived from first principles. Moreover, the rank orderings of SW are not sensitive to changes in the base of logarithm. Here, both indices will be used, since HH is better known in the economic literature while generating results that are consistent with those of SW (Grubb et al., 2005).

10.4 Efficient US and Swiss power generation frontiers in 2003

10.4.1 The data

This study uses time-series data containing annual power generation returns for several technologies, measured in kWh electric power per US dollar.[8] The data cover the years 1981–2003 (USA) and 1985–2003 (Switzerland). Throughout, generation returns comprise fuel cost, cost of current operations and capital user cost.[9] In the case of nuclear power, decommissioning and disposal are also included. The data are adjusted to contain externality surcharges for environmental damage (mainly related to health and global warming). The data on external costs were obtained from the European Commission (2003) for the USA[10] and from Hirschberg and Jakob (1999) for Switzerland. While based on the same

---

[8] For Switzerland, the year 2000 mean value of the Swiss Franc (CHF) exchange rate was used (US Federal Reserve: http://research.stlouisfed.org).

[9] As correctly pointed out by Fabien Roques (personal communication), there are different ways to measure capital user costs, yielding different generation costs. However, the utilities concerned did not provide the background data that would permit calculation of variants of capital user cost.

[10] Since no external cost data for the USA were available, external cost data from the UK were used instead. The UK power industry's generation mix is similar to that of the USA.

methods, these studies contain several externality cost scenarios, ranging from low external costs (optimistic view) to very high costs (conservative view). The conservative estimate will be used throughout. All variables are deflated by the US and Swiss CPI for the USA and Switzerland, respectively, with 2000 serving as the base year (= 100).

Table 10.1 presents the US generation returns for 1995 and 2000, for five categories: oil, coal, gas, nuclear and wind power.[11] Returns range between 9 and 24 kWh/US $ in 2000, with oil attaining the minimum and wind the maximum.

The Swiss dataset contains nuclear,[12] run of river,[13] storage hydro[14] and solar.[15] Three of the four generation technologies are comparable to those of the USA in terms of returns, being in the range 26–57 kWh/US $ in 2000 (Table 10.2). By way of contrast, solar was markedly more expensive in 1995 but have experienced large cost decreases since then, resulting in a steady increase of return.

The historical development of returns in US power generation is shown in Figure 10.2. Oil exhibits large fluctuations in returns, particularly in the aftermath of 9/11. Similar fluctuations can be found for gas, pointing to its strong correlation with oil. By way of contrast, wind and nuclear might have favorable

Table 10.1   US generation returns taking account of external costs (kWh/US $)

Year	Oil	Coal	Gas	Nuclear	Wind
1995	8.87	8.74	16.13	17.34	18.37
2000	9.03	10.22	11.48	22.48	24.18

Table 10.2   Swiss generation returns taking account of external costs (kWh/US $)

Year	Nuclear	Run of river	Storage hydro	Solar
1995	20.14	38.57	17.59	1.24
2000	26.65	57.09	28.80	1.77

[11] Data for oil, coal, gas and nuclear were obtained from the UIC (2005). Wind (State Hawaii, USA (www.state.hi.us) and US Department of Energy (www.energy.gov)). Since the wind data were not available for every year, values for 1983, 1985–1987, 1989–1994, 1996–1999 were generated by cubic spline interpolation.

[12] Data sources: KKL (2005), KKG (2005).

[13] Data source: personal correspondence.

[14] Data source: personal correspondence.

[15] RWE Schott Solar (2005); The average exchange rate of 2000 was used to convert euro cents into US cents (source: US Federal Reserve). RWE Schott Solar data from Germany are used as a proxy for Swiss solar power data, since solar generation technologies in both countries are similar.

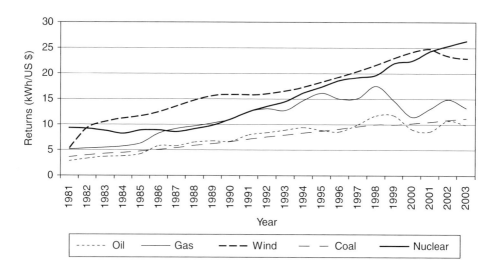

FIGURE 10.2   US returns in power generation (kWh/US $), 1981–2003.

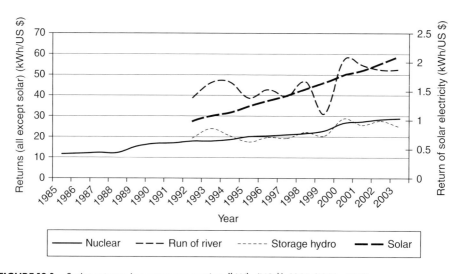

FIGURE 10.3   Swiss returns in power generation (kWh/US $), 1985/1992–2003.

diversification properties thanks to their independence from fluctuations in fossil fuel prices.

In the Swiss dataset, run of river exhibits the strongest fluctuations, particularly in 1999 and 2000 (Figure 10.3). The likely reason is changes in financial transactions between key run of river power suppliers (Axpo, 2002). In contrast, returns of nuclear increase slowly over time, making it a likely candidate for diversification.

10.4.2 Actual mix of power generation as of 2003

In order to have a benchmark against which to hold efficient solutions, actual 2003 shares of current input to US and Swiss power generation are displayed in Table 10.3. The US mix contains 56% coal, 21% nuclear, 18% gas, 3% oil and 2% wind (see Table 10.3(a)). No data were available for Hydro power, which normally makes up for some 7% of total power production in the USA. Still, the ensuing analysis covers more than 90% of US capacity, going beyond earlier studies that were limited to gas, coal and wind (Awerbuch, 2006b) and gas, coal and oil, respectively (Humphreys and McClain, 1998). The actual Swiss power mix (see Table 10.3(b)) consists of 40% nuclear, 32% storage hydro, 24% run of river and 4% solar. However, solar serves as a proxy for (negligible) conventional thermic and renewable sources for which data are unavailable. Again, the data account for more than 90% of capacity.

10.4.3 SURE results for the USA and Switzerland

Recall that SURE seeks to increase the efficiency of estimation by accounting for correlations in unobserved shocks. Table 10.4 provides evidence supporting this

Table 10.3 Actual shares of power generation technologies (2003)

Technology	Shares (%)
(a) USA	
Coal	56
Nuclear	21
Gas	18
Oil	3
Wind	2
(b) Switzerland	
Nuclear	40
Storage hydro	32
Run of river	24
Solar	4

Table 10.4 Partial correlation coefficients, USA (1981–2003)

	Oil	Coal	Gas	Nuclear	Wind
(a) Correlations between returns					
Oil	1	0.9354	0.9524	0.8303	0.9031
Coal	0.9354	1	0.8830	0.9588	0.9711
Gas	0.9524	0.8830	1	0.7503	0.8259
Nuclear	0.8303	0.9588	0.7503	1	0.9169
Wind	0.9031	0.9711	0.8259	0.9169	1
(b) Correlations between $\hat{u}_{i,t}$ residuals from Equation (10.6)					
Oil	1	0.0803	0.2704	0.0988	−0.0860
Coal	0.0803	1	0.7754	−0.4051	−0.4405
Gas	0.2704	0.7754	1	−0.2805	−0.4813
Nuclear	0.0988	−0.4051	−0.2805	1	−0.2265
Wind	−0.0860	−0.4405	−0.4813	−0.2265	1

notion. In Table 10.5, in (a) partial correlation coefficients relating to returns (kWh/ US $) in the USA are shown. The figures indicate strong and positive correlations. For instance, oil and gas exhibit a correlation coefficient of no less than 0.95. In (b) in Table 10.4 are shown the correlations of $\hat{u}_{i,t}$, i.e. the residuals from Equation (10.6), which represent the components due to unobserved shocks. Correlation coefficients drop throughout, turning negative in some cases, but remain substantial. For example, the correlation across the equations for oil and gas is still 0.27.

The evidence for Switzerland is presented in Table 10.5. Here, the most marked correlation is between solar and nuclear, amounting to 0.98 (see (a), Table 10.5). The corresponding correlation between residuals $\hat{u}_{i,t}$ is estimated as 0.39 (see (b), Table 10.5). Correlation coefficients drop as well, but much less than in the case of the USA.

A possible explanation for this difference is that prices for primary energy sources purchased by Swiss utilities, being predominantly domestic, are much more detached from world market developments than their US counterparts.

Table 10.6 displays the SURE regression results for the USA. As can be seen from the column denoted $\overline{R}$, wind has the largest expected return, amounting to 18.01 kWh/US $, while coal has the smallest expected return, at a mere 7.83 kWh/ US $. The standard deviations (SD) of all technologies vary widely, with oil being the least volatile (2.00) and nuclear the most volatile (6.35).

All regressions include a time trend, which, however, turned out to be insignificant for oil and gas. The positive coefficients for trend in the coal, nuclear and

Table 10.5 Partial correlation coefficients, Switzerland (1992–2003)

	Nuclear	Run of river	Storage hydro	Solar
(a) Correlations between returns				
Nuclear	1	0.6421	0.7534	0.9795
Run of river	0.6421	1	0.8522	0.5535
Storage hydro	0.7534	0.8522	1	0.6879
Solar	0.9795	0.5535	0.6879	1
(b) Correlations between $\hat{u}_{i,t}$ residuals from Equation (10.6)				
Nuclear	1	0.4934	0.7420	0.3861
Run of river	0.4934	1	0.7967	0.0205
Storage hydro	0.7420	0.7967	1	−0.0021
Solar	0.3861	0.0205	−0.0021	1

Table 10.6 Results of SURE regression, USA (1981–2003)

	$\overline{R}$	SD	Const.	R_{t-1}	R_{t-2}	R_{t-3}	R_{t-3}	Trend	Obs.	R^2
Oil	8.23	2.00	1.84**	1.00***	−0.77***	0.73**	–	−0.69	19	0.86
Coal	7.83	2.26	0.88***	1.07***	−0.28*	–	–	0.07*	19	0.99
Gas	12.58	2.63	2.99***	1.17***	−1.07***	1.11***	−0.67***	0.18	19	0.90
Nuclear	15.06	6.35	0.65**	0.62***	–	–	–	0.44***	19	0.99
Wind	18.01	4.19	5.39***	1.15***	−0.53	−0.35	–	0.52***	19	0.99

*** significant at 10% level, ** significant at 5% level, * significant at 1% level.

wind regressions indicate that expected returns increase over time. The coefficients of determination, R^2, are comfortably high.

Turning to Switzerland, the $\overline{R}$ column in Table 10.7 shows expected returns for run of river to be maximum with 47.09 kWh/US $, whereas solar only generates 1.47 kWh/US $. Comparing Tables 10.6 and 10.7, nuclear in Switzerland displays both higher expected return and less risk than in the USA. The time trend has a positive and significant coefficient for all generation technologies, showing the strongest increase for run of river (2.11); however, only 45% of the variation can be explained.

Table 10.7   Results of SURE regression, Switzerland (1992–2003)

	$\overline{R}$	SD	Const.	R_{t-1}	R_{t-2}	R_{t-3}	R_{t-4}	Trend	Obs.	R^2
Nuclear	20.29	4.93	4.01**	0.28	–	–	–	0.92***	9	0.94
Run of river	47.09	4.76	8.87	−0.17	0.17	0.13	–	2.11***	9	0.45
Storage hydro	23.73	2.22	−0.50***	0.04	−0.15	–	–	1.85***	9	0.79
Solar	1.47	0.39	0.10	0.10	–	–	–	0.10***	9	0.99

*** significant at 10% level, ** significant at 5% level. * significant at 1% level.

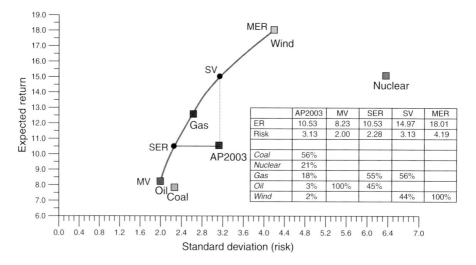

FIGURE 10.4   Efficient frontier for the USA (2003, SURE-based, no constraint, high external costs). AP: actual portfolio; MV: minimum variance; SER: same expected return; SV: same variance; MER: maximum expected return.

10.4.4 Efficient power generation frontiers

10.4.4.1 Efficient frontiers for the USA

The efficient frontier of US power generation technologies is shown in Figure 10.4. With no feasibility constraints applied, a US utility interested in minimizing risk would opt for the MV portfolio, which contains oil exclusively. There, the standard deviation (Risk) is a mere 2.00, 1.13 percentage points less than the actual portfolio (AP2003) with 3.13 (see insert). If the utility is risk neutral, causing it to opt for the MER portfolio, then 100% wind would be efficient, offering an expected return of 18.01, again more than the AP2003 with 10.53. Two intermediate solutions of interest are the SER portfolio and the SV portfolio. For the SER, the benchmark is the 10.53 kWh/US $ achieved by the AP2003, which contains 55% gas and 45% oil. At this mix, the SER offers the same expected return but at a lower risk (down from 3.13 to 2.28). However, it calls for more gas (up from 18 to 55%) and more oil (up from 3 to 45%) but no coal, nuclear or wind. As to the SV, its expected return is 14.97 rather than 10.53 kWh/US $, achieved by changing from 18 to 56% gas and from 2 to 44% wind. This time, coal, nuclear and oil are not part of the efficient portfolio.

However, such unconstrained portfolios are not very realistic. For instance, 44% wind generation in 2003 would not have been technically feasible. Therefore, two additional scenarios are presented, with one constraining the share of wind to a maximum of 5% (Figure 10.5, C1) and the other constraining the shares of coal, oil, nuclear and wind to no more than 60, 10, 25 and 5%, respectively (Figure 10.6, C2), shares that reflect technical feasibility. As expected, the MV_C1 and SER_C1 portfolios of Figure 10.5 contain the same mixes as in Figure 10.4, because wind was not part of the MV and SER efficient portfolios in the first place. However, a look at the SV_C1 and MER_C1 portfolios reveals modifications. The SV_C1 portfolio places a greater weight on gas (78 rather than 55% as in the unconstrained portfolio) and nuclear (17 versus 0%, compared to 13% in the AP2003). The constraint on wind becomes binding at a share of 5%. Interestingly, these constraints cause only minor losses in terms of performance. For example, the SV_C1 portfolio has an expected return amounting to 13.27 as compared to 14.97 kWh/US $ for the unconstrained frontier. Nuclear takes a weight of 95% in the MER_C portfolio, while wind is constrained to its binding share of 5%.

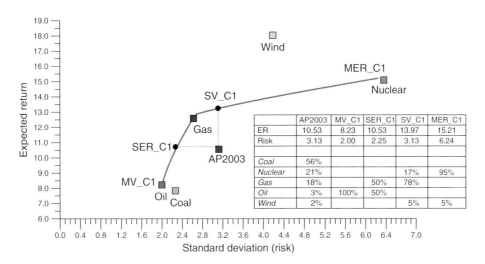

FIGURE 10.5 Efficient frontier for the USA (2003, SURE-based, with constraint, high external costs). Constraint imposed: wind ≤ 5%. AP: actual portfolio; MV: minimum variance; SER: same expected return; SV: same variance; MER: maximum expected return.

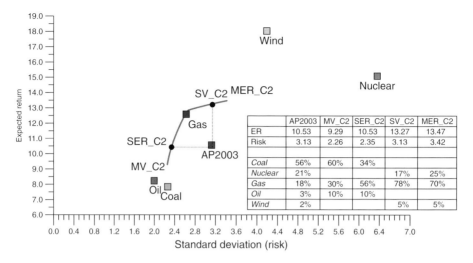

FIGURE 10.6   Efficient frontier for the USA (2003, SURE-based, with constraints, high external costs). Constraints imposed: coal ≤ 60%, nuclear ≤ 25%, oil ≤ 10%, wind ≤ 5%. AP: actual portfolio; MV: minimum variance; SER: same expected return; SV: same variance; MER: maximum expected return.

As expected, expected return is lower and risk higher as in the unconstrained portfolio (15.21 and 6.24 rather than 18.01 and 4.19, respectively).

By way of contrast, Figure 10.6 shows a more diversified mix of generation technologies, due to the imposition of item-wise maximum shares that prevent one single technology from becoming dominant. Portfolio risk increases beyond that of the minimum variance portfolios of Figures 10.4 and 10.5 but still falls short of AP2003.

Focusing on the SV portfolios, a comparison of Figure 10.6 with Figures 10.4 and 10.5 reveals two salient features. First, all portfolios put some weight on gas, with shares ranging between 56 and 78%. This is much more than the 18% of AP2003. Second, maximum expected returns fall the more constraints are applied.

From this section, the following conclusions can be drawn with regard to the USA. A feature common to all scenarios is that a move toward the efficient frontier is possible with an increasing share of wind. This feature is particularly marked in all SV portfolios, where wind takes a share between 5 and 44%, depending on the scenario considered. In all SV portfolios, expected return exceeds that of AP2003 with no increase in risk. Assuming that utilities are rather risk averse, the MV portfolio should be of particular interest, implying a very strong reliance on oil (between 10 and 100%, depending on the scenario considered). The SER portfolios point to gas and oil with combined shares between 66 and 100%. If maximum returns are of interest, a combination of nuclear power and wind appears promising.

However, these results need to be compared with those of other studies. In his US portfolio analysis, Awerbuch (2006b) finds that gas-generated power should play a major role in the SER, SV and MER portfolios, with shares between 45 and

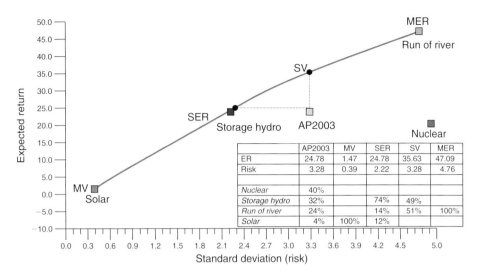

FIGURE 10.7   Efficient frontier for Switzerland (2003, SURE-based, no constraints, high external costs). AP: actual portfolio; MV: minimum variance; SER: same expected return; SV: same variance; MER: maximum expected return.

100%. The same holds true for wind according to his MV and SER portfolios. The present study arrives at similar conclusions, with SER and SV portfolios displaying a share of gas between 55 and 78%.

10.4.4.2 Efficient frontiers for Switzerland

Figure 10.7 displays the set of efficient power generation portfolios for Switzerland, without any constraints imposed. As in the case of the USA (Figures 10.4–10.6), the actual portfolio (AP2003) is located inside the efficient frontier, indicating a good deal of inefficiency (see Section 10.3.1). The SV portfolio serves to increase expected return to 35.63 kWh/US $ (up from 24.78 in the AP2003), while its volatility coincides with the actual portfolio. It implies a mix of 51% (rather than 24%) run of river and 49% (rather than 32%) storage hydro.

Risk-averse utilities opting for the MV portfolio would use 100% solar, reducing risk to 0.39, down from 3.28 in the AP2003. Conversely, the MER portfolio would imply 100% run of river, with expected return of 47.09, twice that of AP2003. By keeping expected returns at 24.78 (the AP2003 value), utilities could reduce risk from 3.28 to 2.22, with a mix containing 74% storage hydro, 14% run of river and 12% solar.

Again, unconstrained portfolios such as these are not realistic (at least not in the short run). For shares of 100% solar (MV) the climate is not sufficiently sunny, and for 100% run of river (MER), the extra hydro resources are lacking (Laufer, et al. 2004). Therefore, run of river, storage hydro and solar are constrained to equal their AP2003 shares, leaving only nuclear to be freely determined.

The corresponding efficient frontier is shown in Figure 10.8. The MV_C portfolio mirrors the technology shares of the AP2003, which are 40% nuclear, 32%

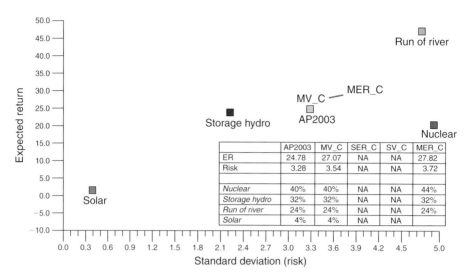

FIGURE 10.8 Efficient frontier for Switzerland (2003, SURE-based, with constraints, high external costs). Constraints imposed: run of river ≤ 24%, storage hydro ≤ 32%, solar ≤ 4%. AP: actual portfolio; MV: minimum variance; SER: same expected return; SV: same variance; MER: maximum expected return; NA: not applicable.

storage hydro, 24% run of river and 4% solar. Interestingly, it exhibits slightly more risk than the AP2003 (3.54 as compared to 3.28), which is due to the use of stabilized correlations in this particular instance (see Section 10.3.2). At the same time, it has 2.3 percentage points more expected return (27.07 versus 24.78), making it an attractive choice. No SER_C and SV_C portfolios can be determined, since the imposed constraints result in a solution set of measure zero. In the absence of risk aversion, the MER_C portfolio would be preferred, implying 44% nuclear, 32% storage hydro and 24% run of river. Not surprisingly, expected return attains a high 27.82, compared to 24.78 kWh/CHF in the AP2003, while risk increases slightly from 3.28 to 3.72.

In all, in the case of Switzerland, portfolios containing solar power serve to reduce risk significantly, as can be seen in Figure 10.7 (MV and SER portfolios) and Figure 10.8 (MV_C portfolio). Conversely, nuclear power helps to maximize expected return, as shown in Figure 10.8 (MV_C and MER_C portfolios). Storage hydro and run of river continue to weigh heavy in both Figure 10.7 (SER and SV portfolios) and Figure 10.8 (MV_C and MER_C portfolios).

10.4.4.3 Comparing efficient power frontiers: a tale of two countries

Both the USA and Switzerland, different as they may be otherwise, share one salient feature with regard to power generation. Their actual portfolio AP2003 definitely falls short of the efficient frontier. Both countries could reduce risk greatly by allocating larger shares to new-renewable technologies. As shown in Figures 10.4–10.6 for the USA, increasing the shares of wind beyond the AP2003 value goes along with higher expected return while risk increases slightly at worst. In Switzerland, all portfolios that contain more solar than AP2003 entail less risk. The SER portfolio of Figure 10.7 even suggests that a mix of 74% storage hydro, 14% run of river and 12% solar achieves 1.06 percentage points less risk while attaining the same expected return as AP2003.

Utilities in both countries are likely to be puzzled by these results. It seems obvious to them that increasing the shares of wind (in the USA) and solar (in Switzerland) must reduce expected returns. Although this notion has intuitive appeal because wind and solar have comparatively low returns, it does not hold true. According to Equation (10.1), $E(R_p)$ admittedly decreases if a below-average return component is added, provided the other shares remain (roughly) constant. However, the transition from the actual portfolio to a point on the efficient frontier causes this condition not to hold any more. For instance, the Swiss share of nuclear drops from 40 to 0% while that of solar increases from 4 to 12% (Figure 10.7, SER portfolio). An important implication is that in dynamic and uncertain environments, the merits of generating technologies must be determined not by evaluating single technologies, but by technology portfolios. This is also the key explanation for the divergence between users' actual choices and efficient choices. In the past, utilities and policy makers have been selecting generating technologies solely on an individual, case-by-case basis, failing to consider their contribution to overall portfolio performance.

On the whole, remaining within the technically feasible and assuming that US utilities are risk averse, it appears that they would have gained by adopting the MV_C2 portfolio by 2003, containing 60% coal, 30% gas and 10% oil. This mix would have reduced volatility by 0.8 percentage points but also expected return by 1.2 points below the AP2003 benchmark. Swiss utilities may be said to act in an efficient manner by adopting the MV_C portfolio, which is identical to the AP2003 (made up of 40% nuclear, 32% storage hydro, 24% run of river and 4% solar).

10.4.5 Supply security

Current concentration values for US and Swiss power generation portfolios are obtained by calculating the SW and HH indices (see Sections 10.3.3 and 10.3.4 for details). Both indices help to determine whether a power generation portfolio is sufficiently diversified in terms of technologies, which also implies diversification in terms of purchases of primary energy sources. In addition, the SR is calculated to identify portfolios with favorable return-to-risk values (see Section 10.3.1).

10.4.5.1 Supply security for the USA
Figure 10.9 provides an overview of all US portfolios that were presented in Section 10.4.4.1, with the AP2003 appearing in the first column. Its SW exceeds 1.00, which corresponds to a reasonably diversified portfolio. However, the HH exceeds 1800, suggesting that generation technologies and therefore purchases of primary sources for US power generation are concentrated. The SR has a fairly low value of 3.4. With the sole exception of MER_C1 (last column; see Figure 10.5), all efficient portfolios offer a higher expected return than AP2003

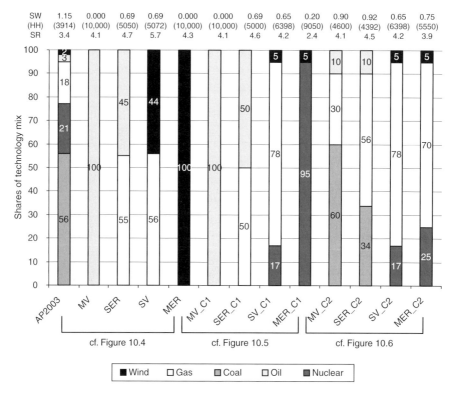

FIGURE 10.9   Shannon–Wiener index (SW), Herfindahl–Hirschman index (HH) and Sharpe ratio (SR) for the USA (2003). AP: actual portfolio; MV: minimum variance; SER: same expected return; SV: same variance; MER: maximum expected return.

for the same amount of risk. As expected, MV, MER and MV_C1 portfolios are heavily concentrated and thus prone to supply disruptions. However, their SRs are high, attaining 4.3 in the case of the MER portfolio. Incorporating short- to medium-term technological constraints, the MV_C2 and SV_C2 portfolios presumably appeal to US utilities. While their SRs are high (4.1 to 4.2), they are not well diversified according to the SW index. This points to a tradeoff between economic efficiency and supply security in US power generation.

10.4.5.2 Supply security for Switzerland

Like the actual portfolio of the USA, AP2003 of Switzerland has an SW in excess of 1.00, indicating a reasonably secure mix of technologies and hence primary energy sources (Figure 10.10). However, the HH exceeds 1800, implying that more diversification of generation technologies (and hence more competition between suppliers of primary energy, *ceteris paribus*) would be beneficial. Arguably the best technology mix for risk-averse utilities and policy makers would be the SV portfolio, because it keeps portfolio volatility constant while limiting the increase

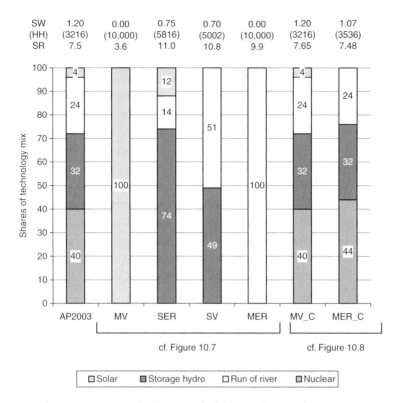

SW	1.20	0.00	0.75	0.70	0.00	1.20	1.07
(HH)	(3216)	(10,000)	(5816)	(5002)	(10,000)	(3216)	(3536)
SR	7.5	3.6	11.0	10.8	9.9	7.65	7.48

FIGURE 10.10   Shannon–Wiener index (SW), Herfindahl–Hirschman index (HH) and Sharpe ratio (SR) for Switzerland (2003). AP: actual portfolio; MV: minimum variance; SER: same expected return; SV: same variance; MER: maximum expected return.

in the HH to 5002 and the drop in SW to 0.70, but raising the SR to a high 10.8 (compared to 3216, 1.20 and 7.5, respectively, under the AP2003). However, the implied shares of run of river of 51% and storage hydro of 49% will not be attainable in the near future. The 12% share of solar that comes with the SER may be considered realistic. Compared to the AP2003, it serves to reduce portfolio volatility while raising SR to 11.0. The real difficulty lies with the concomitant increase of storage hydro's share to no less than 74%, more than double its current value. As evidenced by Laufer et al. (2004), present storage hydro technologies and geographical conditions render such a share unattainable.

All in all, Swiss utilities, policy makers and ultimately consumers face a situation that is very different from their American counterparts'. If they wish to break away from a markedly inefficient AP2003 (relative to domestic potential, not compared to the USA's), they have to bet on storage hydro (SER). This calls for a reduction in run of river capacities (which are alleged to be environmentally friendly, but recall that external costs are accounted for throughout) and scrapping nuclear power entirely. However, the restricted MV_C portfolio is secure according to the SW while containing the same technology shares as AP2003. Risk-averse current users thus seem to be best advised to keep the AP2003 mix. By way of contrast, risk-averse utilities (and consumers) in the USA could increase their SR from 3.4 to almost 4.1 by opting for the SV_C2 portfolio which, however, contains significantly more oil and gas and therefore would jeopardize security of supply.

10.5 Conclusions

The objective of this contribution was to determine the efficient frontiers of electric power generation in the USA and Switzerland, taking into account their implications in terms of security of supply. In contrast to existing portfolio studies, returns are defined as kWh of power per US dollar spent, which amounts to adopting a current user rather than an investor point of view. The observation period covers the years 1981–2003 (USA) and 1985–2003 (Switzerland). Because the error terms of the expected return regressions are correlated, SURE was adopted for estimating the covariance matrix used in determining efficient portfolios.

In the absence of information about the degree of risk aversion of utilities and policy makers in the two countries, the MV, SER, SV and (as a contrast) MER portfolios were singled out for detailed analysis. One could argue that for populations as risk averse as the USA's and especially the Swiss, the minimum variance portfolio is the appropriate one. However, this choice may result in allocations that are too different from the actual ones to be deemed technically feasible in the short to medium run. With realistic maximum shares of 60% coal, 25% nuclear, 10% oil and 5% wind for the USA, the following efficient portfolio of the constrained MV type (MV_C2) was obtained. It contains 60% coal, 30% gas and 10% oil, but no nuclear at all. While volatility declines, both the SW and the HH indices suggest that power generation technologies (and with them, supplies of primary energy sources) are not sufficiently diversified.

The one MV_C portfolio for Switzerland considered comprises 40% nuclear, 32% storage hydro, 24% run of river and 4% solar, which mimics the AP2003. Therefore, Swiss utilities can be said to generate electricity in an efficient manner. In addition, the SW index indicates a degree of diversification that is sufficient to avoid any threat to supply security.

An issue of interest would have been the efficient use of technologies along the load curve. However, this aspect had to be neglected in the present work. First, not all generating technologies can economically contribute to all load segments; for instance, hydro storage is reserved for the peak-load segment for good reasons. Second, the available data do not allow a consistent assignment of cost to load segments, which seems to have been a problem with previous studies as well (Awerbuch, 2003, 2005b, 2006a).

To sum up, depending on the scenario considered, failure to exploit new renewable resources such as wind (in the USA) and solar (in Switzerland) causes expected returns to be below and volatility to be the same as or even in excess of the efficient frontier of power-generating portfolios. In addition, a larger share of nuclear would move both countries closer to their feasible MER portfolio. However, such a move would entail a loss in terms of supply security.

Future research should deal with several aspects that had to be neglected in this work. First, investments in energy technology often must be considered irreversible, raising the issue of their optimal timing which cannot be addressed by Markowitz theory. The appropriate approach in these cases is real options theory (Dixit and Pindyck, 1994), whose prescriptions might differ from those presented here. Second, although defining returns in terms of kilowatt hours generated per US dollar has some appeal, future changes in this quantity, reflecting an investor's point of view, are relevant for planning. The present study thus needs to be complemented by using changes rather than levels of costs and productivities of energy technologies. Third, information about the risk preferences of utilities, policy makers and ultimately citizens with regard to both the costs and prices of final energy and purchases of primary energy inputs would be extremely valuable because this would permit truly optimal (rather than a set of efficient) power portfolios to be determined. The present work takes the first steps toward the attainment of these more ambitious goals.

References

Adegbulugbe, A., Dayo, F. and Gurtler, T. (1989). Optimal structure of the Nigerian energy supply mix. *The Energy Journal*, 10(2), 165–176.

Awerbuch, S. (2004). *Towards a Finance-Oriented Valuation of Conventional and Renewable Energy Sources in Ireland. Perspective from Abroad Series.* Dublin: Sustainable Energy Ireland. www.awerbuch.com

Awerbuch, S. (2005a). *The Role of Wind Generation in Enhancing Scotland's Energy Diversity and Security: A Mean-Variance Portfolio Optimisation of Scotland's Generation Mix.* Scotland: Airtricity. www.airtricity.com

Awerbuch, S. (2005b). *The Cost of Geothermal Energy in the Western US Regions: A Portfolio Based Approach.* Sandia Report SAND2005-5173. Oak Ridge, TN: Sandia National Laboratories. www.awerbuch.com

Awerbuch, S. (2006a). *Wind Provides Competitive Advantage for Scotland: A Report on Scotland's Electricity Mix.* Scotland: Airtricity. www.airtricity.com

Awerbuch, S. (2006b). Portfolio-based electricity generation planning: policy implications for renewables and energy security. *Mitigation and Adaptation Strategies for Global Change*, Vol. 11, pp. 693–710, New York: Springer.

Awerbuch, S. and Berger, M. (2003). *Energy Security and Diversity in the EU: A Mean-Variance Portfolio Approach.* IEA Report No. EET/2003/03. Paris: IEA (February). http://library.iea.org/dbtw-wpd/textbase/papers/2003/port.pdf

Axpo (2002). *Axpo Geschäftsbericht 2002* (Annual Bureau Report 2002). www.axpo.ch

Bar-Lev, D. and Katz, S. (1976). A portfolio approach to fossil fuel procurement in the electric utility industry. *Journal of Finance*, 31(3), 933–947.

Berger, M., Awerbuch, S. and Haas, R. (2003). *Versorgungssicherheit und Diversifizierung der Energieversorgung in der EU* (Security of Supply and Diversification of Energy Supply in the EU). Vienna: Bundesamt für Verkehr, Innovation und Technologie (Federal Office for Transportation, Innovation and Technology).

Brealey, R. and Myers, S. (1994, or any edition). *Principles of Corporate Finance.* New York: McGraw Hill.

Dixit, A. and Pindyck, R. (1994). *Investment under Uncertainty.* Princeton, NJ: Princeton University Press.

Doherty, R., Outhred, H. and O'Malley, M. (2005). *Generation Portfolio Analysis for a Carbon Constrained and Uncertain Future.* Electricity Research Centre, University College Dublin.

European Commission (2003). *External Costs.* Brussels: European Commission.

Fabozzi, F., Gupta, F. and Markowitz, H. (2002). The legacy of modern portfolio theory. *Journal of Investing*, 11(Fall), 7–22.

Gantner, U., Jakob, M. and Hirschberg, S. (2000). *Perspektiven der zukünftigen Energieversorgung in der Schweiz unter Berücksichtigung von nachfrageorientierten Massnahmen.* (Perspectives on the Future Provision of Energy in Switzerland, with Special Emphasis on Demand-Side-Management). Draft, PSI, Switzerland.

Grubb, M., Butler, L. and Twomey, P. (2005). *Diversity and Security in UK Electricity Generation: The Influence of Low Carbon Objectives.* Cambridge Working Papers in Economics. Cambridge: University of Cambridge.

Hirschberg, S. and Jakob, M. (1999). Cost structure of the Swiss electricity generation under consideration of external costs. *SAEE Seminar, Tagungsband*, Bern (11 June).

Humphreys, H. and McClain, K. (1998). Reducing the impacts of energy price volatility through dynamic portfolio selection. *The Energy Journal*, 19(3), 107–131.

Jansen, J., Beurskens, L. and van Tilburg, X. (2006). *Application of Portfolio Analysis to the Dutch Generation Mix.* Dutch Ministry of Economic Affairs (EZ).

KKG (2005). *Annual Report.* Downloaded from: www.kkg.ch (last visited April 2005).

KKL (2005). *Medienkonferenz 20 Jahre KKL, 10. Januar 2005. Portrait – Fakten – Zahlen zu 20 Jahre Kernkraftwerk Leibstadt* (Portrait, Facts and Figures Concerning 20 Years of the Nuclear Plant at Leibstadt). Downloaded from www.kkl.ch (last accessed April 2005).

Krey, B. and Zweifel, P. (2006). *Efficient Electricity Portfolios for Switzerland and the United States.* SOI Working Paper No. 0602. University of Zurich.

Laufer, F., Grötzinger, S., Peter, S. and Schmutz, A. (2004). *Ausbaupotentiale der Wasserkraft* (Potential for Expansion of Hydro Power). Bern: Bundesamt für Energie (Federal Office of Energy).

Markowitz, H. (1952). Portfolio selection. *Journal of Finance*, 7, 77–91.

NEPG (2001). *Report of the National Energy Policy Development Group.* Washington, DC: US Government Printing Office.

Roques, F., Newberry, D., Nuttall, W., Connors, S. and de Neufville, R. (2005). *Valuing Portfolio Diversification for a Utility: Application to a Nuclear Power Investment when Fuel, Electricity, and Carbon Prices are Uncertain.* Draft Research Paper, University of Cambridge.

Roques, F., Nuttall, W., Newberry, D., de Neufville, R. and Connors, S. (2006). Nuclear power: a hedge against uncertain gas and carbon prices? *The Energy Journal*, 27(4), 1–24.

RWE Schott Solar (2005). Data on solar generated electricity. www.rewschottscolar.com

Stirling, A. (1998). On the economics and analysis of diversity. *SPRU Electronic Working Paper Series*, 28.

UIC (2005). *The Economics of Nuclear Power.* Briefing Paper 8 (May). http://www.uic.com.au/nip08.htm

Portfolio Optimization and Utilities' Investments in Liberalized Power Markets[1]

Fabien A. Roques*, David M. Newbery** and William J. Nuttall***

Abstract

Monte Carlo simulations of gas, coal and nuclear plant investment returns are used as inputs of a mean-variance portfolio optimization to identify optimal base-load generation portfolios for large electricity generators in liberalized electricity markets. The impact of fuel, electricity and carbon dioxide price risks, and their degree of correlation on optimal plant portfolios, are studied. High degrees of correlation between gas and electricity prices, as observed in most European markets, reduce gas plant risks and make portfolios dominated by gas plant more attractive. Long-term power purchase contracts and/or a lower cost of capital can rebalance optimal portfolios towards more diversified portfolios with larger shares of nuclear and coal plants.

Key words   Electricity and fuel price risks, fuel mix, mean-variance portfolio theory, Monte Carlo simulation.

Acknowledgements

This work is dedicated to the memory of Shimon Awerbuch whose research and advices have inspired this chapter. The authors would like to thank Shimon Awerbuch, Chris Hall, Paul Twomey, anonymous referees from the Florence School of Regulation and Cambridge University EPRG, as well

---

[1] An earlier and more detailed version of this paper (Roques et al., 2008) can be found on the Cambridge Electricity Policy Research Group webpage, at www.electricitypolicy.org.uk

* International Energy Agency, Economic Analysis Division, Paris, France
** Faculty of Economics, University of Cambridge, Cambridge, UK
*** Judge Business School, University of Cambridge, Cambridge, UK

Analytical Methods for Energy Diversity and Security © 2008 Elsevier Ltd.

as seminar participants at the University of Cambridge, Florence School of Regulation, University of Toulouse, and University Paris Dauphine and University Paris XI for their helpful comments. This work was carried out while the lead author was at the Judge Business School, University of Cambridge, England. Financial support from the British Council and the Cambridge-MIT Institute under the project 045/P 'Promoting Innovation and Productivity in Electricity Markets' is gratefully acknowledged. The views in this paper are those of the authors alone. A large part of the material contained in this paper was orginally published in Energy Economics (2008), Vol 30/4, pp 1831–1849.

11.1 Introduction

The volatility of oil and gas prices over the few past years and the growing mistrust over Russian gas supplies in Europe have raised concerns over security of supplies and revived the debate over the optimal power generation fuel mix. One issue that attracts growing attention is whether liberalized electricity markets encourage investment in an adequately diverse mix of technologies. Do liberalized markets bias investment choices toward some generating technologies? If so, what barriers discourage investors from socially optimal fuel mix choices?

This chapter concentrates on private investors' investment incentives in liberalized electricity markets, where fuel mix diversification is a possible strategy for reducing exposure to electricity, fuel and carbon price risks. The chapter makes two contributions. First, on the methodological side, a two-step simulation approach is introduced, which assesses the impact of both input (fuel) and output [electricity and carbon dioxide (CO_2)] price risks on the returns of different generation technologies. While the existing literature focuses on *plant cost* risk (corresponding to a central planning paradigm), the focus on *plant returns* risk seems more relevant in liberalized markets. Monte Carlo simulations of a discounted cash-flow model of investment in combined cycle gas turbine (CCGT), coal and nuclear plant are first used to compute the distribution of returns and their correlation. Mean-variance portfolio (MVP) theory is then applied to identify the portfolios that maximize investor returns for given risk levels. Second, the simulations give new insights into investor incentives to diversify given fuel, electricity and CO_2 price risks. The results demonstrate the critical impact of the degree of correlation between electricity, fuel and CO_2 prices. It was found that for the correlations between gas, carbon and electricity prices observed in Britain over 2001–2005, gas-fired choices are excessively encouraged, but long-term fixed-price power purchase contracts may offset this bias.

In short, the correlation between electricity, gas and carbon markets makes 'pure' portfolios of gas power plants more attractive than diversified portfolios as gas plants' cash flows are 'self-hedged'. This is consistent with the empirical evidence, as most new power plants built in Britain over 2001–2005 were gas-fired. In contrast, when investors can secure a long-term power purchase agreement, and/or when they can have access to cheaper capital, optimal portfolios contain a more balanced mix of CCGT and coal and/or nuclear plants. These results suggest that liberalized electricity markets characterized by strong correlation between electricity and gas prices (such as Britain) are unlikely to reward fuel mix diversification sufficiently to make private investors' choices align with the socially optimal fuel mix, unless investors can find counter-parties with complementary risk profiles to sign long-term power purchase agreements.

11.2 Diversification in liberalized electricity markets

The old vertically integrated franchise monopoly model under state ownership or cost-of-service regulation was normally able to finance any required capacity in generation. That model occasionally experienced financing difficulties if governments restrained final prices (although that was more of a problem in developing countries) and certainly provided poor incentives for delivering investment in a timely and cost-effective way. Averch and Johnson (1962) demonstrated that regulated utilities might rationally prefer to invest in capital-intensive technologies. Their theoretical prediction is consistent with an emphasis on large coal and nuclear power stations in both the USA and Europe. Moreover, the subordination of utilities to government direction often gave rise to other distortions of investment choices. Many countries directly controlled or influenced the fuel mix through giving protection to 'national' fuels (such as coal or lignite) or the financing of 'national' technologies (such as nuclear) (Newbery and Green, 1996). When examining the alleged biases in technology choice caused by market liberalization, one should remember that public ownership and cost-of-service regulation also introduced biases.

The liberalization of the electricity industry shifted the investment risk burden from consumers to producers. While cost-plus regulation provided investors with prospects of stable returns, in liberalized electricity markets the volatility of electricity, CO_2 and fuel prices present significant risks for an investor. One central issue to the long-term benefit of liberalization of energy markets lies in their ability to deliver sustainable investment signals, without inappropriately biasing investment incentives toward some generating technologies rather than others. That requires identifying the drivers of technology and incentives to diversify in the liberalized industry and the barriers that may prevent investors making socially optimal fuel mix choices.

11.2.1 Fuel mix diversification and corporate strategy

In a liberalized industry, investment decisions are made by individual investors. It is important to understand how the shift from central planning to decentralized investment decision making has impacted investment choices and whether the value of a diverse fuel mix is factored into utilities' corporate strategy.

In the liberalized industry, investments are profit motivated. Utilities can no longer automatically pass on costs to consumers, and have to factor new constraints into the investment decision. When it comes to raising funds to finance a new power project, the impact of this investment on the company financial ratios has to be considered, with the pressure of stakeholders searching for high returns and a quick 'payback' period. In that perspective, the industry investment time-frame has considerably shortened, with power investment being amortized over no longer than 15 years, as debt repayments and power purchase agreements only exceptionally exceed 10–15 years. The opportunity cost of a new power investment is particularly high for capital-intensive technologies such as nuclear power plants.

Deregulation forced utilities to change radically their corporate strategy, and diversification now plays a central role in electricity companies' strategies. However, a closer look at the role of diversification for an electric utility reveals the complexity of that concept. Diversification can indeed apply to generation fuel, but also to plant manufacturers, fuel procurement contracts, plant geographical location, etc. It is not clear to what extent fuel mix diversification would benefit a utility, as the lowered exposure to fuel price risks has to be weighed against the gains of choosing the cheapest technology.

There is no such thing as one business model in the electricity industry, as electricity companies must decide which business segments to enter (generation, distribution, retail) and the degree of vertical or horizontal integration, as well as their geographical scope. This chapter investigates technology diversification from the point of view of a generation company that is not engaged significantly in the upstream and downstream parts of the industry.

11.2.2 The lack of long term financial risk management instruments in the electricity industry

The lack of long term financial risk management instruments is another issue that impacts generation companies' technology and diversification choices, insofar as it may favor technologies that have a degree of 'self-hedging' in the current industry framework. For instance, gas-fired power plants revenue can be expected to be stable in electricity markets such as the UK, which exhibit strong correlation between electricity and gas prices.

When the industry liberalized in the 1990s, market analysts predicted rapid growth in the use of electricity derivatives. However, in the last quarter of 2000, the market for exchange-traded electricity futures and options virtually collapsed in the USA, with knock-on effects in all electricity markets around the world (DOE, 2002).[2] Enron's collapse highlighted the problems of credit risk and default risk in electricity markets. Since 2000, market participants in the USA and to a lesser extent in Europe have become increasingly cautious and have been hedging risks by relying on more traditional utility suppliers and consumers with known physical assets, and by reducing the scope of their derivative products (e.g. moving toward shorter-term forward contracts).

---

[2] By February 2002, the New York Mercantile Exchange (NYMEX) decided to delist all of its futures contracts owing to lack of trading. The Chicago Board of Trade (CBOT) and the Minneapolis Grain Exchange (MGE) also suspended trading in electricity futures.

The US Department of Energy (DOE, 2002) suggests that the failure of exchange-traded electricity derivatives and the lack of liquidity of the over-the-counter (OTC) markets in the USA and in Europe seem to have resulted from problems in the underlying market for electricity itself, such as the lack of competition and regulatory uncertainty. In addition to these structural obstacles, the development of liquid electricity futures markets more than a few months in advance is hindered by the nature of electricity as a commodity, the extreme volatility of prices, the complexity of the existing spot markets and the lack of price transparency (Geman, 2005).

Even with the development of robust competitive markets, however, the use of derivatives to manage electricity price risk will remain difficult, because the simple pricing models used to value derivatives in other energy industries do not work well in the electricity sector (DOE, 2002). This suggests that innovative derivatives that are based on something other than the underlying energy spot price, such as weather derivatives, marketable emissions permits and specialty insurance contracts, will be important for the foreseeable future.[3] As financial markets regain confidence in the electricity industry, and as market participants improve their understanding of the specificities of the electricity industry, more tailored and innovative risk management instruments will emerge. But for the time being, the lack of long-term electricity specific financial risk management products limits the possibilities for generation companies to diversify their risks exposure.

11.2.3 From macroeconomic to microeconomic diversification incentives

Besides the possible 'short-termism' of investors, one might also worry about the ability of decentralized decision making adequately to coordinate individual decisions. Before liberalization, traditional strategic planning emphasized long-term resource allocation (what types of plant, at what locations, etc.), and privately owned utilities' decisions were constrained by public utility commissions.

---

[3]Commonly used electricity derivatives traded in OTC markets include forward price contracts, swaps, options and spark spreads. Several designs for electricity futures also appeared briefly on the NYMEX, CBOT and MGE exchanges before being withdrawn. Although derivatives that focus on price risk *per se* have had mixed success in the electricity industry, three interesting tangential derivatives for managing risk in the industry are also being used: emissions trading, weather derivatives and insurance contracts. SO_2 and NO_x allowance trading has flourished in the USA in recent years and the recently launched European Union (EU) Carbon Emission Trading Scheme is already experiencing large trading volumes. To manage weather risk, some independent power producers have weather adjustments built into their fuel supply contracts. Other large energy companies and power marketers are now using 'weather hedges' in the form of custom OTC contracts that settle on weather statistics. Lastly, to cover the risk from low-probability events such as a plant breakdown, multiple-trigger derivatives and specialty insurance contracts can be used to complement normal derivative products (DOE, 2002).

In state-owned utilities, the link between energy policy and investment choice was more direct, making it easy to influence investment decisions to achieve the desired fuel mix.

Electricity markets may not appropriately signal the need of diversity and flexibility at the macroeconomic level. Consider the case of increased gas dependency, where Britain faces an increased risk of large price increases for imported gas. Plant that uses fuel whose price does not move sympathetically with gas (such as nuclear and wind, and to some extent coal) would be an attractive complement in the portfolio of either a generator or investors holding shares in power companies, and to that extent diversity will be rewarded. However, the macroeconomic risks associated with a large increase in the price of imported gas will not be reflected in the profit of generation from other fuels, and may even be penalized if the macroshock causes an economic downturn and a fall in overall demand. Individual plant choices may therefore not respond to the social risks of increased fuel specialization and reduced diversity.

A perfect market should motivate individual investment decisions leading to the socially optimal fuel mix, but the conditions for this to hold are strong: the usual general equilibrium assumptions of a complete set of spot and forward markets or perfect foresight, price-taking behavior by producers and consumers, risk neutrality (or adequate risk-sharing contracts) and convex production possibilities (Arrow and Debreu, 1954; Debreu, 1959). The lack of informative distant futures markets may lead to a suboptimal degree of diversity. In particular, herd behavior (in which investors observe others' decisions, and assume they are based on superior information that justifies imitating their choices) may encourage investment in one or two dominant technologies, as well as waves of investment leading to boom-and-bust investment cycles (Ford, 1999, 2001; Olsina et al., 2006).

Moreover, imperfections in capital markets may limit the ability of utilities to diversify their risk exposure. Technology diversification does not appear as a primary motivation for cross-participation in utilities' equity. However, alternative diversification strategies to handle fuel risks have developed in response to capital market imperfections. Long-term fixed-price fuel procurement contracts are the most common such strategy, which does not require physical ownership of the production assets. For example, the gas and electricity utility Centrica took steps in 2004 to diversify its power generation portfolio by agreeing its first coal-indexed power purchase deal with International Power. The three-year agreement starting in October 2004 is linked to the monthly average of the API2 coal market index and will see International Power's Rugeley coal station supplying 250 MW of peak electricity to Centrica's British Gas domestic energy business. Centrica owns a number of gas-fired power stations in the UK and also has an off-take agreement with the nuclear generator British Energy. It had been seeking exposure to the coal power market to balance its power generation book and had considered taking an equity stake in a UK coal-fired power plant or off-take agreements from coal stations. The coal-indexed deal will allow Centrica to diversify into the coal power market without the cost of buying aging coal power plant.

11.2.4 Technology diversification and the consumer interest

The standard Arrow–Debreu economic theory of decision making under risk postulates a complete set of competitive risk and futures markets on which all goods and services can be traded now to ensure that investment decisions are efficient (e.g. Arrow and Hahn, 1971). While this is wildly unrealistic, it serves as a useful benchmark for identifying possible market failures. The institutional counterparts to the imagined full set of Arrow–Debreu markets are claims on the profits of companies (i.e. shares in those companies) and futures markets. These can go some way toward offering hedging instruments to share and hence reduce the cost of risk.

In the present context of risky future energy prices, consumers would (if they were well informed) wish to hedge such risks. To be more precise, high gas prices that translate into high electricity prices will harm consumers who buy electricity. High gas prices are likely to lead to high profits in gas-producing companies. One natural hedge would be for consumers to hold shares in companies that specialized in producing gas (or buying gas on long-term fixed-price contracts). Another hedge would be for consumers to hold shares in a specialized nuclear power generating company that would earn higher profits when selling at higher electricity prices. For instance, from the time when British Energy was relisted on 17 January 2005 to the end of 2005, British Energy shares have exhibited a 91% correlation to the one-year forward electricity price in the UK (Figure 11.1).

In both cases the extra profits from the shares could offset the extra costs of electricity. Note that holding shares in a gas-fired electricity company would be no use at all, as higher gas and electricity prices would leave their profits more or less unchanged. Finally, consumers could theoretically hedge directly by

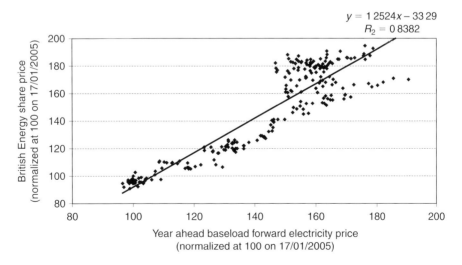

FIGURE 11.1   Correlation between British Energy share price and base-load forward annual power price in 2005.

buying futures in gas (or more directly in electricity). These would pay off in high gas/electricity states of the world and compensate for the high electricity prices.

However, only an extremely small number of commodities is competitively supplied and both storable and homogeneous enough to support liquid futures markets. Oil products and gas satisfy these conditions in some jurisdictions (notably the USA) where there are futures markets, but gas is not sufficiently competitively traded in most of Europe to offer reliable hedges or sustain liquid futures markets. Electricity (whose value can vary each hour and across space) is not sufficiently homogeneous for really low-cost liquid futures markets to take off, although some products (base and peak seasonal and annual) are traded in some markets (though with relatively low turnover/physical traded volume).

Shares in companies also offer some insurance, but few companies are sufficiently specialized to offer pure hedges, perhaps because their value as such is attenuated by the information costs facing individual consumers. For many purposes, these information and transaction costs give an advantage to portfolio companies (or mutual funds for small investors), and as such remove the hedging options needed to reflect consumer demand for diversity.

All of this suggests that utilities are likely to have to bear much of the cost of risk in their investment decisions, unless they can find counter-parties with complementary risk attitudes. Again the simple story above is illuminating: risks that profit companies may harm consumers and vice versa, giving rise to the prospect of profitable exchange (as the Centrica example shows). A supply company selling electricity at fixed prices to final consumers is one natural counter-party to a generation company selling electricity at variable prices but burning fuel whose price is uncorrelated with the electricity price. Vertical integration between such companies avoids the need for contracts or cross-ownership of shares and is common in these markets. Energy-intensive consumers such as pulp and paper manufacturers may be willing to equity-finance nuclear power plants for similar reasons (and do so in Finland) (IEA, 2006; Roques et al., 2006b). One test of how well such risk markets or their surrogates might work is whether a nuclear power company's shares are seen as complementary or substitutable in consumers' portfolios or whether ignorance and/or information costs give an advantage to energy companies in constructing physical rather than financial portfolios of plants.

This suggests that utilities are likely to have to bear much of the cost of risk in their investment decisions, unless they can find willing counter-parties with complementary risk attitudes. The working assumption in the rest of this chapter is that ignorance of risks and/or information costs give an advantage to energy companies in constructing physical rather than financial portfolios of plants, and that therefore the best case for plant diversity is probably within the portfolio of large well-capitalized energy companies, rather than stand-alone share-issuing specialized companies (which theoretically ought to offer more adaptable portfolio hedging options for consumers). The rest of the chapter explores the potential of MVP theory as an analytical framework for private investors to value technology and fuel mix diversity in liberalized electricity markets.

11.3 Using mean-variance portfolio theory to identify optimal generation portfolios

MVP theory, based on Markowitz's (1952) seminal work, was initially developed for financial securities and has found wide application in the financial industry.[4] As power markets have liberalized, so interest in quantifying and managing market risks has grown. The MVP method can be applied to determine the optimal portfolio of generation plants either for a country or for a particular company.

Although applications of MVP to optimizing generation portfolios are growing, the existing literature has so far adopted a *social welfare maximization* perspective, aiming to minimize generation cost for each risk level, concentrating on the *production cost* in the face of risky fossil fuel prices.[5] Based on projected unit costs and volatility covariation patterns, these studies determined 'efficient' (Pareto optimal) portfolios of generating assets. Thus, Awerbuch and Berger (2003) state 'our analysis is cost-based, since from a societal perspective, generating costs and risks are properly minimized [...]. Since the analysis and the expected portfolio returns are cost-based, variations in electricity market prices are not relevant.' But in liberalized markets, the objective function is different, as portfolios are chosen to maximize financial returns to these investing generators, given portfolio risk levels, which will also depend on electricity and emission permit price risks. This chapter is the first application of MVP theory taking the perspective of large electricity generators in liberalized electricity markets.

Of course, other risks may also be important and the MVP approach can be extended to include them. Thus, recent developments take account of operation and maintenance risks as well as construction time risk (Awerbuch and Berger, 2003, Jansen et al., 2006). More challenging is to allow for differing capacity factors resulting from variations in the merit order as fuel prices vary. Revenue from ancillary services may also be important, while the intermittency of renewable resources would warrant more sophisticated electricity market modelling. This chapter concentrates on base-load technologies (coal, nuclear and CCGT plant) and does not offer guidance on the choice of mid-merit and peak load plant. Finally, this chapter, as well as the existing literature, is static, and a useful extension for future research would be to develop a multiperiod analysis to determine optimal trajectories for rebalancing portfolios from the base year to the target year, taking account of different vintages of generation plant (Kleindorfer and Li, 2005).

---

[4] See, for example, Elton and Gruber (1994) and Fabozzi et al. (2002) for a recent review of the developments of Portfolio theory. The MVP approach has been developed in the context of financial markets to include transaction costs, margin loans for security purchases and short sales, and unsecured portfolio debt and its relationship to the probability of insolvency to the investor (e.g. Pogue, 1970; Chen et al., 1971). Such extensions appear less relevant for portfolios of real assets such as power plants.

[5] Most studies define portfolio return as the reciprocal of unit generating cost (reciprocal of cost per kWh) and price risk in terms of price volatility per holding period (per year), but Jansen et al. (2006) argue that such an approach has several pitfalls and that for transparency, it is better to use directly a simple cost frontier rather than a return frontier.

On the methodological side, this chapter uses simulation to explore a wide range of degrees of correlation between electricity, fuel and CO_2 prices based on central estimates derived from empirical data, in contrast to purely empirical papers, whose results are very sensitive to the methodology used to compute the correlation matrix between the different risky inputs. High cross-correlations cause problems of multicollinearity when deriving optimal portfolios directly from these risk estimates. For such empirical studies, a number of improvements could be made on the methodology used to derive the correlation matrix between the different risky parameters.[6] Humphreys and McClain (1998) and Krey and Zweifel (2006) present two different approaches to refine the econometric estimation of the correlation matrix to obtain time-invariant matrices as an input to the determination of efficient electricity-generating energy portfolios.[7] The simulations reported here consider a wide range of degrees of correlation around the empirical base case, so such refinements in the econometric estimates appear less critical.

11.3.1 A two-step simulation framework with portfolio optimization

MVP theory requires a method for deriving the mean and variance of any portfolio. This is accomplished in two steps. The first step consists of running Monte Carlo simulations of a discounted cash-flow model to estimate the mean and variances of the distribution of net present value (NPV) of each technology. The second step takes these and their associated correlations to determine the efficient risk–return frontier.

The discounted cash-flow model represents three base-load technologies (CCGT, coal and nuclear plant) currently available for new build in Britain. The cost and technical parameters presented in Table 11.1 are derived from a variety of sources (Deutch et al., 2003; IEA/NEA, 2005; IEA, 2006).[8] The construction time lag (measured from the moment of the actual start of construction) is five years in the case of nuclear, four years in the case of coal, and only two years in the case of the CCGT plant. The capital costs ('overnight cost' and 'O&M incremental cost') are much higher for the nuclear plant, and to a lesser extent for the coal plant, than for the CCGT plant, while the converse is true for fuel costs. Nuclear plant

---

[6]The pioneering MVP studies relied on simple ordinary least squares (OLS) econometric regressions to compute covariance matrices; the robustness of such approaches could be improved by computing some further statistical tests including autocorrelation, endogeneity and multicollinearity).

[7]Humphreys and McClain (1998) introduce a time-varying covariance matrix computed using so-called generalized autoregressive conditional heteroscedastic (GARCH) models to filter out systematic changes in volatility in response to price shocks. Krey and Zweifel (2006) use seemingly unrelated regression estimation in order to take into account all of the common shocks (e.g. weather related) impinging on the generation costs from different energy sources.

[8]All the costs are expressed in real 2005 British Pounds.

Table 11.1   Cost and technical parameters

Parameter	Unit	Nuclear	Coal	CCGT
Technical parameters				
Net capacity	MWe	1000	1000	1000
Capacity factor	%	85	85	85
Heat rate	Btu/kWh	10 400	8600	7000
Carbon intensity	kg-C/mmBtu	0	25.8	14.5
Construction period	years	5	4	2
Plant life	years	40	30	20
Cost parameters				
Overnight cost	£/kWe	1140	740	285
Incremental capital costs	£/kWe/year	11.4	8.6	3.4
Fuel costs	£/mmBtu	See distribution parameters		
Real fuel escalation rate	%	0.5	0.5	1.2
Fixed O&M	£/kWe/year	36	13	9
Variable O&M	£/MWh	0.23	1.93	0.3
O&M real escalation rate	%	0.5	0.5	0.5
Nuclear waste fee	£/MWh	0.6	0	0
Financing parameters				
Real discount rate	%	10	10	10
Marginal corporate tax	%	30	30	30
Regulatory actions				
Carbon tax	£/tC	See distribution parameters		
Carbon price escalation rate	%	1	1	1
Revenues				
Electricity price	£/MWh	See distribution parameters		
Electricity price escalation rate	%	0.5	0.5	0.5

NGCC: natural gas combined cycle; O&M: operation and maintenance.

incurs a 'nuclear waste fee' to cover the cost of decommissioning and nuclear waste treatment. The three plants are assumed to operate base-load with an average annual capacity availability factor of 85%.[9] The cost of CO_2 emission permits in the European Emission Trading Scheme is represented by a carbon tax, so prices will be quoted per tonne of carbon (£/tC) rather than the permit price, which is given in €/tonne CO_2. Real post-tax weighted average costs of capital (WACC) of 5% and 10% are used.[10] Plant lifetimes of 20, 30 and 40 years for gas, coal and nuclear plants, respectively, represent the capital recovery period.

The model concentrates on market risks and does not consider technical or operational risks (e.g. construction cost overruns, plant availability).[11] Fuel, electricity and CO_2 prices are represented by normally distributed random variables, whose cross-correlation and standard deviation base estimates are derived from UK historical time series of quarter-ahead fuel and power prices from January 2001 to August 2005, and quarter-ahead CO_2 allowance price data from 2005 (see Roques et al., 2008, for a more detailed description of these empirical data).[12] The distribution parameters and correlation coefficients between the different market prices are shown in Tables 11.2 and 11.3 for the base case. When modelling commodity prices, it is important to distinguish price risk from price variability, which corresponds to usual daily and seasonal fluctuation patterns (Geman, 2005). The focus is here on price risk. Fuel and CO_2 allowances are assumed to be bought and sold on spot or forward markets, or through contracts indexed on the spot market price, thereby subjecting generators to quarterly price volatility.[13] The model does not account for long-term fixed-price gas procurement contracts, which are unusual in liberalized electricity markets, where most gas plant investment projects are financed with flexible gas intake contracts (e.g. through tolling arrangements).[14] The forward prices of CO_2 (second period, from January 2006 to September 2007) have varied from 15 to 25 €/tonne CO_2 with a mean of

[9]CCGT, and to a lesser extent coal plant, cycle up and down such that actual average capacity factors are usually lower. The average availability capacity factor for gas and coal plant is assumed to capture part of the value associated with plant operational flexibility; see Roques (2007) for a more detailed valuation of such flexibility.

[10]The International Energy Agency generating costs estimates also consider the two cases of 5% and 10% discount rates (IEA/NEA, 2005).

[11]Note that this two-step methodology can also be used to take into account other non-market risks, e.g. operating or construction costs uncertainty, or performance risk. See Roques et al. (2006a) for an example of Monte Carlo simulation incorporating these other risks.

[12]A drawback of using historical data for estimating portfolio variance is that the method does not provide a measure of absolute worst loss.

[13]Generators do to some extent buy gas, coal and nuclear fuel on forward markets, and sell power on forward markets. As forward and spot markets are strongly correlated, no distinction is made here between sales on spot and forward markets.

[14]See Wiser et al. (2004) for an empirical study of gas-plant fuel purchase agreements in the USA and Roques (2007) for a discussion of how these affect power plant risk exposure and valuation.

Table 11.2 Empirical prices correlation coefficients, January 2001 to September 2005

Correlation coefficient	Base electricity price	Gas price	Coal price	CO_2 price
Base electricity price	1			
Gas price	0.89	1		
Coal price	0.56	0.77	1	
CO_2 price[a]	0.73	0.45	−0.46	1

[a]November 2004 to August 2005.

Table 11.3 Monte Carlo simulation risk distribution inputs

Normal distribution parameters	Technology	Mean	SD
Cost parameters			
Fuel cost (£/mmBtu)	Nuclear	0.35	0.1
	CCGT	3.3	1.0
	Coal	1.3	0.6
Carbon tax (£/tC)	All	40	10
Revenue			
Electricity price (£/MWh)	All	40	10

CCGT: combined cycle gas turbine.
Note: These distribution parameters are based on UK empirical values from 2001–2005.

£49/tonne carbon (tC, 3.67 t CO_2 = 1 tC) and a standard deviation of £10/tC. The parameters in Table 11.3 are based on earlier estimates and may underestimate both the future mean and possibly also the standard deviation of carbon prices.

If, as here, the price distributions are assumed to be normally distributed, then the simulated distributions of plant returns emerge with distributions very close to normal, as required by MVP analysis. In fact, the actual fuel and electricity price distributions are positively skewed, and arguably better approximated by log-normal distributions. This relies on the argument that disregarding moments higher than variance will generally not affect portfolio choice (Bodie et al., 2002, p. 174). The likely direction of errors of fitting normal distributions is unclear: the model does not rule out negative returns (when input costs exceed output prices) which would in practice be avoided by closing the plant until output prices recovered, and so may amplify the downside risk, but it ignores the fatter upside tails that imperfectly correlated but log-normal prices would generate. Future research may investigate the effects of a dispatch model with log-normal input and output prices on the distribution and cross-correlation of plant returns.

11.3.2 Net present value Monte Carlo simulation results

The NPV of an investment in the three technologies was simulated for three case studies and using two different discount rates (100,000 runs for each simulation). The results are expressed in £ million for a 1000 MW plant, but can be interpreted as £/kW of capacity (the form in which the capacity costs are normally represented, as in Table 11.1). The results are presented using the highest 10% discount rate, which is more representative of investment financing conditions in liberalized electricity markets; a sensitivity analysis of the results to a lower (5%) cost of capital is presented in Section 11.3.3. In the first illustrative case study, electricity, fuel and CO_2 prices are risky but the correlation between them is set to zero. This is a benchmark case study that would correspond to hypothetical isolated fuel, electricity and CO_2 markets. In the second case study, electricity, fuel and CO_2 prices are risky and the correlation coefficients are set to the values of Tables 11.2 and 11.3. In the third case study, electricity prices are fixed while fuel and CO_2 prices are risky. This case study corresponds to investments for an electricity generation company which has a significant portion of its output contracted over the long-term through a fixed-price power purchase agreement (indexed to the retail price level). The financing of the new nuclear power plant in Finland is an extreme example of such long-term power purchase agreement: the shareholders will have access to electricity at production costs during the full life of the plant in proportion to their share (IEA, 2006).

Figure 11.2 shows the NPV probability distributions of the different technologies in the three case studies, for a commercial 10% discount rate. In all three case studies, the expected net present value (ENPV) of the coal and nuclear plants does not change much and is slightly negative (about −£70/kW and −£40/kW, or −9% and −4% of the overnight costs, respectively), while the ENPV of a CCGT plant is positive (about £130/kW, 46% of overnight cost), as shown in Table 11.4.[15] The spread of the three NPV probability distributions varies greatly in the three case studies, depicting very different risk–return profiles. In the first case study without correlation, the spread (or SD) of the three NPV distributions is relatively similar, with the nuclear and coal plant appearing less risky than the CCGT plant. In the second case study, the correlation between electricity, gas and CO_2 prices gives rise to an interesting phenomenon, as the standard deviation of the NPV distribution of the CCGT plant is smaller than that of the coal

---

[15] It should be noted that these negative central NPV estimates correspond to cost and price assumptions in the UK over the period 2001–2005, characterized by relatively low gas and electricity prices by historical standards. All plants have a positive ENPV for a lower discount rate of 8% or 5% as presented in Section 11.3.3. Besides, lower prices for coal in countries such as the USA or some countries in continental Europe over the same period would yield different results and make coal more competitive relative to the other technologies.

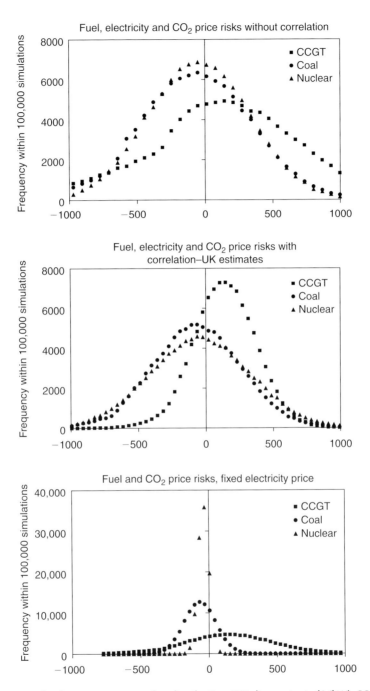

FIGURE 11.2   Single plants net present value distribution, 10% discount rate (£/kW). CCGT: combined cycle gas turbine.

Table 11.4 Single plants net present value distribution statistics, 10% discount rate (£/kW)

Scenario	1st scenario: no correlation between fuel/CO_2/electricity			2nd scenario: with correlations between fuel/CO_2/electricity			3rd scenario: fixed electricity price (PPA)		
Statistics	CCGT	Coal	Nuclear	CCGT	Coal	Nuclear	CCGT	Coal	Nuclear
Mean	111	−76	−42	139	−73	−43	134	−68	−41
SD	586	426	378	233	336	377	331	116	39
Minimum	−2699	−2310	−1990	−1042	−1706	−1872	−1782	−593	−208
Maximum	2447	1749	1698	1118	1462	1694	1530	451	120
Range	5146	4059	3688	2159	3169	3566	3312	1044	329

PPA: power purchase agreement; CCGT: combined cycle gas turbine.

Table 11.5 Correlation coefficients between the three technologies net present values, 10% discount rate

Correlation of returns	CCGT/Nuclear	CCGT/Coal	Coal/Nuclear
Fixed electricity price	0.002	0.118	0.007
No correlation electricity/gas/CO_2 prices	0.797	0.789	0.953
With correlation electricity/gas/CO_2 prices	0.594	0.596	0.959

CCGT: combined cycle gas turbine.

and nuclear plants. In the third case study, with risky fuel and CO_2 prices but a fixed electricity price, the CCGT, and to a lesser extent the coal plant, have much higher standard deviations, and are therefore much more likely to make a loss.

The correlation between the NPV probability distributions of the three technologies is relatively high in the first two case studies, in which fuel, CO_2 and electricity prices are risky (Table 11.5). In contrast, in the third case study with fixed electricity price, the returns of the three technologies are only slightly positively correlated, since fuel price risks are relatively uncorrelated. In addition, comparing the first two case studies, the correlations between plant returns fall (except between coal and nuclear plant) after allowing for the high correlations between electricity, fuel and CO_2 prices of Table 11.2. Looking at the technologies themselves, the returns of the coal and nuclear plants are generally more correlated than the returns of the CCGT and nuclear plants, or the returns of the CCGT and coal plants.

11.4 Optimal base-load generation portfolios in liberalized electricity markets

The returns, risks and correlation data from the Monte Carlo simulations are now used as inputs to identify the optimal portfolios of the three base-load technologies, using MVP theory. MVP theory prescribes not a single optimal portfolio combination, but a range of efficient choices, represented by the *risk–return efficient frontier* in the graph of portfolio return against portfolio standard deviation (SD). Investors will choose a risk–return combination based on their own preferences and risk aversion. The expected return $E(r_p)$ of portfolio P containing N assets i [expected return, $E(r_i)$, SD, σ_i] in proportion X_i is just the weighted average of the expected returns of the N assets:

$$E(r_p) = \sum_{i=1} X_i E(r_i)$$

The portfolio standard deviation σ_p is defined by:

$$\sigma_p = \sqrt{\sum_{i=1}^{NP} X_i^2 \sigma_i^2 + \sum_{i=1}^{N} \sum_{\substack{j=1 \\ i \neq j}}^{N} X_i X_j \rho_{ij} \sigma_i \sigma_j}$$

where ρ_{ij} represents the correlation between the returns r_i and r_j of the two assets.[16]

Whether only two of the three technologies available are efficient or whether all three lie on the frontier will depend on their characteristics. A standard model of the tradeoff between risk and return is to use a mean-variance utility function, given by

$$U = E(r_p) - \frac{1}{2} \lambda \text{var}(r_p)$$

where U is the power generator's utility, $E(r_p)$ is the expected return (NPV) of portfolio p, $\text{var}(r_p)$ is the variance of the return, and λ is the coefficient of risk aversion.[17] This describes a convex increasing function in the graph of mean against standard deviation.

11.4.1 The impact of correlation between fuel, carbon dioxide and electricity prices

Figure 11.3 shows the feasible portfolios of nuclear, coal and CCGT plants in the first two case studies. To clarify the graph, only the frontier delineating the feasible combinations of pairs of plants is shown. In both case studies, fuel, CO_2 and electricity prices are risky. The first case study (the curved triangle on the right-hand side) corresponds to the hypothetical case in which there is no correlation between electricity, fuel and CO_2 prices, as if the electricity, fuel and CO_2 markets were independent. In this case study, a CCGT offers the greatest return, but is more risky than a nuclear or coal plant. The efficient frontier corresponds to the convex line with circle markers (the lowest standard deviation for any given mean), consisting of combinations of nuclear and CCGT plants only. There are no coal plants in the efficient portfolio, as a coal plant has both lower returns and higher risks than a nuclear plant. In this hypothetical case in which the electricity, fuel and CO_2 markets are independent, diversification strategies according to MVP theory would therefore induce investors to invest in a mix of CCGT and nuclear plants. The greater the risk aversion of investors, the more nuclear plants there would be in the optimal portfolio.

In practise, electricity, fuel and CO_2 prices are correlated in liberalized electricity markets. The correlation between fuel and electricity prices is the result of a

---

[16] Lagrange multipliers can be used to compute the efficient frontier (Bar-Lev and Katz, 1976). Optimization procedures are also available and practical (Awerbuch and Berger, 2003). In such optimization procedures, the program calculates all possible portfolio combinations and finds the efficient frontier using an iterative approach (Kwan, 2001).

[17] See Neuhoff and De Vries (2004) and Green (2004) for a discussion of empirical estimates of risk aversion of generators and suppliers in the electricity sector.

complex set of phenomena, including the fuel used by the plants which have the highest marginal costs of production and are therefore clearing the market, but also other factors such as the terms and duration of fuel procurement contracts, the operational dispatch strategies of electric companies holding portfolios of diverse generation technologies, and the behavior of traders on electricity and fuel markets. In Britain, for instance, the correlation coefficients between gas and electricity prices have been relatively high and fluctuating from year to year in the recent past: from 40% to more than 90% from 2001 to 2005. Gas-fired plants were often the marginal price-setting plants in the British electricity market, which led to very strong correlation between base electricity and gas prices (Roques et al., 2008).

Figure 11.3 also shows the second case study using the data of Tables 11.2 and 11.3 based on the empirically observed average correlations between electricity, fuel and CO_2 prices in the British market from 2001 to 2005. The left-hand triangle of lines delineates the set of possible portfolios: its boundary made of pairs of technologies and the interior, of all three. Introducing correlation dramatically decreases the riskiness of the CCGT technology and slightly lowers the riskiness of the coal technology, such that a nuclear plant becomes the most risky investment. The optimal portfolio for an investor in this case study is to invest only in CCGTs, as any other portfolio would both reduce returns and increase risk. This dramatic impact of the empirical correlation between electricity, fuel and CO_2 prices is consistent with the observed behavior of investors in the British market, who have invested heavily in CCGTs during the past decade, and have not seemed to value highly fuel mix diversity.[18]

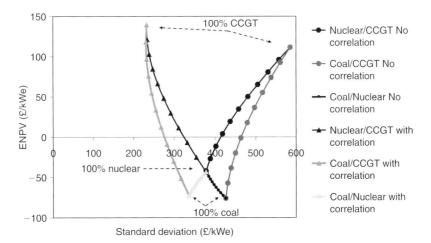

FIGURE 11.3   Feasible portfolios of nuclear, coal and combined cycle gas turbine (CCGT) plants with and without empirical correlation between electricity, fuel and CO_2 prices (10% discount rate). ENPV: expected net present value.

---

[18] Some companies already owned coal plant and so gas represented a diversification, but the new independent power producers concentrated on gas alone.

Figure 11.4 extends the previous analysis and shows the composition of optimal portfolios for degrees of correlation between electricity, fuel and CO_2 prices ranging from 0% to 100% and for different degrees of risk aversion. Green (2004, 2007) uses Grinold's (1996) 'grapes for wine' technique to compute estimates of the value of the coefficient of risk aversion λ from stock-market data on *ex post* returns and their variability. He finds evidence using the FT All-share index mean return in the UK from 1955 to 2000 that investors' risk aversion is likely to be in a relatively low range, with values of λ lower than 0.1.[19] When taking into account the risk aversion of plant investors, pure portfolios of CCGT plant are optimal for such low coefficients of risk aversion ($\lambda \leq 0.1$), whatever the degree of correlation between fuel, electricity and CO_2 prices. As the risk aversion coefficient increases, mixed portfolios become more attractive to investors. For medium risk aversion coefficients ($\lambda = 10$), Figure 11.3 shows that optimal portfolios consist almost entirely of CCGT plant for correlation coefficients greater than 70%, possibly with some coal and nuclear plants at the margin. As the degree of correlation between fuel, electricity and CO_2 prices decreases, mixed portfolios of CCGT, coal and nuclear plant become optimal. For an average degree of correlation of 50%, the optimal portfolio is composed of a majority of CCGT plant (60%), with 20% of coal and 20% of nuclear plant. For degrees of correlation lower than 50%, the share of CCGT plant in the optimal portfolio decreases and the share of nuclear plant increases concurrently. Such low degrees of correlation are unlikely within liberalized markets, but could be interpreted from the perspective of generators investing in separate and isolated markets (e.g. Europe and Latin America). Pure portfolios of nuclear plants become optimal when considering diversification of plant returns across different, separate markets. For very high degrees of risk aversion ($\lambda = 1000$ in Figure 11.4), the patterns are similar, with a greater proportion of mixed portfolios for any given degree of assumed correlation between fuel, electricity and CO_2 prices, and a significant shift toward more coal plant.

11.4.2 The impact of long-term fixed-price power purchase agreements on optimal generation portfolios

High correlations between the most costly fuel and the resulting electricity price reduce the risk of the marginal price-setting plant, and hence raise the risk facing nuclear power, which might be hedged by a long-term power purchase agreement. Figure 11.5 compares generation portfolios in the second and third case studies (the latter corresponding to the case in which electricity prices are fixed thanks to a long-term contract). Figure 11.5 shows that the sets of optimal portfolios

---

[19] The value of λ (the coefficient of absolute risk aversion) is not dimensionless, in contrast to the coefficient of relative risk aversion, R. The relationship between them is that $\lambda = R/W$, where W is a suitable measure of wealth. As R is more likely to be stable than λ, and as Green's estimate was based on a trader's wealth exposure, rather than a generating company, the appropriate value of λ here should be less than Green's value, and possibly considerably less.

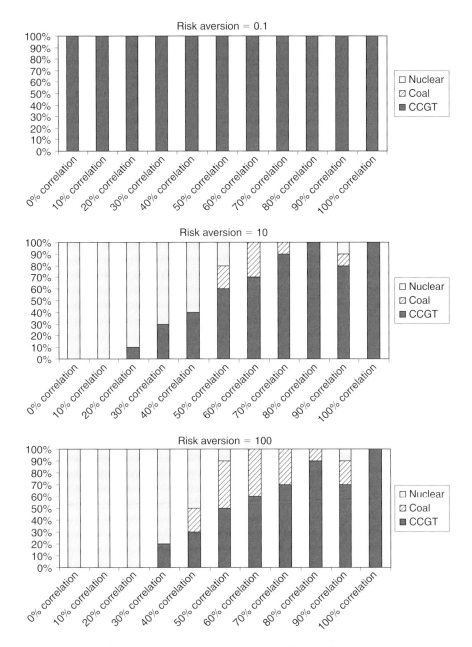

FIGURE 11.4 Optimal generation mixes for low, medium and high risk aversion (10% discount rate). CCGT: combined cycle gas turbine.

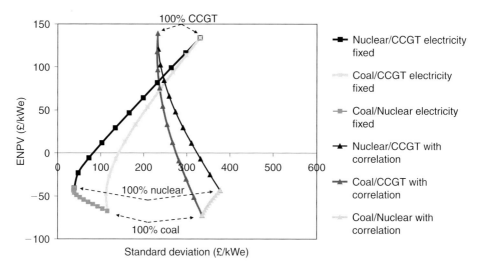

FIGURE 11.5 Efficient frontier for portfolios of nuclear, coal and combined cycle gas turbine (CCGT) plants with fixed and risky electricity prices (10% discount rate). ENPV: expected net present value.

in the two case studies are very different. When the electricity company sells its output on spot markets or with contracts that are indexed on the spot market, and is thereby exposed to annual electricity price risk, the efficient frontier is the same as the left side of Figure 11.3, and again the optimal portfolio has only CCGT. In contrast, when the electricity price is fixed by a long-term power purchase agreement, efficient portfolios lie on the efficient frontier represented by the line with square markers, consisting of combinations of CCGT and nuclear plants in various proportions, depending on the investor's risk aversion. The more risk averse the investor, the more nuclear (or coal) plants in the optimal portfolio.

The difference between the optimal MVP portfolios with and without long-term power purchase agreements points toward a critical issue with regard to generating companies' investment risk management strategies. Utilities are likely to have to bear much of the cost of risk in their investment decisions, unless they can find counter-parties with complementary risk attitudes to sign long-term contracts. Such contracts are, however, quite rare in liberalized markets. The best case for plant diversity is probably within the portfolio of large well-capitalized energy companies that also have supply (retail) businesses. The second case study may therefore be more realistic for merchant investors in liberalized electricity markets, and is consistent with the observed dominance of portfolios of CCGTs in Britain, particularly for new entrants who lack downstream supply businesses. The third case study shows, nevertheless, that if generation companies can find counter-parties with complementary risk attitudes, or if they are guaranteed a stable revenue stream (through for instance feed-in tariffs), then diversifying away from CCGTs by investing in nuclear and/or coal becomes the optimal strategy.

11.4.3 The impact of the cost of capital on optimal generation portfolios

In theory, according to the Capital Asset Pricing Model, the cost of capital should change with the portfolio as each portfolio corresponds to a different degree of risk exposure. For practical reasons, using a constant discount rate (e.g. the company cost of capital) is common for investment valuation, particularly when the impact of risk is already taken into account through Monte Carlo simulation. The results were presented using a 10% average post-tax weighted average cost of capital, which appears realistic for investment evaluation in liberalized electricity markets. However, power investment projects are often underpinned by specific contractual arrangements to transfer some of the construction, operational or market risks away from the plant investor toward other stakeholders. For instance, 'turn key' contracts for part or the whole plant, such as the one between AREVA and TVO for the nuclear plant under construction in Finland, significantly lower the risk exposure of the plant investor and should allow access to cheaper debt financing. Governments might also want to intervene to facilitate access to cheaper financing by underwriting some of the risks associated with a specific technology.[20] This section presents a sensitivity analysis of the main results to a lower (5%) discount rate, which can be interpreted as a case in which plant investors have managed to shift away onto other stakeholders major project risks, thereby obtaining favorable financing conditions.

With a 5% discount rate, the ENPV of a nuclear plant is much higher than that of a coal or CCGT plant, which are similar. The relative riskiness of the three technologies does not change significantly, with nuclear being less risky than gas and coal when only cost (gas price and CO_2 price) risk is taken into account, and CCGT plant becoming less risky than nuclear plant when both electricity price and cost risks are taken into account, particularly for high degrees of correlation as in the British market. The higher ENPV of nuclear plant implies that, in contrast to the 10% discount rate case, optimal portfolios when fuel, electricity and CO_2 prices are highly correlated do not contain exclusively CCGTs. Figure 11.6 shows that any combination of nuclear, CCGT and possibly coal plant is efficient, depending on the risk aversion of investors. Optimal portfolios when generators can obtain a long-term fixed-price power purchase agreement contain a majority of nuclear plant.

Figure 11.7 shows the composition of optimal portfolios for degrees of correlation between electricity, fuel and CO_2 prices ranging from 0% to 100% and for different degrees of risk aversion. When taking into account the risk aversion of plant investors, pure portfolios of nuclear plant are optimal for low coefficients of risk aversion characterizing investors' behavior in the UK stock market over 1955–2000 ($\lambda \leq 0.1$), contrary to the 10% commercial discount rate case presented

---

[20] For instance, the US 2005 Energy Act, besides a production tax credit, has provisions to compensate investors for additional expense related to delays in the regulatory licensing process. It also provides loan guarantees for up to 80% of the project cost for the first 6 GW of nuclear plants built, which will allow investors to increase project leverage and to have access to debt at a lower interest rate, all of this lowering the weighted average cost of capital significantly (IEA, 2006).

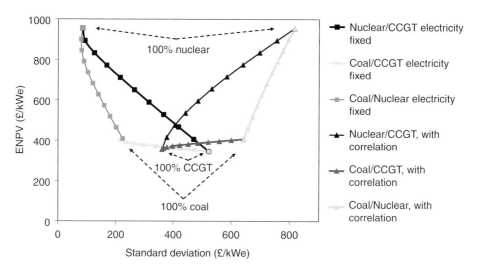

FIGURE 11.6   Efficient frontier for portfolios of nuclear, coal and combined cycle gas turbine (CCGT) plants with fixed and risky electricity prices (5% discount rate). ENPV: expected net present value.

earlier in which pure portfolios of CCGT plant were dominant. For higher coefficients of risk aversion, Figure 11.7 shows that optimal portfolios become more similar to the higher discount rate case, with mixed portfolios of nuclear and CCGT plant becoming more attractive to investors, the share of CCGT being the greater the higher the correlation between fuel, electricity and CO_2 prices. The higher the risk aversion coefficient ($\lambda = 10$ in Figure 11.7), the more coal plants take over gas plants in the optimal portfolio.

11.5 Conclusion and policy implications

This chapter has used simulation techniques and portfolio optimization to study private investors' diversification incentives in liberalized electricity markets. The main contribution on the methodological side is the introduction of a two-step simulation approach to assess the impact of both plant input (fuel) and output (electricity and CO_2) price risks on the return of different base-load generation technologies. In contrast to the existing literature which used empirical data and focused on *plant cost* risk, this chapter relies on Monte Carlo simulations of a discounted cash-flow model of investment (in CCGT, coal, and nuclear plant) to compute the *plant return* risks and their correlation. These then serve as inputs to a portfolio optimization using MVP theory.

The chapter demonstrated the usefulness of this new theoretical approach by studying optimal portfolios in three case studies, using central parameter estimates derived from historical electricity, fuel and carbon prices data from Britain over 2001–2005. These case studies gave new insights into diversification

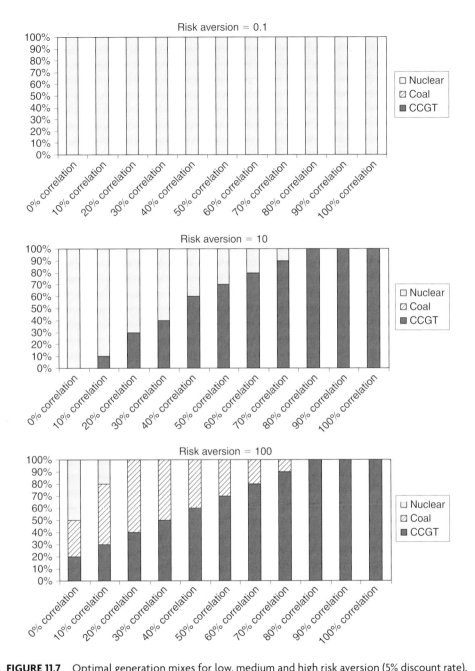

FIGURE 11.7 Optimal generation mixes for low, medium and high risk aversion (5% discount rate). CCGT: combined cycle gas turbine.

incentives for power investors in liberalized markets. The simulations demonstrated in particular the critical impact of the degree of correlation between electricity, fuel and CO_2 prices. If electricity, gas and CO_2 prices were to move independently, optimal generation portfolios would contain a mix of gas and coal or nuclear plants. However, as electricity, gas and CO_2 prices are highly correlated in liberalized markets such as in Britain, optimal portfolios for a private company contain mostly CCGT plants, possibly with some coal and nuclear plants if the investor is very risk averse in regard to risk aversion estimates based on UK stock market historical data (for a commercial discount rate of 10%). This appears consistent with the empirical evidence which shows that most new power plants built in Britain in the past decade have been CCGTs.

The dominance of CCGT in optimal generation portfolios can thus be traced back to the high degrees of correlation observed in many liberalized markets between electricity and gas prices, which reduce the risk of investment in this technology. The correlation between electricity and gas prices warrants further research, particularly with regard to dispatching issues and the impact of carbon emission allowances on the marginal technology in the market (e.g. Newbery, 2006). When investing in a gas-fired plant, investors take into account not only the expected returns of this investment, but also the positive externality effect of this investment on the correlation between electricity and gas markets. In a relatively isolated electricity market with little interconnection capacity, such as in England and Wales, the more generators invest in gas plant, the more closely correlated the electricity price with the gas price, principally because open-cycle gas turbines often set the marginal electricity price, but also because the increasing share of gas-fired plant in the fuel mix leads to greater integration of electricity and gas markets. For an electric company, investing in an additional gas plant in such market has therefore an 'externality value' as it increases the correlation between electricity and gas prices, thereby not only reducing the volatility of the returns of the new gas plant investment, but also reducing the volatility of the returns of the other gas units that the electricity company already operates. The simulations demonstrated that this externality effect outweighs the risk-reducing benefits of diversifying in other technologies, which suffer the opposite negative externality (especially coal plants) of reducing the correlation between electricity and gas prices.

This chapter also examined optimal generation portfolios when investors can secure a long-term fixed-price power purchase agreement, and showed that in that case optimal portfolios would contain a more balanced mix of nuclear and CCGT plant. While finding counter-parties with complementary risk attitudes may be difficult for investors in current liberalized electricity markets, recent experience with new nuclear build in Finland suggests that such long-term arrangements may interest some specific industrial consumers. Similarly, optimal portfolios when investors have access to cheaper capital (through the transfer of some of the project risks to other parties or through support from the government) are more diversified, with a larger share of nuclear and coal-fired plants.

These simulation findings can be used to inform the current policy debate about the role of government and regulators in electricity markets, as they show that if a generating company were granted a stable source of revenue through institutional or market changes (e.g. long-term capacity contract or a capacity payment), or through other support mechanisms such as a feed-in tariff or credit guarantees for nuclear, then optimal generation portfolios would integrate some share of nuclear power generation. Perhaps more importantly, these results suggest that liberalized electricity markets characterized by strong correlation between electricity and gas prices such as in Britain may not reward fuel mix diversification enough to make private investors' choices socially optimal. These findings raise questions as to whether and how policy makers and regulators should modify the market framework, given the macroeconomic and security of supply benefits of a diverse fuel mix (Stirling, 1994, 1998, 2001; Awerbuch and Berger, 2003; Awerbuch, 2004). Finding an instrument that corrects the potential market failure while not unduly distorting the market is a challenge. Whether integration between supply and generation companies will solve the problem remains to be adequately tested.

References

Arrow, K. and Debreu, G. (1954). Existence of equilibrium for a competitive economy. *Econometrica, Journal of the Econometric Society*, 22(3), 265–290.

Arrow, K.J. and Hahn, F.H. (1971). *General Competetive Analysis*. San Francisco, CA: Holden-Day.

Averch, H. and Johnson, L. (1962). Behavior of the firm under regulatory constraint. *American Economic Review*, 52, 1053–1069.

Awerbuch, S. (2004). *Towards a Finance-Oriented Valuation of Conventional and Renewable Energy Sources in Ireland. Perspective from Abroad Series*. Dublin: Sustainable Energy Ireland.

Awerbuch, S. and Berger, M. (2003). *Energy Security and Diversity in the EU: A Mean-Variance Portfolio Approach*. IEA Research Paper. Paris: IEA (February). www.iea.org/techno/renew/port.pdf

Bar-Lev, D. and Katz, S. (1976). A portfolio approach to fossil fuel procurement in the electric utility industry. *Journal of Finance*, 31(3), 933–947.

Bodie, Z., Kane, A. and Marcus, A. J. (2002). *Investments*, 5th edn.. New York: McGraw-Hill.

Chen, A., Franck, H., Jen, C. and Zionts, S. (1971). The optimal portfolio revision policy. *The Journal of Business*, 44(1), 51–61.

Debreu, G. (1959). *Theory of Value: An Axiomatic Analysis of Economic Equilibrium*. New York: Wiley.

Department of Energy (2002). *Derivatives and Risk Management in the Petroleum, Natural Gas, and Electricity Industries*. Report downloadable from www.eia.doe.gov/oiaf/servicerpt/derivative/chapter4.html

Deutch J., Moniz, E., Ansolabehere, S., Driscoll, M., Gray, P., Holdren, J., Joskow, P., Lester, R. and Todreas, N. (2003). *The Future Of Nuclear Power. An MIT Interdisciplinary Study*. http://web.mit.edu/nuclearpower

Elton, E. and Gruber, M. (1994). *Modern Portfolio Theory and Investment Analysis*. New York: Wiley.

Fabozzi, F., Gupta, F. and Markowitz, H. (2002). The legacy of modern portfolio theory. *Journal of Investing, Institutional Investor*, 11(Fall), 7–22.

Ford, A. (1999). Cycles in competitive electricity markets: a simulation study of the western United States. *Energy Policy*, 27(11), 637–658.

Ford, A. (2001). Waiting for the boom: a simulation study of plant construction in California. *Energy Policy*, 29, 847–869.

Geman, H. (2005). *Commodities and Commodity Derivatives – Modelling and Pricing For Agriculturals, Metals and Energy*. New York: John Wiley and Sons.

Green, R. (2004). *Retail Competition and Electricity Contracts*. Cambridge Working Papers in Economics CWPE 0406.

Green, R. (2007). *Carbon Tax or Carbon Permits: The Impact on Generators' Risks*. Working Paper, Mimeo.

Grinold, R. (1996). Domestic grapes from imported wine. *Journal of Portfolio Management*, 26 (Special Issue, December).

Humphreys, H. and McClain, K. (1998). Reducing the impacts of energy price volatility through dynamic portfolio selection. *The Energy Journal*, 19(3), 107–131.

International Energy Agency (2006). *World Energy Outlook 2006*. Paris: OECD/IEA.

International Energy Agency/Nuclear Energy Agency (2005). *Projected Costs of Generating Electricity, 2005 Update*. Paris: OECD.

Jansen, J., Beurskens, L. and van Tilburg, X. (2006). *Application of Portfolio Analysis to the Dutch Generating Mix*. ECN Report C-05-100 (February).

Kleindorfer, P. and Li, L. (2005). Multi-period VaR-constrained portfolio optimisation with applications to the electric power sector. *The Energy Journal*, 26(1), 1–26.

Krey, B. and Zweifel, P. (2006). *Efficient Electricity Portfolios for Switzerland and the United States*. University of Zurich, SOI Working Paper No. 0602. http://www.soi.unizh.ch/research/wp/wp0602.pdf

Kwan, C. (2001). Portfolio analysis using spreadsheet tools. *Journal of Applied Finance*, 1, 70–81.

Markowitz, H. (1952). Portfolio selection. *Journal of Finance*, 7(1), 77–91.

Neuhoff, K. and De Vries, L. (2004). Insufficient incentives for investment in electricity generations. *Utilities Policy*, 12, 253–267.

Newbery, D. (2006). *Climate Change Policy and its Effect on Market Power in the Gas Market*. Electricity Policy Research Group Working Paper 0510 (February). www.electricitypolicy.org.uk

Newbery, D. and Green, R. (1996). Regulation, public ownership, and privatisation of the English electricity industry. In Gilbert, R. J. and Kahn, E. P. (Eds), *International Comparisons of Electricity Regulation*. Cambridge: Cambridge University Press, pp. 25–81.

Olsina, F., Garces, F. and Haubrich, H.-J. (2006). Modelling long-term dynamics of electricity markets. *Energy Policy*, 34, 1411–1433.

Pogue, G. (1970). An extension of the Markowitz portfolio selection model to include variable transactions' costs, short sales, leverage policies and taxes. *The Journal of Finance*, 25(5), 1005–1027.

Roques, F. (2007). *Managing Fuel and Electricity Price Risks through Operating Flexibility and Contractual Arrangements: Impact on Power Projects Valuation*. EPRG Working Paper 07/XX, University of Cambridge. www.electricitypolicy.org.uk

Roques, F., Newbery, D. and Nuttall, W. (2006a). *Using Monte Carlo Simulation to Assess the Impact of Risks and Managerial Flexibility on Different Generation Technologies*. EPRG Working Paper 06/19, University of Cambridge. www.electricitypolicy.org.uk

Roques, F., Newbery, D. and Nuttall, W. (2008). *Fuel Mix Diversification Incentives in Liberalised Electricity Markets: A Mean-Variance Portfolio Theory Approach*. Energy Economics, 30(4), 1831–1849.

Roques, F., Newbery, D., Nuttall, W., de Neufville, R. and Connors, S. (2006b). Nuclear power: a hedge against uncertain gas and carbon prices?. *The Energy Journal*, 27(4), 1–24.

Stirling, A. (1994). Diversity and Ignorance in electricity supply investment. Addressing the solution rather than the problem. *Energy Policy*, 22, 195–216.

Stirling, A. (1998). On the Economics and Analysis of Diversity. *SPRU Electronic Working Paper* No. 28, October 1998; http://www.sussex.ac.uk/spru/publications/imprint/sewps/sewp28/sewp28.html

Stirling, A. (2001). Science and precaution in the appraisal of electricity supply options. *Journal of Hazardous Materials*, 86, 55–75.

Wiser, R., Bachrach, D., Bolinger, M. and Golove, W. (2004). Comparing the risk profiles of renewable and natural gas-fired electricity contracts. *Renewable and Sustainable Energy Reviews*, 8, 335–363.

CHAPTER 12

Risk Management in a Competitive Electricity Market

Min Liu* and **Felix F. Wu****

Abstract

In a competitive electricity market, it is necessary and important to develop an appropriate risk management scheme for trade with full utilization of the multimarket environment in order to maximize participants' benefits and minimize the corresponding risks. Based on analyses of trading environments and risks in the electricity market, a layered framework of risk management for electric energy trading is proposed. Simulation results confirmed that trading among multiple markets is helpful to reduce the complete risk, and value at risk provides a useful approach to judge whether the formed risk-control scheme is acceptable.

Key Words: Mean-variance portfolio theory, risk management, value at risk.

Acknowledgements

This work has been supported by the Research Grant Council, Hong Kong SAR, China, through grant HKU7174/04E, and Guizhou University, Guizhou, China, under Grant GUT2004-014.

12.1 Introduction

Global deregulation in the electrical power industry has introduced the concept of a competitive electricity market. In this new environment, electricity is traded in the same way as other commodities. However, electricity prices are substantially more volatile than any other commodity price since electricity cannot be stored and its transmission is limited by physical and reliability constraints.

*Faculty of Electrical Engineering, Guizhou University, Guiyang, Guizhou, PR China
**Department of Electrical and Electronic Engineering, The University of Hong Kong, Hong Kong

Analytical Methods for Energy Diversity and Security © 2008 Elsevier Ltd.

Confronted with this severe price volatility, market participants need to find ways to protect their benefits (quantified in profits in this chapter), i.e. to manage risks involved in the market.

Risk refers to the possibility of suffering harm or loss, danger or hazard. Risks result from uncertainty. In the electricity market, a trader's profit is influenced by many uncertain factors, including unit outage, other traders' bidding strategies, congestion in transmission and demand change. These uncertainties bring about risks in electricity pricing and delivery. From a mathematics point of view, a trader's profit is a random variable. According to the modern theory of choice under uncertainty (Biswas, 1997), the expected profit is an indication of expected profitability, while the variance or the standard deviation of the profit can be used as an indication of the risk involved.

Risk management is the process of achieving a desired profit, taking the risks into consideration, through a particular strategy. In the financial field, there are two means to control risk. One is through risk financing by using hedging to offset losses that can occur and the other is through risk reduction using diversification to reduce exposure to risks. Instruments for risk management include forward contracts, futures contracts, options and swaps (Luenberger, 1997). Forward contracts are agreements to buy or sell an agreed amount of the commodity at a specified price at a designated time. Futures contracts are standardized forward contracts that are traded on exchange and no physical delivery is necessary. Options are contracts that provide the holder with the right but not the obligation to buy or sell the commodity at a designated time at specified price. A swap contract is an agreement between two parties to exchange a series of cash flows generated by underlying assets without physical transfer of the commodity between the buyer and seller. Hedging is the use of these financial instruments with the payoff patterns to offset the market risks. Diversification is engaging in a wide variety of markets so that the exposure to the risk of any particular market is limited. Applying this concept to energy trading in an electricity market, diversification means trading electric energy through different physical trading approaches.[1] In the energy market, both physical trading approaches (e.g. spot market, contract market) and financial trading approaches (e.g. futures, options and swaps) are available. A combination of these trading approaches is defined as a portfolio and the corresponding risk-control methodology is called portfolio optimization. A commonly adopted measure for risk assessment, i.e. assessing risk exposure of financial portfolios, is the value at risk (VaR) (Jorion, 1997).

Various aspects of risk management have been applied to electricity markets (Liu et al., 2006). For example, different forward contracts that can provide hedging to the risk of spot prices for market participants are proposed in Kaye et al. (1990) and Gedra (1994). The usefulness of the application of futures contracts in an electricity market is demonstrated in Collins (2002) and Tanlapco et al. (2002), and valuation of different contracts is considered in Bjorgan et al. (2000) and

[1] Physical trading approach refers to the trading approach in which actual physical energy is traded, while a financial trading approach only involves financial settlement, and no physical delivery is necessary.

Deng (2000). Monte Carlo simulation and decision analysis have been applied to find the optimal contract combination (Sheble, 1999; Siddiqi, 2000; Vehviläinen and Keppo, 2003). Approaches of portfolio optimization in mean-variance portfolio theory have been adopted in the trading scheduling for a generation company (Genco) (Liu and Wu, 2006). VaR has been applied to risk assessment in electricity markets (Dahlgren et al., 2003). Concepts from financial option theory have been used in the valuation of generation assets (Deng et al., 1999).

This chapter addressed the problem of establishing a framework for risk management in a multimarket environment, i.e. how to make an optimal trading schedule from the point of view of risk control and make an assessment on this trading schedule. The methodology of mean-variance portfolio theory is applied to risk control and VaR to risk assessment on the associated trading portfolio. The following explanation is made from the point of view of a Genco. The chapter first introduces the background of a competitive electricity market, then gives an overview about the framework of the risk management, explains the methodologies of the risk control and risk assessment, gives a numerical example to demonstrate the proposed approach, and finally draws a brief conclusion.

12.2 Electricity markets and pricing systems

In the electricity market, energy-trading markets can be divided into two categories: physical markets and financial markets. In the physical market, energy is physically traded, whereas financial markets only operate as hedging instruments and no physical energy transactions are involved.

Most of the electricity markets provide two types of physical market: the spot market and the contract market. The term 'spot market' in the electricity market typically refers to a market in which trades cover a short period in the very near future. This chapter adopts Federal Energy Regulatory Commission (FERC)-USA definition in its standard market design (US FERC, 2002) that all the energy traded in the real-time and day-ahead market is spot energy. From a Genco's point of view, selling energy in the spot market means submitting a bid (price and quantity) to the exchange (Power Pool/ISO), with one of two alternative results: (1) the exchange accepts the Genco's bid and pays the Genco the market clearing price (MCP) for its actual energy output; or (2) the exchange rejects the Genco's bid, i.e. the Genco sells nothing in the spot market. The MCP depends on everybody's bids, as well as the load demand, and is therefore uncertain. The risk of the price fluctuation is therefore the most important risk in the spot market.

In the contract market, a Genco trades energy by way of signing contacts, which are referred to as physical forward contracts, with its counter-parties (e.g. energy consumers). Specific details such as trading quantity (MW), trading duration (hours), trading price ($/MWh) and delivery point are bilaterally negotiated between Gencos and consumers or their agents. Bilateral contracts are signed before the actual trading period. In other words, trading quantity and price are set in advance. The main risk in the contract market is the congestion charge. Congestion charges depend on the specific pricing system of an electricity market.

There are three pricing systems currently adopted in the electricity market: uniform marginal pricing, zonal pricing and locational marginal pricing (LMP) (Ma et al., 2003). In a market with a uniform marginal pricing system, only one price is used for ex post settlement for each trading interval. A Genco can make certain of its revenue by signing bilateral contracts with its customers at fixed prices. In other words, there is no risk of a congestion charge in a uniform pricing market. In a zonal pricing market or LMP market, bilateral contracts face the risk of congestion charge since the marginal prices will vary from zones or locations when there is a congestion between the zones (zonal pricing market) or in the transaction system (LMP market). The congestion charge is the product of the price difference between the zones or locations and the trading amount involved.

In financial markets, several financial instruments are provided to offset the particular sources of risk in the electricity market. For example, futures contracts, options and swaps are used to hedge the risk of the price volatility in the spot market. Financial transmission rights (FTRs)[2] are adopted to protect market participants from the risk of congestion charge in the contract market.

12.3 Overview of the framework

In order to provide a clear hierarchy of the risk management process, this chapter proposes to establish a risk management framework in which four steps are involved: (1) determination of the trading objective of a Genco; (2) identification to the associated trading constraints such as trading environments, market rules and trading horizons; (3) translating the objectives and constrains into risk-control strategies, and making a trading schedule/trading portfolio under a specific strategy; and (4) risk assessment on the formed trading portfolio. The risk management process is completed if the assessment result is acceptable to the Genco. Otherwise, modifications to the risk-control strategy and the corresponding trading portfolio are needed.

A Genco's objective, in a competitive electricity market, is characterized by the benefit–risk tradeoff between the expected benefits that the Genco wants (benefit requirements) and how much risk it is willing to assume (risk tolerance). Different benefit requirements (e.g. variable or constant) and risk tolerance (e.g. conservative or variable) result in different objectives. Rather than enumerating all the benefit–risk combinations, which is an impossible task, Gencos' objectives may be divided roughly into three types: normal conservative, more conservative and less conservative. A Genco with a normal conservative objective would like to accept variable benefits (as high as possible) provided that the corresponding

---

[2] FTRs (Lyons et al., 2000) are contracts that exist between a market participant – in fact, any individual or organization – and the system operator. FTRs do not entitle their holders to an exclusive right to use the transmission system, but entitle them to be paid the transmission price on a given path (multiplied by the number of rights the owner has) or, in a nodal market, the price difference between two nodes.

risk may be reduced with risk-control instruments. A more conservative Genco is more concerned about the risk than the benefit and, therefore, requires expected benefit as close to constant as possible. A less conservative Genco focuses its attention on transactions with potentially high benefits rather than trying to search for the optimal benefit–risk profile for the entire portfolio.

Trading constraints include trading environments, market rules and trading horizons. Trading environments refer to the types of physical and financial trading approaches provided by the specific electricity market (e.g. spot market, forward contract market, futures market, FTRs market). Market rules vary from one market to another. From a risk management point of view, two aspects of the market rules are concerned. The first concern is the pricing method adopted in the spot market, i.e. uniform pricing, zonal pricing, LMP or others. The second concern is the specific rules on the trading proportions of each trading market, i.e. the maximum and minimum trading proportions of each trading market. For example, some markets require at least 80% of a Genco's energy to be traded through forward contract market and the remainder in the spot market. The trading horizon can be divided into three levels: (1) short-term trading schedule (e.g. weekly or monthly schedule); (2) mid-term trading schedule (e.g. quarterly or annually schedule); and (3) long-term trading schedule (e.g. a schedule for several years).

Risk control has two levels. The first level is diversification through portfolio optimization, i.e. trying to find a portfolio with a reasonable tradeoff between the expected benefit and risk. The second level is hedging specific risks with specific financial instruments. Different Gencos with different objectives and different constraints would adopt different risk-control strategies. This is discussed in detail below.

Risk assessment is done to give a picture of the risk of a trading portfolio and to let a decision maker decide intuitively whether the trading schedule is acceptable. VaR is adopted to value the trading portfolio in the following text.

12.4 Risk control

12.4.1 General case: risk-control strategy for a normal conservative Genco

Assume that all existing physical and financial trading approaches are available in an electricity market. The risk-control strategy of a normal conservative Genco includes two aspects: (1) diversification among multiple physical trading markets; and (2) hedging with specific financial instruments. That is, physical energy is allocated between spot markets and contract markets, while specific risks of the spot market (i.e. spot-price risk) and contract market (i.e. congestion-charge risk) are hedged with specific financial instruments. Detailed risk-control schemes are subject to the specific pricing method adopted in the spot market.[3]

---

[3] Owing to limitations of space, this chapter only gives the associated conclusions about risk-control schemes in different pricing markets. See Liu (2004) for more explanation and demonstrations.

In a uniform pricing market, a Genco would sign contracts with customers who offer the highest price without considering congestion cost. As for the spot market, the price risk can be hedged with financial instruments such as futures, options and swaps. Among these instruments, futures are suitable for a mid-term trading schedule since they are generally traded monthly and up to 18 months. Options can be traded in the exchange or over-the-counter (OTC)[4] market and therefore can be used in short-term, mid-term and long-term trading schedules. However, it tends to be more expensive than futures and swaps owing to the premium payment. Swaps are OTC derivatives and therefore suitable for short-term, mid-term and long-term trading schedules provided that counter-parties are available.

In a zonal pricing/LMP market, a Genco would sign bilateral contracts with customers located in different pricing areas, as well as trade energy in the spot market, aiming to reduce the total risk of the trading portfolio. Spot-price risk can be hedged with futures, options or swaps, while congestion-charge risk can be hedged with FTRs.

Making a trading schedule/portfolio refers to the determination of energy allocation ratios of the markets including both physical trading markets (i.e. spot market and forward contract market) and financial trading markets (i.e. futures market and FTRs market, etc.). There are three steps to achieving an optimal trading schedule. First, calculate the optimal hedge ratio[5] for each physical transaction; then calculate the optimal energy allocation ratio to each physical trading approach; and finally, calculate the optimal allocation ratios to each financial trading approach.

Step 1: Hedging with financial instruments

For each physical trading approach, assuming that all energy is traded in this approach, calculate the optimal hedge ratio denoted by x_i^* (i is the index of the physical trading approach and the associated financial hedging market). Mathematically, the optimal hedge amount can be achieved by minimizing the variance of the profit on the hedged physical trading approach with respect to the trading amount in the financial hedging market. For example, for the transaction in the spot market, if futures contracts are used to hedge the spot-price risk, the optimal hedge ratio can be obtained by minimizing the total risk[6] with respect to the amount traded in the futures market. That is, suppose that a Genco sells α MWh energy in futures market at futures price $f(\$/MWh)$ before the beginning of the trading period. On the date of delivery (i.e. the beginning of the trading period) the Genco settles the futures contracts by buying back α MWh energy in the futures market at the futures price $f^*(\$/MWh)$. Let π_O be the profit on the spot market. If the transaction costs of futures contracts are ignored, the Genco's profit

---

[4]OTC is a kind of derivatives market in which non-standard products (e.g. contracts) are traded. Trades on the OTC market are negotiated directly with dealers.

[5]Hedge ratio refers to the ratio of the energy quantity traded in the financial hedging instrument to the quantity traded in the underlying physical trading approach.

[6]Here, the total risk is the risk of trading in a physical trading market (i.e. spot market) taking the corresponding hedge effect (i.e. hedging with futures contracts) into consideration.

from both spot market and futures market is π_N, where $\pi_N = \pi_O + \alpha(f - f^*)$. The optimal quantity of energy sold in the futures market can be achieved by minimizing the total risk, $\text{Var}(\pi_N)$, with respect to α, i.e.

$$\text{Min}_{\alpha} \ \text{Var}(\pi_N) = \text{Var}(\pi_O) + \alpha^2 \text{Var}(f^*) - 2\alpha \cdot \text{Cov}(\pi_O, f^*)$$

where $\text{Var}(\pi_O)$, $\text{Var}(\pi_N)$ and $\text{Var}(f^*)$ are variance of π_O, π_N and f^*, respectively, and $\text{Cov}(\pi_O, f^*)$ is the covariance between π_O and f^*. Solving this optimization problem results in the optimal selling quantity α^*, where $\alpha^* = \text{Cov}(\pi_O, f^*)/\text{Var}(f^*)$. The optimal hedge ratio is $x^* = \alpha^*/E$, where E is the quantity of energy sold in the spot market.

Step 2: Energy allocation among physical trading approaches

The risk preference of a risk-averse Genco can be described with a utility function which combines the benefit (expected profit) and risk (variance of profit) into a simple relation, e.g. $U(\pi) = E(\pi) - B \cdot \text{Var}(\pi)$ (Liu and Wu, 2006), where $U(\pi)$ is the utility value, and B is the risk penalty factor which indicates the extent that a Genco 'penalizes' the expected profit considering the risk of obtaining the corresponding profit.[7] According to the utility theory (Biswas, 1997), a trading portfolio with the highest utility value is preferred. The trading objective of a normal conservative Genco, i.e. maximizing benefit and minimizing the associated risk, is then achieved by maximizing the utility function.

For each hedged trade (with the optimal hedge ratio x_i^*) determined from step 1, calculate its expected profit and risk (i.e. $E(\pi_i)$, $\text{Var}(\pi_i)$). For any two trades that have been hedged, calculate their covariance (i.e. σ_{ij} where i, j are the indexes of trading approaches). The optimal energy allocation among different hedged trades can be achieved by maximizing the Genco's utility value with respect to the proportions allocated to the trades. Assuming there are n physical trading approaches available in the electricity market, this optimization problem can be described as follows:

$$\text{Max}_{w_i} \ U(w_1, \ldots, w_i, \ldots, w_n) = \sum_{i=1}^{n} w_i E(\pi_i) - B \sum_{i=1}^{n} \sum_{j=1}^{n} w_i w_j \sigma_{ij} \tag{12.1}$$

$$\text{s.t.} \ \sum_{i=1}^{n} w_i = 1$$

$$w_i^{\text{Min}} \leq w_i \leq w_i^{\text{Max}}$$

where w_i is the proportion allocated to the ith trade, and w_i^{Min} and w_i^{Max} are the upper limit and lower limit of the trading proportion for the ith physical trading approach, respectively, which are specified by a specific electricity market.

---

[7] The value of B can be calculated using the formula $B = A/2C$ (Liu and Wu, 2006), where C is the production cost of a Genco, and A is an index of the decision maker's risk aversion. The moderate value of A is 3; $A > 3$ means more risk averse and $A < 3$ indicates less risk averse (Bodie et al., 1999).

Solutions to this optimization problem are the optimal allocation ratios to the physical trading approaches denoted by w_i^* $(i = 1 \sim n)$.

Step 3: Optimal hedging proportions

The proportion of total energy traded in a financial instrument market aiming at hedging the risk of the ith physical trading approach is called the optimal hedging proportion for the ith physical trading approach. This ratio is calculated as $w_i^* x_i^*$.

To summarize, the optimal proportion of total energy allocated to the physical trading market is $w_i^* (i = 1 \sim n)$, and the optimal proportion of total energy allocated to the financial trading market is $w_i^* x_i^* (i = 1 \sim n)$.

12.4.2 Discussion: risk-control strategies for more conservative Gencos and less conservative Gencos

A more conservative Genco would like to trade physical energy in both spot and contract markets. Spot-price risk is hedged with swaps since it can be negotiated bilaterally and no transaction cost is involved. For the contract trade, the Genco would sign contracts with customers who offer the highest price in the uniform pricing market since no congestion cost is charged to the bilateral transaction. In a zonal pricing/LMP market, the Genco would sign contracts with local customers. The reason for this is that contracts signed with non-local customers face potential congestion charges, and it is uncertain whether the corresponding financial transmission rights can be obtained and completely hedge the congestion risk.

A less conservative Genco, in a uniform pricing market, would like to trade physical energy only in the spot market and hedge the spot-price risk with options. Purchasing a put option with a physical sale of electric energy to the spot market lets the Genco avoid the risk of lower prices and benefit from any increase in spot price, although a premium is needed. However, in the zonal pricing/LMP market, the Genco would like to trade energy in the spot market and with customers who offer the highest contract price in the form of forward contracts. It prefers to hedge the spot-price risk and congestion-charge risk with options and FTRs, respectively.

12.5 Risk assessment

12.5.1 Risk assessment technique

Value at risk (VaR) is a risk management concept developed and promoted in the banking industry to provide a common measurement for the risk exposure of financial portfolios. It is defined, in the financial literature, as a monetary value that the portfolio will lose less than that amount over a specified period of time with a specified probability. For example, a one-day 95% VaR of $500,000 indicates that the portfolio is expected to lose an amount less than $500,000 on 95 days out of 100 days.

There are numerous methods to calculate VaR, which use different assumptions and techniques. Since VaR calculations are very sensitive to assumptions and data, quantitative results will differ when the same techniques are applied using different assumptions or different datasets. Jorion (1997) distinguishes four separate routes to measuring VaR: delta-normal method, historical simulation, stress-testing method and the Monte Carlo approach.

12.5.2 Application of value at risk in trading scheduling

In this chapter, the VaR of a trading portfolio is defined as the expected minimum profit (a monetary value) of the portfolio over a target horizon within a given confidence interval. The target horizon is the trading horizon. The confidence level depends on the extent of the Genco's risk aversion. Normally, a Genco with moderate risk aversion adopts the 95% confidence level, a more risk-averse Genco may require a 99% confidence level and a less risk-averse Genco could use a 92.5% confidence level.

In the most general form, VaR can be derived from the probability distribution of the future portfolio value $f(\pi)$, where π denotes the profit on the trading portfolio. At a given confidence level c, we wish to find the lowest possible realization $\hat{\pi}$ such that the probability of exceeding this value is c:

$$c = \int_{\hat{\pi}}^{\infty} f(\pi)\, d\pi$$

or such that the probability of a value lower than $\hat{\pi}$, $p = \text{Prob}(\pi \leq \hat{\pi})$, is $1 - c$:

$$1 - c = \int_{-\infty}^{\hat{\pi}} f(\pi)\, d\pi = \text{Prob}(\pi \leq \hat{\pi}) = p$$

The number $\hat{\pi}$ is called the sample quantile of the distribution. In the simulation approaches to measuring VaR such as the historical simulation, stress testing and Monte Carlo simulation, this sample quantile can be derived from the simulation results of samples. For example, suppose 100 samples are used to simulate the expected profit on a trading portfolio. Simulation results are ranged from the highest profit to the lowest profit. If the confidence level is 95%, then $\hat{\pi}$ is equal to the 95th simulation result in the simulation result list.

In this chapter, VaR is used to measure the risk extent of the scheduled trading portfolio, aiming at providing a rough figure that helps the decision maker to judge whether the scheduled portfolio is acceptable. Therefore, a relatively rough but simplified method may be used to calculate the value of VaR, i.e. the delta-normal method. Suppose the prices of electricity and fuel are normally distributed. The profit on the trading portfolio, which is a linear combination of normal random variables, is then also normally distributed. Under this condition, the VaR figure can be derived directly from the portfolio's standard deviation, using a multiplicative factor that depends on the confidence level.

First, the general normal distribution $f(\pi)$ is translated into a standard normal distribution $\phi(u)$, where u has mean of zero and standard deviation of unity. That is, $\hat{\pi}$ is associated with a standard normal deviate α ($\alpha > 0$) by setting

$$-\alpha = \frac{\hat{\pi} - \mu}{\sigma}$$

where $\mu = E(\pi^*) = \sum_{i=1}^{n} w_i^* E(\pi_i)$, $\sigma = \sqrt{Var(\pi^*)} = \sqrt{\sum_{i=1}^{n}\sum_{j=1}^{n} w_i^* w_j^* \sigma_{ij}}$. w_i^* is the optimal

allocation ratio of the ith physical trading approach, which is the calculation result in the process of risk control. It is equivalent to set

$$1 - c = \int_{-\infty}^{\hat{\pi}} f(\pi)d\pi = \int_{-\infty}^{-\alpha} \phi(u)du = p$$

Thus, the problem of finding a VaR is equivalent to finding the deviate α such that the area to the left of it is equal to $1 - c$. It is made possible by turning to tables of the cumulative stand normal distribution function, which is the area to the left of a standard normal variable with value equal to c. To find the VaR of a standard normal variable, select the desired confidence level in the table, say 95%. This corresponds to a value of $\alpha = 1.65$. Then the VaR of the portfolio (i.e. the cutoff profit $\hat{\pi}$) can be calculated as follows:

$$VaR = \hat{\pi} = \mu - \alpha\sigma \tag{12.2}$$

12.6 Example

The PJM electricity market is an LMP market with several pricing zones such as PSEG, PECO, PENELEC, etc. (see PJM website). Suppose a Genco is located in PENELEC and is normal conservative. The Genco is making a monthly trading plan of August, i.e. determining the trading amount/trading ratio of each market (e.g. spot market, contract market). Before applying the methodologies proposed in this chapter, the profit characteristics (i.e. expected value, variance and covariance) of each trading approach are clarified first in the following subsection.

12.6.1 Profit characteristics

Suppose that there are M trading intervals during the planning period (one trading interval can be one hour, one day, one week, one month or even one year, depending on the planning horizontal). The trading time for each trading interval is t (hour). The following notation will be used: i, j is the index of the trading area or pricing node; k is the index of the trading interval; $\lambda_{i,k}^{B}$ is the kth trading interval's electricity contract price signed with customers of area i; $\lambda_{i,k}^{S}$ is the kth trading interval's electricity spot price of area i; λ_{k}^{F} is the kth trading interval's fuel spot price; π_i is profit on the ith trade, where i is 0 denotes local contract, i is $1 \sim n$ denotes non-local contract and i is $n + 1$ denotes spot transaction; e_k is the

kth trading interval's trading energy; $E(\cdot)$ is expectation; $\text{Var}(\cdot)$ is variance; $\sigma_{i,j}$ is covariance between profits on transaction i and j; and b is heat rate (or consumption coefficient) of a unit.

Assume that the Genco's production exhibits constant returns to scale. Production cost is a function of energy output and fuel price, i.e. $c(\cdot) = be\lambda^F$. For the local contract and spot transaction, the associated cost only involves the production cost; for the non-local contract, the associated costs include congestion charge as well as production cost. In general, congestion charges should be paid by the associated bilateral transaction. But who (Gencos or energy purchasers) should pay what percentage of the involved congestion charges depends on the specific market rules. In this chapter, a factor β ($0 \le \beta \le 1$), is used to denote the payment proportion of the Genco. According to the methodologies of probability, the expectation, variance and covariance of profits on each transaction can be derived as follows:[8]

$$E(\pi_0) = \sum_{k=1}^{M} \left[\lambda_{0,k}^B - bE\left(\lambda_k^F\right) \right] \cdot e_k \tag{12.3}$$

$$\text{Var}(\pi_0) = \sum_{k=1}^{M} (be_k)^2 \cdot \text{Var}\left(\lambda_k^F\right) \tag{12.4}$$

$$E(\pi_0) = \sum_{k=1}^{M} \left[\lambda_{i,k}^B + \beta E\left(\lambda_{0,k}^S\right) - \beta E\left(\lambda_{i,k}^S\right) - bE\left(\lambda_k^F\right) \right] \cdot e_k \quad (i = 1 \sim n) \tag{12.5}$$

$$\text{Var}(\pi_i) = \sum_{k=1}^{M} e_k^2 \begin{bmatrix} \beta^2 \text{Var}\left(\lambda_{0,k}^S\right) + \beta^2 \text{Var}\left(\lambda_{i,k}^S\right) + b^2 \text{Var}\left(\lambda_k^F\right) \\ - 2\beta^2 \text{Cov}\left(\lambda_{0,k}^S, \lambda_{i,k}^S\right) - 2b\beta \text{Cov}\left(\lambda_{0,k}^S, \lambda_k^F\right) \\ + 2b\beta \text{Cov}\left(\lambda_{i,k}^S, \lambda_k^F\right) \end{bmatrix} \quad (i = 1 \sim n) \tag{12.6}$$

$$E(\pi_{n+1}) = \sum_{k=1}^{M} \left[E\left(\lambda_{0,k}^S\right) - bE\left(\lambda_k^F\right) \right] \cdot e_k \tag{12.7}$$

$$\text{Var}(\pi_{n+1}) = \sum_{k=1}^{M} e_k^2 \left[\text{Var}\left(\lambda_{0,k}^S\right) + b^2 \text{Var}\left(\lambda_k^F\right) - 2b\text{Cov}\left(\lambda_{0,k}^S, \lambda_k^F\right) \right] \tag{12.8}$$

$$\sigma_{0i} = -\sum b\beta e_k^2 \text{Cov}\left(\lambda_k^F, \lambda_{0,k}^S\right) + \sum b\beta e_k^2 \text{Cov}\left(\lambda_k^F, \lambda_{i,k}^S\right) + \sum b^2 e_k^2 \text{Var}\left(\lambda_k^F\right) \quad (i = 1 \sim n) \tag{12.9}$$

$$\sigma_{0,n+1} = -\sum be_k^2 \text{Cov}\left(\lambda_k^F, \lambda_{0,k}^S\right) + \sum b^2 e_k^2 \text{Var}\left(\lambda_k^F\right) \tag{12.10}$$

[8] See Liu and Wu (2006) for details.

$$\sigma_{i,n+1} = \sum e_k^2 \left[\begin{array}{c} \beta \mathrm{Var}\left(\lambda_{0,k}^S\right) - \beta \mathrm{Cov}\left(\lambda_{i,k}^S, \lambda_{0,k}^S\right) - b\mathrm{Cov}\left(\lambda_k^F, \lambda_{0,k}^S\right) \\ - b\beta \mathrm{Cov}\left(\lambda_k^F, \lambda_{0,k}^S\right) + b\beta \mathrm{Cov}\left(\lambda_k^F, \lambda_{i,k}^S\right) + b^2 \mathrm{Var}\left(\lambda_k^F\right) \end{array} \right] \qquad (12.11)$$

$$(i = 1 \sim n)$$

$$\sigma_{i,j} = \sum e_k^2 \left[\begin{array}{c} \beta^2 \mathrm{Var}\left(\lambda_{0,k}^S\right) - \beta^2 \mathrm{Cov}\left(\lambda_{i,k}^S, \lambda_{0,k}^S\right) - 2b\beta \mathrm{Cov}\left(\lambda_k^F, \lambda_{0,k}^S\right) \\ - \beta^2 \mathrm{Cov}\left(\lambda_{j,k}^S, \lambda_{0,k}^S\right) + \beta^2 \mathrm{Cov}\left(\lambda_{i,k}^S, \lambda_{j,k}^S\right) \\ + b\beta \mathrm{Cov}\left(\lambda_k^F, \lambda_{i,k}^S\right) - b\beta \mathrm{Cov}\left(\lambda_k^F, \lambda_{j,k}^S\right) + b^2 \mathrm{Var}\left(\lambda_k^F\right) \end{array} \right] \qquad (12.12)$$

$$(i, j = 1 \sim n)$$

The statistics of prices involved in the above equations, i.e. $E(\lambda_{i,k}^S)$, $\mathrm{Var}(\lambda_{i,k}^S)$, $E(\lambda_k^F)$, $\mathrm{Var}(\lambda_k^F)$, $\mathrm{Cov}(\lambda_{i,k}^S, \lambda_{j,k}^S)$ and $\mathrm{Cov}(\lambda_{i,k}^S, \lambda_k^F)$, can be estimated based on historical data according to the statistical method.

12.6.2 Simulation results

The following numerical simulation is performed based on the historical data of daily electricity prices in the PJM electricity market (see PJM website). The statistical characteristics (i.e. expectation, variance and covariance) of prices in each trading interval (i.e. one day, 24 hours) of ten pricing zones are calculated. The average value (i.e. the statistical characteristics of monthly price) is shown in Table 12.1. The Genco owns a coal-fired generation unit with 600 MW capacity and 8.9 MBtu/MWh heat rates (Wood and Wollenberg, 1996). The coal price is given as $\lambda_k^F = 2.0\$/MBtu$ ($k = 1\sim31$). The unit is located in PENELEC.

If the Genco only trades its energy in the spot market, with Equations (12.7) and (12.8), the expected profit and corresponding variance are calculated, and are shown in the second column of Table 12.2. If the Genco is normal conservative, i.e. $c = 95\%$, the VaR is calculated as $6,481,300 (see Table 12.2). The simulation results indicate that, in the spot market, the expected profit is $10,202,000; within

Table 12.1 Statistical characteristics of prices in the PJM market (August)

Zone	Expectation of price ($/MWh)	SD of price (%)	Zone	Expectation of price ($/MWh)	SD of price (%)
PSEG	47.738	84.554	METED	46.715	84.643
PECO	46.351	88.662	PEPCO	47.097	79.188
PPL	44.301	87.376	AECO	49.058	86.245
BGE	47.157	84.294	DPL	48.148	91.941
JCPL	46.044	87.095	PENELEC	40.653	69.177

95% confidence level, the minimum profit is $6,481,300 during the trading month. If the Genco thinks that the VaR is lower and would like to reduce the risk of spot transaction, it can control the risk by trading energy in both the spot market and the contract market.

Suppose the Genco can sign bilateral contracts with consumers located in different pricing zones with the contract prices shown in Table 12.3. Assume that the Genco pays all the congestion charge involved in the bilateral transaction, i.e. $\beta = 1$. The upper and lower limits of the trading proportion of each trading market are 100% and 0%, respectively, i.e. $w_i^{max} = 1$, $w_i^{min} = 0$. The index of the Genco's risk aversion is set to 3 (i.e. $A = 3$) since the Genco is normal conservative. Then the optimal allocation ratios w_i^* can be obtained by solving problem (1) and are shown in Table 12.4; the VaR, calculated with Equation (12.2), is shown in the third column of Table 12.2. Under this trading portfolio (i.e. PECO 19.6%, BGE 32.6%, PENELEC 14.1% and spot market 33.7%), the one-month 95% VaR is $9,033,900. That is, at the same confidence level, the minimum profit of the trading portfolio is increased by 39.4% compared to that of the spot trading. Of course, the values of allocation ratio and VaR depend on the price of each bilateral contract. For example, if the consumers located in DPL would like to offer a higher price, say 47.8 $/MWh, the allocation ratio to the contract signed with DPL consumers increases and other allocation ratios change accordingly (Table 12.5). The expected profit and the associated risk also change, as shown in the fourth column of Table 12.2. Trading portfolios 1 and 2 both demonstrate the effect of

Table 12.2 Simulation results

Characteristic	Spot trading	Trading portfolio (1)	Trading portfolio (2)
$E(\pi^*)$ ($)	1.0202×10^7	1.0055×10^7	1.0063×10^7
$Var(\pi^*)$ (^{2})	5.084×10^{12}	3.8263×10^{11}	4.0435×10^{11}
VaR ($) (c = 95%)	6.4813×10^6	9.0339×10^6	9.0135×10^6

Table 12.3 Contract prices

Location of consumers	Contract price ($/MWh)	Location of consumers	Contract price ($/MWh)
PSEG	47.2	METED	46.2
PECO	45.9	PEPCO	46.6
PPL	43.8	AECO	48.6
BGE	46.7	DPL	47.6
JCPL	45.5	PENELEC	40.0

Table 12.4 Allocation ratio of each transaction (1)

Transaction	Spot	Bilateral contract									
		PSEG	PECO	PPL	BGE	JCPL	METED	PEPCO	AECO	DPL	PENELEC
Ratio	0.337	0	0.196	0	0.326	0	0	0	0	0	0.141

Table 12.5 Allocation ratio of each transaction (2)

Transaction	Spot	Bilateral contract									
		PSEG	PECO	PPL	BGE	JCPL	METED	PEPCO	AECO	DPL	PENELEC
Ratio	0.341	0	0	0	0.294	0	0	0	0	0.201	0.163

diversification, i.e. trading among multiple trading approaches can reduce the risk of the complete portfolio.

If the minimum profit with 95% confidence level of trading portfolio 1 or 2 is still not accepted by the Genco, modification to the risk-control strategy is needed. For example, hedging can be adopted to hedge the specific risks of spot transaction and non-local contract. First, spot-price risk could be hedged with available financial instrument such as futures, options and swaps. Physical energy is then allocated among the hedged spot transaction, local contract and non-local contract. The risk of the modified trading portfolio is expected to be reduced. The risk management process is completed if the corresponding VaR is acceptable. Otherwise, hedging the risk of non-local contract (i.e. congestion-charge risk) with FTRs can be considered.

12.7 Conclusion

This chapter developed an overall framework of risk management for Gencos' trading in a competitive electricity market. First, a Genco's objective and trading constraints are identified. Then the identified objective and constraints are translated into a reasonable and feasible risk-control strategy under which a specific trading schedule could be made. Finally, the formed trading portfolio is assessed with the standard risk measurement technique VaR. If the risk assessment result is not acceptable, the risk-control strategy and associated trading schedule are readjusted until the Genco accepts it.

A risk-control strategy varies with different trading objectives and constraints. With trading constraints assumed to be general, this chapter discussed the risk-control strategies for Gencos with different objectives. Simply stated, a normal conservative Genco would like to control risk with all available trading approaches, i.e. trading physical energy between the spot market and contract market, hedging spot-price risk with futures or swaps according to the specific trading horizon, and hedging the congestion-charge risk with financial transmission rights. A more conservative Genco prefers to control risk through diversification, i.e. trading energy between spot and contract markets. A less conservative Genco tends to trade physical energy in spot markets only and hedge spot-price risk with options.

The risk-management process of a Genco with normal conservative objectives was demonstrated based on the historical data of electricity prices in the PJM market. Simulation results confirmed that diversification, i.e. trading among multiple physical approaches, is helpful to reduce the complete trading risk, and VaR provides a useful approach to judge whether the formed trading portfolio is acceptable.

To summarize, the proposed framework of risk management provides a clear hierarchy of the risk management process, which should help a Genco to identify its objective and achieve an optimal trading portfolio in markets involving risks. It is also applicable to other market participants such as energy purchasers, with little modification.

References

Biswas, T. (1997). *Decision-Making Under Uncertainty*. New York: St Martin's Press.

Bjorgan, R., Song, H., Liu, C. C. and Dahlgren, R. (2000). Pricing flexible electricity contracts. *IEEE Transactions on Power Systems*, 15, 477–482.

Bodie, Z., Kane, A. and Marcus, A. J. (1999). *Investments*, 4th edn. Chicago, IL: Irwin/McGraw-Hill.

Collins, R. A. (2002). The economics of electricity hedging and a proposed modification for the futures contract for electricity. *IEEE Transactions on Power Systems*, 17, 100–107.

Dahlgren, R., Liu, C. C. and Lawarree, J. (2003). Risk assessment in energy trading. *IEEE Transactions on Power Systems*, 18, 503–511.

Deng, S. J. (2000). Pricing electricity derivatives under alternative stochastic spot price models. In *Proceedings of the 33rd Annual Hawaii International Conference on System Sciences*, pp. 1313–1322.

Deng, S. J., Johnson, B. and Sogomonian, A. (1999). Spark spread options and the valuation of electricity generation assets. In *Proceedings of the 32nd Annual Hawaii International Conference on System Sciences*, Vol. 3, p. 3027.

FERC (2002). *Notice of Proposed Rulemaking*. United States of America Federal Energy Regulatory Commission Docket No. RM01-12-000.

Gedra, T. W. (1994). Optional forward contracts for electric power markets. *IEEE Transactions on Power Systems*, 9, 1766–1773.

Jorion, P. (1997). *Value at Risk*. New York: McGraw-Hill.

Kaye, R. J., Outhred, H. R. and Bannister, C. H. (1990). Forward contracts for the operation of an electricity industry under spot pricing. *IEEE Transactions on Power Systems*, 5, 46–52.

Liu, M. (2004). *Energy Allocation with Risk Management in Electricity Markets*. Ph.D. Thesis, University of Hong Kong.

Liu, M., Wu, F. F. and Ni, Y. X. (2006). A survey on risk management in electricity markets. In *Proceedings of 2006 IEEE Power Engineering Society General Meeting*, Montreal.

Liu, M. and Wu, F. F. (2006). Managing price risk in a multi-market environment. *IEEE Transactions on Power Systems*, 21, 1512–1519.

Luenberger, D. G. (1997). *Investment Science*. New York: Oxford University Press.

Lyons, K., Fraser, H. and Parmesano, H. (2000). An introduction to financial transmission rights. *The Electricity Journal*, 13, 31–37.

Ma, X. W., Sun, D. I. and Cheung, K. W. (2003). Evolution toward standardized market design. *IEEE Transactions on Power Systems*, 18, 460–469.

PJM. website: www.pjm.com

Sheble, G. B. (1999). Decision analysis tools for GENCO dispatchers. *IEEE Transactions on Power Systems*, 14, 745–750.

Siddiqi, S. N. (2000). Project valuation and power portfolio management in a competitive market. *IEEE Transactions on Power Systems*, 15, 116–121.

Tanlapco, E., Lawarree, J. and Liu, C. C. (2002). Hedging with futures contracts in a deregulated electricity industry. *IEEE Transactions on Power Systems*, 17, 577–582.

Vehviläinen, I. and Keppo, J. (2003). Managing electricity market price risk. *European Journal of Operational Research*, 145, 136–147.

Wood, A. J. and Wollenberg, B. F. (1996). *Power Generation, Operation, and Control*, 2nd edn. New York: Wiley & Sons. p. 24.

CHAPTER 13

Application of Mean-Variance Analysis to Locational Value of Generation Assets

Serhiy Kotsan * and **Stratford Douglas****

Abstract

This chapter addresses the problem of optimal investment in generation based on mean-variance portfolio analysis. It is assumed the investor can freely create a portfolio of shares in generation located on buses of the electrical network. Investors are risk averse, and seek to minimize the variance of the weighted average locational marginal price (LMP) in their portfolio, and to maximize its expected value. Simulations are conducted using a standard IEEE 68-bus network that resembles the New York–New England system, and LMPs calculated in accordance with the PJM methodology for a fully optimal AC power flow solution. Results indicate that the network topology is a crucial determinant of the investment decision as line congestion makes it difficult to deliver power to certain nodes at system peak load. Determining those nodes is an important task for an investor in generation as well as the transmission system operator.

Acknowledgements

This research was sponsored in part by a US DOE/EPSCoR WV State Implementation Award.

13.1 Introduction

The emergence of centralized physical markets in electric power has led to complexities and new opportunities for managing risks in both the operation and finance of power generation facilities. In power markets such as PJM,

*Market Monitoring Unit, New York Independent System Operator New York, USA
**Department of Economics, West Virginia University, Morgantown, WV, USA

Analytical Methods for Energy Diversity and Security © 2008 Elsevier Ltd.
978-0-08-056887-4 All rights reserved.

compensation of generators is based on locational marginal cost prices (LMPs) calculated at each node on the system. The main factor that causes dispersion of the nodal prices of electricity is network congestion, which requires expensive generators to be dispatched before cheaper ones. An important task for the system operator is to identify the locations of chronically congested lines, and to determine the best places to add new generators or transmission system enhancements.

Identifying congested lines requires numerical simulations owing to the non-linearity and complexity of the full AC power flow problem (Bergen, 2000). In addition, demand for electricity is a highly stochastic variable that depends on weather, development of industrial units, etc. Resulting fluctuations in nodal electricity prices make it hard to estimate the ideal location of additional generators or transmission lines. The non-linear nature of the physical power flows causes price peaks to appear in different places on the network as the load changes. It is even possible for the high price of electricity in one area to be eliminated by an increase in demand in another area. Evidence of this can be found in the PJM (2004) data in the form of negative prices. Negative prices imply that an increase in consumption of power can relieve congestion and save production costs.

When planning additional generating units, it is important to consider both the nodal price and its variance. The price of electricity is a highly volatile variable. It can jump from \$30/MWh to more than \$1000 during peak demand.

Mean-variance portfolio theory, developed by Harry Markowitz (1952), is widely used by financial economists to determine the investment portfolios that produce efficient outcomes under various economic conditions. In the framework of investment planning policy it can be used to identify where in the network additional generation capacity should be installed, or alternatively, which buses will be most affected by network congestion as the load increases.

There have been several academic studies on mean-variance analysis for the deregulated electricity market in both the economics and the engineering literature. Economists usually focus the attention of the investor on the cost of electricity generation. For example, Awerbuch and Berger (2003) look at the value of diversification of investment among various production technologies such as nuclear, fossil fuel or green power. An efficient generating portfolio minimizes expected costs of electricity for a given risk of fuel cost increase. This approach is useful in the long run (more than a year) since it does not consider line congestion, network topology and production constraints. Yu (2003) analyzed a spatial mean-variance model, with its spatial nature captured using the correlation of prices in geographically separated markets. Overall, the economic literature applies mean-variance analysis to assess the diversification of risk on the cost side. Optimal frontier analysis can be used to help determine the optimal investment shares of various production technologies. The optimal investment share allocation across technologies will provide the lowest expected price for the electricity for a given level of fluctuations in the cost of fuel.

Revenue in the power market comes from nodal prices, paid at each node. Hogan (1992) showed that the LMP pricing mechanism should provide proper incentives for efficient distribution of power production among generators. High volatility of LMPs is often driven by network congestion, and LMP volatility generally far exceeds the volatility of fuel costs.

In contrast, the engineering literature emphasizes the importance of full AC load flow analysis, subject to network topology, voltage constraints, transmission losses, etc., and does not emphasize the use of financial tools in investment policy. One exception is Denton et al. (2003), who suggest construction of an efficient frontier to determine intermediate-term market risk over a period of one month to a year.

The mean-variance approach helps to identify the nodes with the highest expected revenues and lowest risk, taking into consideration the correlation among prices at different nodes. The simulations presented here are designed to determine the most attractive buses for future investment. They are conducted on the standard IEEE 68-bus test network with generators, high-voltage lines and loads that represent a simplified New York–New England power system.

13.2 Simulation methodology

The goals of modelling are to simulate a competitive power auction using an electrical engineering network model and then to examine correlations among nodal prices and construct the efficient frontier from a financial perspective. The model uses a 16-generator system with 86 transmission lines (see Appendix B) and is a simplified representation of the existing power system in the New York–New England area. The following six steps represent the simulation algorithm:

1. Simulate a load increase.
2. Calculate the optimal power flow (OPF) for each given load.
3. Determine the LMP at each node for each given load.
4. Calculate the LMP mean and variance–covariance matrix.
5. Estimate the mean-variance efficient frontier.
6. Determine optimal investment shares at each bus for the representative agent.

The first three steps represent the 'engineering' part of the research. Each simulated load scenario is satisfied in the least expensive way. Production costs are minimized when generators sell power on the competitive market, such as the PJM power auction. Both cost minimization and nodal prices are calculated for the full AC load flow model following the PJM algorithm.

Steps 4–6 represent the financial part of the research. A mean-variance frontier is constructed for the nodal prices calculated for the stochastic load increase. The representative agent selects a portfolio on the mean-variance frontier, depending on his risk aversion. That portfolio yields the highest expected nodal prices for the given level of risk. Put another way, the investor will be interested in adding generators to the buses that pay the most for the power, at lowest possible risk.

Each of the steps is explained in detail below.

13.2.1 Load simulation scenario

A random walk with drift is used to simulate a 10% increase in load over the T simulation periods:

$$\text{LOAD}_t^i = \text{LOAD}_0^i + \alpha_i t + \varepsilon_t$$

where $t = 1, 2 \ldots T$, LOAD_0^i = initial load at the i, $\alpha_i = \dfrac{\Delta \text{LOAD}^i}{\Delta T} = \dfrac{0.1 * \Delta \text{LOAD}^i}{T - 1}$ slope of the drift, and $\varepsilon \sim N(0, 0.001 * \text{LOAD}^i)$.

Figure 13.1 demonstrates load simulation example for the 2.53 MW load at one of the buses over 30 periods. The straight line represents the trend line without shocks. The load at each bus of the system was simulated independently using the same technique.

13.2.2 Linear programming optimal power flow and locational marginal price

For purposes of this simulation, it is assumed that power is sold competitively on the power auction maintained by PJM. Hogan (1992) shows that with sufficient competition among generators a power auction will yield an efficient dispatch. The LMP is the market clearing price at each node of the power network that yields the lowest possible cost of generation (efficient solution). In this chapter, the power auction was simulated in reverse order. First, the lowest possible

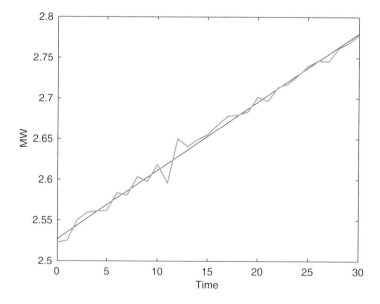

FIGURE 13.1 Example of load fluctuation over time ($T = 30$).

generation costs were found by solution of the OPF, and then the nodal prices (LMP) for the efficient allocation were calculated.

The PJM (2004) methodology was used to calculate LMPs in this chapter. This approach takes into account line congestion and the marginal costs of generators, but neglects line losses.

The OPF was solved for load simulations at each point of time, and LMPs representing the system operator's willingness to pay per unit of power at each bus were calculated.

13.2.3 Constructing the efficient frontier and finding optimum investment buses

In classical portfolio theory, maximizing the expected return for a specified level of risk is a standard problem (Markowitz, 1952). Simply put, portfolio optimization is the search for a vector of investment portfolio shares that satisfies all constraints and provides the minimum total variance at any given level of return or, equivalently, the maximum return for any level of variance. There are typically many such vectors, and their risk/return metrics allow construction of the 'efficient frontier' of this problem space.

The efficient frontier approach requires calculating the mean and variance of return for each portfolio. In this case, the portfolio consists of ownership shares of nodes on the system. Returns at each node are calculated from the LMPs observed through time. The mean LMP at each bus is calculated by taking the arithmetic average of its LMP over time. Calculating portfolio return variance required construction of the variance–covariance matrix between LMPs.

13.2.4 Portfolio selection

The efficient frontier for the investor represents the set of investment portfolios that have highest expected LMP for each given variance. The choice of a particular portfolio on the frontier depends on the investor's degree of risk aversion. This level of aversion to risk can be characterized by defining the investor's indifference curve. Constant absolute risk aversion (CARA) preferences $U = -e^{-w\gamma}$ are typically used in financial theory (Grossman and Stiglitz, 1980). U is investor utility and w is wealth. The parameter γ represents the investor's risk aversion. Typical risk aversion coefficients range between 2 and 8, with the higher number representing less tolerance to risk. The same CARA utility function was used to simulate the representative investor in the power network. Note that $V = E(w) - (\gamma/2) \text{Var}(w)$ captures the investor's tradeoff between risk and return. Also, maximizing V is equivalent to maximizing U (Grossman and Stiglitz, 1980).

The example below emphasizes the importance of the mean-variance analysis. Table 13.1 demonstrates three different assets. Each asset has a payoff that occurs in one of two possible states of the world. For simplicity it is assumed that each state is equally likely to happen. Next, the expected return and risk (variance) of each asset are calculated.

Table 13.1 Example where a portfolio of two assets is preferred to the higher return, lower risk asset

	State 1	State 2	μ	σ^2
Asset 1	−7	7	0	49
Asset 2	−1	5	2	9
Asset 3	8	13	10.5	6.25

At first glance it seems that an investor should spend all his money purchasing asset 3, which is the asset with the highest expected return and lowest variance. However, because the variance of a portfolio's return also depends on the covariances among all of the returns, it may be possible to design a portfolio investment from assets 1 and 2 that the investor will prefer over spending all his money on asset 3. To illustrate, assume that assets 1 and 2 are perfectly negatively correlated, but asset 3 is not correlated with any other asset, $\rho_{12} = -1$, $\rho_{13} = \rho_{23} = 0$.

When the investment dollar is split (0.3, 0.7) among asset 1 and asset 2, respectively, the expected return of the portfolio is $E(R_p) = 0.3 \times 0 + 0.7 \times 2 = 1.4$, and the portfolio variance is $\mathrm{Var}(R_p) = 0.3^2 \times 49 - 2 \times 0.3 \times 0.7 \times 7 \times 3 + 0.7^2 \times 9 = 0$. Thus, two relatively high-risk assets can be combined into a portfolio with a much smaller level of risk. In this particular example, the portfolio of assets 1 and 2 is risk free. An investor with a CARA utility function will prefer this portfolio over a portfolio containing only asset 3 when his risk aversion parameter $\gamma > 2.912$.[1]

This example demonstrates that for investment planning it is important to consider the correlations of returns among a portfolio of assets located at all nodes in the system, rather than analyzing returns at a single node in isolation.

13.3 Simulation results

Figure 13.2 demonstrates the outcome of the simulations for the 68-bus New York–New England IEEE system. The efficient frontier is calculated for the LMPs using an OPF algorithm in the increasing load scenario (steps 1–5). The highlighted area on the efficient frontier is a set of portfolios that would be chosen by representative investors with risk aversion coefficient in the range from 2 to 8 of the CARA utility function. Those portfolios represent sets of buses that will have highest expected LMPs after the anticipated 10% load increase. In other words, the efficient portfolio identifies combinations of nodes that will have the highest average price at each level of price risk. Investment shares for this portfolio can be interpreted as corresponding to the shares that each node contributes for the expected peaking LMP, or the percentage share of investment at each node. In this network, locational prices on buses 3, 33 and 65 (see Figure. 13.B1

---

[1]That is, when $\infty = 2.912$, $U(R_p) = U(\text{asset 3})$.

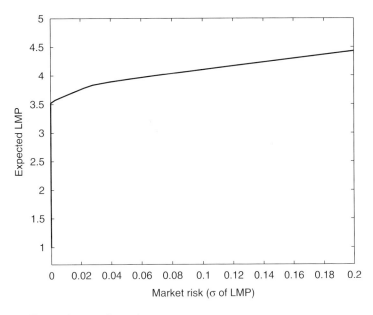

FIGURE 13.2   Efficient frontier for 68-bus power system.

in Appendix B) will form the highest expected LMP. The mean-variance technique allows the system operator or investor to determine the combinations of high-priced system nodes with the lowest possible variation of price. For example, the optimal investment share is around 99% in buses 3 and 33, which are located in the congested part of the network. Congestion requires out-of-merit dispatch, which means that more costly generators are dispatched ahead of less costly ones, which raises the nodal price. Therefore, the identified buses are seen as attractive sites for installing additional generation capacity. From a transmission planner's point of view, the mean-variance analysis identifies these nodes as needing relief from transmission congestion.

It is important to mention that the expected peaking price determined in the scope of this chapter assumes a competitive market, and hence does take into account the potential with holding productive capacity, exercising market power or any other form of generator gaming. Hence, the predicted peak may be substantially lower than the actual price experienced in a real system that is subject to market power manipulation.

13.4 Adding generators

The identification of optimal nodes for new generation in the previous section assumed a static situation in which the addition of the new generation does not affect the distribution of LMPs. A more sophisticated analysis would look at the dynamics of the market with new entry. For example, a system operator may

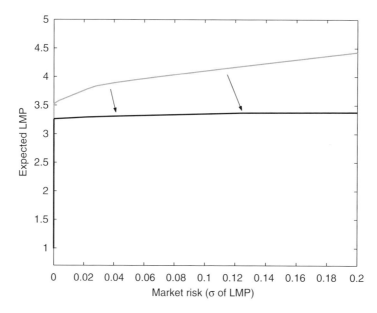

FIGURE 13.3   Shift of the efficient frontier after adding generators to buses 3 and 33.

wish to plan future expansion of generation in such a way as to minimize LMP increases caused by network congestion. One way to do so is to install additional generators on the buses with high predicted LMPs.

The effect of adding generators can be measured by constructing the new mean-variance frontier with the new generation installed and comparing it with the one found with no additional generation. This makes it easy to compare different plans for adding generators to the network. When one or more generators are added, the mean-variance curve will be shifted downward owing to lower prices at each node at all times.

These effects were calculated when additional generators were added to buses 3 and 33 and all the simulations steps were repeated. The resulting mean-variance curve, together with the original one, is shown in Figure 13.3.

13.5 Optimal investment strategy

The efficient frontier represents the highest expected price of electricity for any given level of risk as the system load increases. This will maximize the objective function of the investor in generation that is small relative to the total output of the system. Therefore, this investment strategy can maximize the expected profit of the small power producer (e.g. wind turbine).

In contrast, some market participants, such as regulators or system operators, may wish to minimize expected price at each node. In particular, when a system operator plans to add a large generator to the network, the objective may

Table 13.2 Example demonstrating an increase in the locational marginal price while production costs are decreasing

1. Initial condition				2. Adding a generator, no network congestion				3. Adding a generator to the congested network			
	MC	MW	TC		MC	MW	TC		MC	MW	TC
Gen. 1	$20	Idle	$0	Gen. 1	$20	Idle	$0	Gen. 1	$20	1	$20
Gen. 2	$15	2	$30	Gen. 2	$15	1	$15	Gen. 2	$15	Offline	$0
Gen. 3	$10	5	$50	Gen. 3	$10	5	$50	Gen. 3	$10	5	$50
Gen. 4	$10	6	$60	Gen. 4	$10	6	$60	Gen. 4	$10	6	$60
				Gen. 5	$5	1	$5	Gen. 5	$5	1	$5

be to satisfy increasing demand, reduce network congestion and thus yield lower electricity prices. A substantial increase in productive capacity will cause a downward shift of the efficient frontier in return-variance space, which would serve the system operator's interest.

Figure 13.3 represents a set of efficient frontiers achieved by installing a generator at various nodes of the network. Adding a generator to bus 4 produces the almost horizontal frontier located below. First, it implies a low expected price throughout the network. Second, a flat frontier implies a relatively constant price for any level of load variation, implying market stability and low congestion. The efficient frontier after adding a generator to bus 4 is located below the one obtained by adding two generators (at buses 3 and 33) to the network.

This result implies that a larger capacity cannot be installed at buses 3 and 33 owing to network topology, resulting in congestion and high prices. Finally, large-scale generation capacity may be sponsored by a profit-seeking firm. Such a firm will want the highest possible location of the efficient frontier. The highest expected LMP after the efficient frontier shift will maximize the company's profit (in the framework of constant marginal costs).

At first sight, it may seem impossible for per-unit production costs to fall while market clearing prices are rising in a competitive market. After all, a decrease in production costs occurs when a generator is dispatched whose production cost is lower than that of the most expensive generator currently operating on the network, but a competitive market price is set by the marginal generator. Certainly, system costs could fall and leave LMP unchanged if an inframarginal generator becomes more efficient, but how can the addition of a lower cost generator increase the LMP?

The addition of a lower cost inframarginal generator can increase LMPs if it creates congestion in the system that causes a higher cost generator to be dispatched out of merit order. This situation is illustrated in Table 13.2. In the initial condition, generator 2 is the marginal generator and the system price is $20.

If small but low-cost generator 5 is added to the system and there is no congestion, the system price stays constant, but the total cost of generation falls. If, however, the addition of generator 5 creates congestion that precludes generator 2 from producing, high-cost generator 1 is dispatched instead, driving the price to $20 everywhere on the system (except for generator 2's node).

A profit-maximizing investor who understands the effect of transmission system congestion would try to locate a new plant in such a way that it will increase the system marginal cost in the way illustrated in Table 13.2.

13.6 Conclusion

This chapter demonstrated a mean-variance analysis of a simulated system located on the standard IEEE 68-bus test system. The load was allowed to increase stochastically by 10%. OPF was calculated for the given load fluctuations and LMPs were calculated using the PJM calculation methodology. The process allowed nodes to be identified that were most attractive for new generation investment, assuming that the new generating facility's output is negligible compared to the overall system, and it does not change the level of transmission congestion.

When the production capacity of the new entrant is substantial it may change the pattern of network congestion and therefore the nodal prices of the system. The effects of such a substantial addition to generation were analyzed from the point of view of a system operator who plans the new generation with a desire to minimize network congestion and consequently keep nodal prices low. In that case adding a

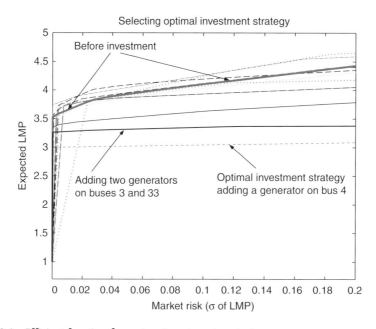

FIGURE 13.4 Efficient frontiers for various investment projects.

large-scale generator to bus 4 will yield the lowest expected price at all nodes and a flat efficient frontier (Figure 13.4). When the owner of a large new generator is responsible for the investment decision making, there will be an incentive to locate it on the system in such a way as to raise nodal prices as much as possible. Thus, even a low-cost plant can result in higher electricity prices, if strategically located.

Mean-variance analysis can be used by the system operator and investors to calculate the risk and returns from investments at various nodes in the system with stochastic load fluctuations. This approach takes into consideration both nodal price variance and covariance with the prices on the other nodes. The analysis becomes more complex when dynamic effects of the new investment on LMPs are taken into consideration, but these considerations may be important to the profitability of the new investment.

References

Awerbuch, S. and Berger, M. (2003). *Applying Portfolio Theory to EU Electricity Planning and Policy Making*. IEA/EET Working Paper. Paris: IEA.

Bergen, A. R. (2000). *Power System Analysis*. Englewood Cliffs, NJ: Prentice Hall. Chapters 5–6.

Denton, M., Palmer, A., Masiello, R. and Skantze, P. (2003). Managing market risk in energy. *IEEE Transactions on Power Systems*, 18(2), 494–502.

Grossman, S. and Stiglitz, J. (1980). On the impossibility of informationally efficient markets. *American Economic Review*, 70, 393–408.

Hogan, W. (1992). Markets in real electric networks require reactive prices. *The Energy Journal*, 14(3), 211–242.

Markowitz, H. (1952). Portfolio selection. *Journal of Finance*, 7, 77–91.

PJM website (2004). http://www.pjm.com

Yu, Z. (2003). A spatial mean-variance MIP model for energy market risk analysis. *Energy Economics*, 25, 255–268.

Appendix A

Formal derivation of the portfolio frontier

$$\min_{\{s\}} \frac{1}{2} s' \sum s$$
$$s.t. \quad \boldsymbol{\mu}' s = \mu_p$$
$$\mathbf{1}' s = 1$$

where $\Sigma = [\sigma_{ij}]$ represents the $n \times n$ variance–covariance matrix of LMP at each bus, where $\sigma_{ij} = \sigma_i^2$, $\boldsymbol{\mu}$ is a column vector of the expected returns μ_i, and s represents a column vector of the portfolio shares s_i, such that $\sum_{i=1}^{n} s_i = 1$.

Thus portfolio variance is minimized subject to a given expected portfolio return μ_p and given that all portfolio shares add up to 1.

Appendix B

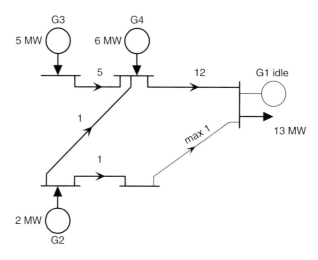

FIGURE 13 B1   Power diagram demonstrating power flows before adding new generator.

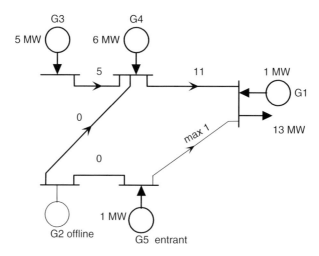

FIGURE 13 B2   Power diagram demonstrating power flows after entrance of the new generator.

CHAPTER 14

Risk, Embodied Technical Change and Irreversible Investment Decisions in UK Electricity Production

An Optimum Technology Portfolio Approach

Adriaan van Zon[*] and **Sabine Fuss**[**]

Abstract

UK climate change policy has long been concerned with the transition to a more sustainable energy mix, both environmentally and in terms of energy security. Electricity producers have to handle the uncertainties surrounding investment decisions for new capacity. This chapter will focus on just two of these sources: the volatility of fuel prices and uncertainty concerning technological progress itself in a context of embodied technical change and irreversible investment. Technological uncertainty in combination with high capital costs per MW of installed capacity is likely to deter investors from irreversibly committing resources to the adoption of renewable technologies on a larger scale, even though they have to accept a higher degree of fuel price risk by doing so. By selecting a portfolio of technologies, risk-averse producers can effectively hedge the uncertainties mentioned above. An extended version of the model developed in van Zon and Fuss (2005) is used, which combines a clay–clay vintage framework with elements from financial portfolio theory and thus captures both dynamic investment aspects and the irreversibilities associated with large sunk costs. Using the extended model, several characteristics of present UK policy are implemented to illustrate the principles involved. The reduction of risk is accompanied by an increase in total costs. For increasing risk aversion, investors are willing to adopt nuclear energy relatively early. Moreover, the embodiment of technical change, in combination with the expectation of a future switch toward another technology, can reduce current investment in that technology (while temporarily increasing current investment in competing technologies). This enables

*UNU-Merit/University of Maastricht, Maastricht, the Netherlands,
**International Institute of Applied Systems Analysis, Laxenburg, Austria

Analytical Methods for Energy Diversity and Security © 2008 Elsevier Ltd.
978-0-08-056887-4 All rights reserved.

rational but risk-averse investors to maximize productivity gains by waiting for ongoing technical change to materialize until they plan to switch and subsequently invest more heavily in the most recent vintages.

14.1 Introduction

During the 1990s the UK electricity mix, which up to that time had been based mainly on coal, became more diversified, as the share of gas in fuels used for power generation started to rise. However, by 2004 still less than 4% of all electricity produced in the UK came from techniques based on renewable energy, while the fraction of electricity coming from oil-fired generators was even smaller. Figure 14.1 illustrates the composition of the current electricity mix in the UK, which consists mainly of coal, gas and nuclear energy.

In the face of the large contribution of power generation to overall carbon dioxide (CO_2) emissions, UK policy makers have set out clear goals for emissions reductions in the Energy White Papers of 2003 and 2007. In 2007, the UK adopted the target to reduce greenhouse gas emissions by 30% by 2020 compared to 1990 levels, with a view to reducing them by 60–80% by 2050, using two instruments: (1) capping and trading, which means that a cap is set for emissions determining the number of permits subsequently traded [the European Union (EU) Emissions Trading Scheme (ETS)] and (2) promoting renewable energy through the Renewables Obligation (RO), which requires electricity retailers eventually to acquire at least 10% of their electricity from renewable sources.[1] At the same time, these measures will serve another important goal of the UK, namely less dependence on fossil fuels from abroad and thereby enhanced energy security.

Investors in power generation equipment face a number of uncertainties. These range from volatile fuel prices over uncertainties regarding government interventions to uncertainty about the development of renewable technologies. However, by using a portfolio of technologies, they may try to avoid downswings in the aggregate return to investment by adjusting their portfolios through investment and disinvestment at the margin. In this chapter, this is modelled by extending the van Zon and Fuss (2005) clay–clay vintage portfolio framework by adding emissions and a number of other features to the basic model as described in van Zon and Fuss (2006). This enables several important aspects of UK climate change policy to be implemented.

---

[1] Currently at a level of 6.7%, this percentage has recently been updated to reach 15.4% by 2015 in the new Energy White Paper and to remain at that level until 2027, with an option to raise the level up to 20% if necessary (DTI, 2007a). The levels of envisaged CO_2 savings from the RO may seem low relative to the total level of required reductions. However, there is little scope for additional large-scale hydropower systems as suitable sites are scarce. Likewise, the resources for some biomass technologies are limited, e.g. for landfill gas, and importing crops for biomass-fired electricity generation is often not economically viable, while at the same time such imports would also cause emissions related to transport that will reduce net potential savings.

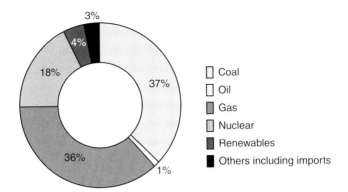

FIGURE 14.1   2006 UK electricity generation mix. Source: DTI, 2007a.

The model uses a rather limited dataset, where data were assembled from different sources, i.e. Anderson and Winne (2004), EIA (2005), DTI (2007b), and some sources had to be aggregated. However, data on technical change and fuel price growth projections are often only suggestive and even though they serve the purpose of illustrating this new method rather nicely, the authors want to warn against mistaking the investment patterns computed with this framework for an exact prediction. Still, preliminary findings are roughly in line with current investment outlooks. The finding that nuclear energy may serve as a bridge to a more sustainable electricity mix, for example, is in line with DTI (2007a) suggestions.

The rest of the chapter first discusses the current literature about investment under uncertainty and portfolio applications to investment in the electricity sector to define the present contribution more precisely. Section 14.3 provides a brief overview of the model. Section 14.4 presents the results and addresses their potential policy implications, while Section 14.5 summarizes and concludes.

14.2 Literature review

The model consists of an optimum portfolio investment approach combined with a two-dimensional clay–clay vintage model. The first of these two dimensions refers to the kind of basic technology that is used to produce electricity, e.g. coal-fired plants or gas-fired plants. The second dimension refers to quality/productivity differences between various generations within these basic technologies, due to embodied technical change. The clay–clay assumption implies that investment is (largely) irreversible. Embodied technical progress may turn the vintage that was cutting edge by the time it was installed obsolete owing to the arrival of a new vintage, a process generally known as 'creative destruction' (Schumpeter, 1942; Aghion and Howitt, 1992).

Portfolio theory dates back to Markowitz (1952). Initially thought to determine optimal portfolios of financial assets, two decades later the first applications to 'real' assets were developed to analyze the energy sector. In the field of energy generation, Bar-Lev and Katz (1976) find that in the USA electric utilities are more or less efficiently diversified in each region, even though their results show that utilities tend to combine high rates of expected return with high levels of risk. This willingness to accept more risk is attributed to the large number of regulations, which the authors claim to be a threat to investors, so that they feel forced to take on higher risks and behave in a less risk-averse way than would otherwise have been the case.

Humphreys and McLain (1998) refine the approach by Bar-Lev and Katz (1976) by building a generalized autoregressive conditional heteroscedastic (GARCH)-type model, which allows the covariance matrix to be systematically updated over time as new events occur. This leads to the result that in times of, for instance, oil price shocks (like the spike during the Gulf War), a diversification away from oil-intensive generation equipment is found to be efficient. Moreover, even though the electric utilities in the USA were operating at a minimum variance position by the end of the 1990s, overall energy consumption was far from efficient during that time, according to the authors.

Awerbuch and Berger (2003) conclude in a similar study, but with a static setting, that the existing portfolio of EU power-generating technologies is suboptimal and therefore inefficient, i.e. there are portfolios combining lower risks with higher returns. Furthermore, they find that fuel price risks have important implications for energy security. Awerbuch and Sauter (2006), for example, point out that the fluctuations in the prices of fossil fuels depress macroeconomic activity as measured by gross domestic product (GDP) growth, employment and inflation. Energy portfolios that have high shares of fossil fuel energy technologies are more vulnerable to this volatility and therefore reduce energy security.[2] This is taken up in Awerbuch (2006), who focuses on issues of energy security, making a case for a higher proportion of renewable energy to be chosen for the overall energy mix. He argues that if a country or a region starts from a portfolio based on fossil fuels, adding renewable energy carriers will reduce the risk (measured by year-to-year cost volatility). Awerbuch (2006) shows that this is possible by reshuffling the portfolio, bringing the sector to a point where lower generating cost is combined with the same level of risk as before, implying that the starting mix could not have been an efficient portfolio. Empirically, he finds that compared to the performance of the EU and the USA, Mexico and other developing countries have or plan to have much less efficient electricity generation portfolios.[3]

---

[2] Another way in which to view energy insecurity is to associate it with sudden disruptions in the supply of fossil fuels as an input to the production process, which could also be captured by price spikes.

[3] From the calculations it is seen that this result might (at least partially) be due to the plan to include a substantial portion of oil-based energy in the generation portfolio, which lowers the return at the same level of risk compared to the efficient frontier. This decision may have to do with strategic and political considerations, as much as with the fact that existing equipment may not easily be scrapped and, in addition, be costly to replace with renewable alternatives.

Whereas these applications of mean-variance optimum portfolio theory in the style of Markowitz (1952) neglect the possible fact that the distributions of uncertain variables could be distributed non-normally and that fat tails could lead to larger losses than conveyed by the use of mean-variance optimization, other studies have incorporated different types of risk measures that do take these possibilities into account. Fortin et al. (2008), for example, employ the conditional value at risk (CVaR)[4] to find an optimal portfolio of electricity generating technologies.

This chapter illustrates the working of a vintage portfolio model. In order to keep matters as simple as possible, the focus is only on capital and fuel costs; however, other cost items such as maintenance and CO_2 costs[5] and their associated uncertainties could in principle have been included as well, thus increasing both the practical relevance and computational burden of the setting, but at the same time reducing its transparency. For the moment, the aim is for greater transparency, as the possibilities offered by the vintage portfolio modelling setup are still being explored.

The particular contribution of this model is that it integrates portfolio considerations with a vintage setting incorporating the irreversibility of investment decisions, and uncertainties not only in factor prices, but also technical change itself in a setting of embodied technical change. The mean-variance portfolio selection methods used in previous studies are inherently static in that they compute optimum portfolios at specific points in time. However, the investor can exploit more opportunities by not only diversifying across technologies, but also by reshuffling the portfolio over time. Adopting a vintage framework not only allows for embodied technical change as stated above, but also enables the investor to choose between different 'versions' of the same general production technology. Investors can thus control the cost and risk characteristics of their entire electricity generation portfolio through both horizontal and vertical differentiation through investment in different 'technology families'[6] and in different vintages within those families. Another strand of literature that takes the irreversibility of investment explicitly into account is real options theory, which typically finds a negative relationship (through an enhanced value of waiting) between investment and uncertainty.[7] Unfortunately, a real options approach considering the simultaneous heterogeneity in the family and the vintage dimension is unfeasible for the present purposes.

Vintage modelling arose in the late 1950s. The basic idea is that the potential of *technical change* can only be realized by first incorporating the new idea in a

---

[4]See Rockafellar and Uryasev (2002), for example, for a discussion of the risk measures involved.
[5]See Roques et al. (2006) for a combination of CO_2 and other factor prices in a static (non-vintage) portfolio setting.
[6]A technology family is defined by the type of fuel that is used to generate electricity.
[7]There have been numerous applications of real options to electricity planning. These contributions go back as early as Pindyck (1991, 1993) and a section in Dixit and Pindyck (1994). A study by Madlener et al. (2005) combines elements from real options theory with a vintage setting.

piece of machinery and subsequently using that machinery to produce output. Because technical change is therefore *embodied* in individual pieces of machinery and equipment, vintage models emphasize the fact that complementary invest- ment has to take place in order to realize the productivity promises of new ideas.[8] The main focus of a vintage model, therefore, is on the *diffusion* of techni- cal change. The embodiment of technical change results in a capital stock that is heterogeneous in terms of the unit operating cost associated with individual vintages.

Depending on the vintage model type, the arrival of superior technologies may render the old ones obsolete, leading to the economic scrapping of inferior equipment. It is through investment and disinvestment at both ends of the vin- tage spectrum that the average productivity characteristics of the capital stock can gradually be changed.

There are different types of vintage model, ranging from 'putty–putty' (Solow, 1960; Phelps, 1962), through 'putty–clay' (Johansen, 1959; Salter, 1960; Phelps, 1963), to 'clay–clay' models (Kaldor and Mirrlees, 1962; Solow et al., 1966).[9] Even 'putty–semi-putty' models exist (Fuss, 1978). However, a clay–clay model applied to individual technology families is best suited for the present purposes, as it implies the absence of alternative choices of production tech- niques within a family, and the absence of possibilities to change factor propor- tions (including the fuel output ratio) ex post. A clay–clay model therefore also reflects the irreversibility of investment decisions ex post.

Since entrepreneurs 'know' beforehand that they will be constrained ex post by the investment choices they make ex ante, the irreversibility of investment ex post also means that changes in factor prices and technology have to be fore- cast ex ante. For example, a rise in fuel costs not properly foreseen results in an economic lifetime that is too short (the economic lifetime of a vintage is equal to the duration of the period over which it would be most profitable to operate that vintage; see Malcolmson, 1975). And so, because of the irreversibility of invest- ment ex post, one runs the risk of being stuck for a long time with the negative cost-consequences of erroneous factor price and technology forecasts ex ante. It seems then, that the incorporation of risk aversion in ex ante investment decision making is especially important, since ex post vintage portfolio adjustments are limited relative to those in financial portfolios.

---

[8] There is also disembodied technical change in most vintage models that is associated not with particular vintages as such, but with the overall organization of the production pro- cess instead. To keep matters as simple as possible, disembodied technical change is disre- garded here.

[9] The metaphors 'putty' and 'clay' refer to different assumptions about factor proportions: if vintages are 'putty', proportions are still flexible, while for 'clay' vintage proportions cannot be changed. The somewhat far-fetched words 'putty' and '(hard-baked) clay' come from the world of pottery, and they refer to the fact that the characteristics of a 'pot' are intrinsically different before and after firing the pot. Before firing, the pot can still change shape (i.e. one can still change factor proportions), but after firing a pot, i.e. turning it into hard-baked clay, its shape can no longer be changed without breaking it.

From a policy point of view too, the irreversibility of investment is important, as it implies that society's trust in technical change to solve some of its problems, e.g. global warming, may involve high investment costs. Moreover, adjustments may take a long time to complete, as the effective pace of embodied technical change depends on the rate of investment. Thus, either one would be forced to bear very large (just-in-time) adjustment costs or, from a risk diversification point of view, one would have to promote investment in new energy-saving technologies sooner rather than later.

With respect to the economic scrapping of old equipment, the most general rule is formulated by Malcolmson (1975) and states that an existing vintage should be used up to the point in time when its variable unit cost exceeds the total unit cost of the latest vintage, since by replacing old capacity with new capacity, one would 'save' the difference between total unit cost on the newest vintage and unit variable cost on the old ones. A 'modified' Malcolmson scrapping rule is used in this chapter, implicitly comparing the variance-adjusted contributions of new and old vintages to total variance-adjusted costs in order to decide whether to keep old vintages operational or to replace them with the newest ones.

The basic results obtained are qualitatively somewhat different from those of others, as they are in between the standard predictions of portfolio and real options theory. More specifically, price and technological volatility have intrinsically different effects on investment: price volatility generates 'standard' portfolio outcomes in that it makes the concerned asset less attractive, whereas technological volatility does not; in fact, producers postpone investment in technologies exhibiting lower degrees of technological uncertainty. Portfolio theory predicts that low-risk assets command a larger portfolio share immediately. In real options models, higher volatility usually increases the value of waiting and investing later. The findings in this chapter and in van Zon and Fuss (2005, 2006) are partly in contrast with this: whereas in real options theory the option value of waiting and keeping the investment opportunity open falls with a decrease in the variance, the (implicit) option value here is adjusted for the benefits that can be realized through the cumulativeness of technical change, where the latter can more than outweigh the immediate gains from lower variance (van Zon and Fuss, 2005).

14.3 The vintage portfolio model

14.3.1 Model outline

In electricity production it makes sense to distinguish between production capacity and actual production, since production capacity must be sufficient to absorb peaks in demand. Actual production will on average be lower than capacity production. So we have to make simultaneous choices regarding (dis-)investment in production capacity (i.e. the capital-output ratio, where the capital stock itself is subject to an exponential rate of physical decay) and the actual use of that capacity (i.e. not capacity output but actual output).[10] The first choice affects capital costs

---

[10]Capacity is measured in MW, and capital costs themselves are measured per unit of initially installed capacity.

(measured per MW of installed capacity), while the second choice affects fuel costs and their contribution to total cost variance.

It follows that generation capacity in electricity production has three different dimensions that need to be taken into account: the dimension of the technology family it belongs to, that of the vintage within the family it belongs to and that of time itself.[11] Technical change is embodied, and has a direct effect on unit fuel and capital requirements. It is assumed that the proportional rates at which these unit requirements fall (i.e. there is fuel-saving and capital-saving embodied technical change at exogenously determined rates) may vary around a given mean. The variance of technical change and the variance in the rates of growth of fuel and investment prices give rise to variance in costs at the portfolio level. The portfolio variance can therefore be controlled to some extent by changing the composition of the electricity production portfolio both in capacity terms and in actual production terms.

The decision how much to invest per technology family as well as decisions regarding the timing of investment are assumed to be irreversible ex post. Moreover, it is assumed that producers want to minimize the present value, further called PV, of the cost of obtaining and operating the entire electricity production portfolio over a fixed planning period of given length. PV – or the forecast of present value of total portfolio cost – is the discounted sum over time of expected input costs for each technology family and vintage. It thus consists of payments both for the capacity and for the operation of the vintages used for production.

In order to calculate the variance of the PV of total cost, several simplifying assumptions are made that enable a linearization of the expected total portfolio cost in combination with the forecasting errors of all individual cost items that make up total portfolio cost to be obtained. These forecasting errors in turn allow an approximation of the portfolio variance to be obtained.[12]

The formulation of an intertemporal investment plan concerning the various electricity production technologies is now but one aspect of the bigger 'plan' of installing new equipment, discarding economically inefficient or particularly 'risky' old equipment, and formulating consistent plans for the actual usage of available equipment over a given planning horizon, which covers 30 years in this case. The bigger plan itself can be obtained from the assumption that electricity producers are risk averse, and hence want to minimize a weighted sum of the expected total portfolio costs over the given planning period and its corresponding expected variance pertaining to that period, given the feasibility of that intertemporal portfolio implicitly defined by a number of additional constraints that will be discussed in more detail in Section 14.3.3 below.

---

[11] Time is important because of the technical decay that (exogenously) reduces generation capacity over time.

[12] A more extensive discussion of the features of the model can be found in van Zon and Fuss (2006).

14.3.2 Vintage portfolios versus standard mean-variance portfolios

Even though a first impression of the contribution of this work compared to previous work has already been made in Section 14.2, it is important to devote some more discussion to the precise differences between the standard mean-variance portfolio (MVP) approach and the new vintage portfolio framework at this point. Both the MVP and the vintage portfolio approach take into account that the existence of uncertainty in terms of variances and covariances of individual assets makes diversification worthwhile. Indeed, this analysis will show that the vintage portfolio approach produces results that are largely consistent with those of a static MVP approach. However, the vintage approach adds a definite time and quality dimension to the problem, so diversification over time and vintages becomes an issue – or rather an opportunity to exploit by the rational investor. This is why the decrease in investment into a technology in response to lower technological uncertainty[13] seems at odds with the standard theory at first glance. However, such behavior is in fact perfectly rational because a temporary increase in investment into a substitute technology enables the investor to invest less into the now more secure technology at first, so that at the time that the planned increase in investment in the now more secure technology comes, one can benefit to the fullest extent possible from the cumulative nature of technical change, which has continued at a more secure rate up to the time that the new technology will be invested in.[14]

This example shows that the primary advantage of the vintage portfolio approach is that it has an explicitly temporal dimension, whereas standard MVPs are inherently myopic when applied to real assets.[15] Furthermore, vintage portfolios take the irreversibility associated with the large sunk cost of electricity-related investments into account by allowing for changes to the portfolio only at the margin through the installation of modern vintages. Major changes in the portfolio composition may therefore take a considerable amount of time, which explains why renewables have been observed to spread rather slowly, even though their performance has been increasing at a pace where a quicker transition may have been expected. This is a sort of temporary lock-in effect, where the value of holding on to existing capacity is enhanced by irreversibility, uncertainty and a lack of (technology switching) incentives. By contrast, standard MVPs emerging from finance feature (virtually) zero transaction costs for adding and removing assets to and from a portfolio and thus do not capture the irreversibility of investment relevant in an electricity production setting.

---

[13] This statement refers to a technological uncertainty experiment, the outcomes of which will be presented in Section 14.4.4.

[14] Recall that since technical change is embodied, productivity improvements on existing machinery and equipment stop after the moment of installation of that equipment.

[15] In finance this is different, and many advances have been made since Markowitz (1952) in making MVPs dynamic.

Another characteristic that is new in the vintage portfolio approach is the indirect valuation of the opportunity to wait or to bring investment forward in time, which can be substantial in the face of underlying dynamics and uncertainties and the evolution thereof. This is like an implicit option value: once the option to invest is realized there is no value to it any more, but as long as it is kept open, its value is increased by a number of factors such as ongoing technical change improving the newest vintages' productivity or cost of installation. In standard MVP frameworks investments are optimized at the point in time under consideration and thus no value is computed for the option to exercise investment at a different point in time and correspondingly to compose the current portfolio differently.

These advantages of vintage portfolios over standard MVPs, but also the similarities and consistency between the two approaches, will become more apparent in the experiments conducted in the following section.

14.3.3 Modelling details

14.3.3.1 The clay–clay vintage setup

The variables K^f, Y^f, X^f and F^f represent the (vintage) level of investment, capacity output, actual output and fuel consumption per technology family, respectively. The model allows only for embodied capital- and fuel-saving technical change at a proportional rate with a given expected value and a given (expected) variance of that rate. For the development of the volume of capital associated with each vintage, we postulate

$$K_{v,t}^f = e^{\delta^f \cdot (t-v)} \cdot K_{v,v}^f \qquad (14.1)$$

where $K_{v,t}^f$ measures the amount of capital still left at time t of a vintage belonging to technology family f and that was installed at time v, with $v \leq t$. In Equation (14.1), δ^f is the (constant) exponential rate of physical decay associated with vintages belonging to family f.

For capacity output associated with a vintage we have

$$Y_{v,t}^f = \frac{K_{v,t}^f}{\kappa_v^f} \qquad (14.2)$$

In Equation (14.2) κ_v^f is the capital-output ratio associated with a vintage of family f that was installed at time v. As it is assumed that there is no ex post disembodied technical change, κ_v^f only depends on v. However, embodied (capital- and fuel-saving) technical change takes place at a given expected proportional rate and with a given expected variance of that rate. Therefore

$$\kappa_v^f = \kappa_0^f \cdot e^{\hat{\kappa}^f \cdot v} \qquad (14.3)$$

where $\hat{\kappa}^f$ is the expected proportional rate of change of the capital-output ratio.[16] By analogy, we postulate for the fuel-output ratio φ_v^f that

$$\varphi_v^f = \varphi_0^f \cdot e^{\hat{\varphi}^f \cdot v} \tag{14.4}$$

where φ_v^f is the expected proportional rate of change of the fuel-output ratio.[17] Hence, for fuel consumption per vintage belonging to family f we must have

$$F_{v,t}^f = \varphi_v^f \cdot X_{v,t}^f \tag{14.5}$$

Equation (14.2) can be used to find the 'demand' for capital per vintage in function of the level of installed capacity (in 'capital' terms):

$$K_{v,t}^f = \kappa_v^f \cdot Y_{v,t}^f \tag{14.6}$$

14.3.3.2 Portfolio selection

Given the technological setting described above, there are now two problems to solve. The first one is the problem of how much to invest per technology family, given its specific characteristics. The second problem is the timing of investment. Since investment is irreversible ex post (i.e. capital costs are sunk), the investment planning process should involve both forward-looking expectations and a measure of risk aversion to accommodate this irreversibility. Therefore, it is assumed that producers minimize the weighted sum of the expected present value of total cost[18] and the variance of that cost by carefully choosing a composition of their vintage portfolio in both the family dimension and the vintage/productivity dimension, because as rational, risk-averse investors they would be willing to reduce risks by spreading investments both over technologies and over time. However, as the model uses a planning period with fixed length, the irreversibility of investment would provide a bias against investment in capital-intensive technologies at the end of the planning period. Hence, irreversibility is taken to mean 'ex post clay during the planning period,' rather than 'ex post clay for all times.' The latter is implemented by noting that in principle the value of investment should be equal to the present value of interest and depreciation charges on investment (see Van Zon and Fuss, 2006, for more details on the calculations). So, in order to make the relative contribution of capital costs to total costs during

---

[16] A negative/positive value of this rate therefore reflects capital-saving/-using technical change.

[17] See footnote 16.

[18] The focus is on costs, even though electricity producers also suffer from volatile electricity prices in liberalized markets. However, since this model is interested in the aggregate electricity mix of a whole country, the authors think that the assumption that all individual power plant owners are price takers is warranted.

the planning period comparable between vintages that are installed at different points in time during the planning period, it is simply assumed that the relevant capital costs are actually the present value of the interest and depreciation charges associated with a particular vintage that are incurred until the end of the planning period.[19]

In order to calculate the portfolio variance of the present value of buying and using the vintage portfolio, we first describe how capital and fuel costs are expected to develop over time and what the corresponding variance of these expectations will be. The PV of capital and fuel costs for all technology families f over a planning period with length θ is given by

$$PV = \sum_f \sum_{t=0}^{\theta} e^{-\rho t} \cdot \left(\Psi_{t,\theta}^f \cdot P_t^f \cdot \kappa_t^f \cdot Y_t^f + \sum_{v=0}^{t} Q_t^f \cdot \varphi_v^f \cdot X_{v,t}^f \right) \qquad (14.7)$$

In Equation (14.7), ρ is the given and constant rate of discount, whereas $\Psi_{t,\theta}^f = 1 - \left((1 - \delta^f)/(1 + \rho) \right)^{\theta-t+1}$ reflects the share of initial investment outlays that can be regarded as the discounted[20] flows of factor payments (i.e. interest and depreciation charges) for the years from t until the end of the planning period. P_t^f is the cost of a unit of investment of a vintage belonging to family f at the time of its installation t, with $0 \leq t \leq \theta$. In Equation (14.7), depreciation charges are valued at historic cost-prices, rather than at replacement value.[21] κ_t^f is the capital/capacity-output ratio associated with the vintage belonging to family f, which is installed at time t. Since there is no disembodied technical change ex post by assumption, the capital-output ratio does not change once a vintage has been installed. Q_t^f is the user price of a unit of fuel f used at time t. The price of fuels does not depend on the time of installation of a vintage v for which it is used. Hence, for all vintages v, Q only depends on t. φ_v^f is the corresponding fuel-output ratio. Y_v^f is the total capacity of vintage v at its time of

---

[19]This is covered by the Ψ term in Equation (14.7), which represents the share of present value of capital costs 'generated' within the planning period as a fraction of the present value of total capital costs.

[20]These flows during the period $t \ldots \theta$ are discounted back until time t. The term $e^{-\rho \cdot t}$ then takes account of further discounting costs until the beginning of the planning period, i.e. time zero. Note that for an infinitely long horizon, the share would approach a value of 1, whereas for a very short horizon, the shortest possible being 0 for investment taking place in the first year after the planning period, the share is equal to zero. So, for t approaching θ, the share is falling toward zero. For further details see van Zon and Fuss (2006).

[21]Note that a change in investment prices then affects only the marginal vintage in a technology family, as opposed to changing fuel prices that would affect production on all vintages in a technology family at the same time. So valuation at historic cost-prices introduces a qualitative difference between capital and fuel costs that would vanish in part if capital were valued at replacement costs. Of course, there would still be the qualitative difference arising from capital costs being associated with capacity installed and fuel costs with capacity used.

installation. That amount will decrease owing to technical decay. The latter will therefore constrain actual output on a vintage v at time t, i.e. $X_{v,t}^{f}$, in accordance with $X_{v,t}^{f} \leq e^{-\delta^{f} \cdot (t-v)} \cdot Y_{v}^{f}$.

In order to calculate the variance of the PV of total cost as given by Equation (14.7), several simplifying assumptions have been made. The first one is that the (constant) discount rate also reflects the required internal net rate of return on investment. The second one is that forecasting errors are serially uncorrelated, and that (co)variances of the growth rates of fuel and investment prices, but also of the rates of fuel-saving and capital-saving technical change, are constant. In that case, it should be noted that for constant expected values of the growth rates of prices and capital and fuel coefficients, a first order approximation of Equation (14.7) is given by

$$PV \approx \sum_{f} \sum_{t=0}^{\theta} e^{-\rho \cdot t} \cdot \Psi_{t,\theta}^{f} \cdot \tilde{P}_{t}^{f} \cdot \tilde{\kappa}_{t}^{f} \cdot Y_{t}^{f} \cdot (1 + S_{t}^{f,\hat{P}} + S_{t}^{f,\hat{\kappa}}) +$$

$$\sum_{f} \sum_{t=0}^{\theta} \sum_{v=0}^{t} e^{-\rho \cdot t} \cdot \tilde{Q}_{t}^{f} \cdot \tilde{\varphi}_{v}^{f} \cdot X_{v,t} \cdot (1 + S_{t}^{f,\hat{Q}} + S_{v}^{f,\hat{\varphi}}) \qquad (14.8)$$

where $S_{t}^{f,\hat{P}} = \sum_{j=0}^{t} \varepsilon_{j}^{\hat{P}^{f}}$, $S_{t}^{f,\hat{\kappa}} = \sum_{j=0}^{t} \varepsilon_{j}^{\hat{\kappa}^{f}}$, $S_{t}^{f,\hat{Q}} = \sum_{j=0}^{t} \varepsilon_{j}^{\hat{Q}^{f}}$ and $S_{v}^{f,\hat{\varphi}} = \sum_{j=0}^{v} \varepsilon_{j}^{\hat{\varphi}^{f}}$, and where ε_{j}^{x} is the forecasting error associated with variable x for time j.[22] Moreover, in Equation (14.8), $\hat{P}^{f}$ and $\hat{Q}^{f}$ are the expected growth rates of investment prices and fuel prices for technology family f. $\hat{\kappa}^{f}$ and $\hat{\varphi}^{f}$ are the expected rates of embodied capital- and fuel-*using* technical change.[23] All forecasting errors ε_{j}^{x} are assumed to have zero expectation. Note the subscript v in $S_{v}^{f,\hat{\varphi}}$. The other sums of error terms all depend just on t.

Equation (14.8) can now be used to calculate the (approximated) expected forecasting error in the present value of total capital and fuel costs.[24] Given that Z is the set of stochastic variables, i.e. $Z = \{P, \kappa, Q, \varphi\}$, while z1 and z2 are 'running' elements of this set, the expectation of its squared value will be equal to the total variance of the PV, which in turn is given by

$$\text{var}(PV) = \sum_{t1=0}^{\theta} \sum_{t2=0}^{\theta} \sum_{f1} \sum_{f2} \sum_{z1 \in z} \sum_{z2 \in z} \min(t1, t2) \cdot m_{t1}^{f1,z1} \cdot \sigma_{f1,z1}^{f2,z2} \cdot m_{t2}^{f2,z2} \qquad (14.9)$$

where $\min(t1, t2)$ represents the minimum of $t1$ and $t2$. $\sigma_{f1,z1}^{f2,z2}$ is the covariance between the growth rates of the different stochastic variables z1 and z2 for technology families $f1$ and $f2$. The 'terms' $m_{t1}^{f1,z1}$ and $m_{t2}^{f2,z2}$ are defined in terms of

---

[22] Variables with a tilde represent their expected values.
[23] See footnote 16.
[24] One can obtain the latter forecasting error by subtracting the expected PV [obtained from Equation (14.8) by setting all S terms equal to zero] from Equation (14.8) itself.

the actual control variables of the problem, i.e. investment in individual vintages of different technology families and the corresponding production plans for those vintages.[25]

14.3.3.3 Additional features and constraints

In order to increase the degree of realism of the model, several new features are included in the van Zon and Fuss (2005) framework. First of all, it is not just prices and technical change that are uncertain in this model, but also demand is no longer assumed to be known with certainty. The risk of facing higher demand than expected is captured by the introduction of two demand scenarios, a low demand scenario (which is the standard extrapolation of known trends) and a high demand scenario (which exceeds the expectation of unchanged growth in demand). Weighing both scenarios by their probabilities, the optimum value of investment that needs to be undertaken to be able to meet demand in all circumstances can be determined. Second, additional capacity may not be installed without bounds as mentioned above. In the case of hydroelectric utilities, for example, the UK has almost reached maximum installable capacity, i.e. there are not enough suitable sites left to realize additional investments. Therefore, investment is constrained here by the estimates for maximum installable capacity. Third, there are differences across technology families with respect to their load characteristics. While coal-fired turbines, for example, have capacity factors of 80% and more, wind energy and solar techniques depend on external circumstances that do not allow them to produce electricity continuously. Fourth, a distinction is made between base-load and peak-load technologies, where typically coal, nuclear and renewables are used for base-load production, whereas gas can be used to meet peaks in demand. Fifth, the UK government has expressed interest in producing at least some output using renewable fuels. This is introduced as an explicit constraint in the model. Finally, as environmental concerns pertain, for an important part, to (cumulative) CO_2 emissions, these are also included in the model. This will show what the introduction of emission caps would mean for the technology composition of the electricity production portfolio and the timing of investment.[26]

14.3.3.4 The 'ultimate' objective function

For a given demand scenario, s, it is assumed that producers want to minimize a weighted sum of the expected PV of their total production cost and its corresponding variance:

$$\Phi^s = PV^s + \lambda \cdot \mathrm{var}(PV^s) \tag{14.10}$$

---

[25] In the actual calculations it is assumed that all covariances are equal to zero, as there are relatively few data available to measure these covariances, and the immediate purpose here is to illustrate the working of the model. This has the added bonus of considerably speeding up the calculations. This is implemented by requiring that $z2 = z1$ and $f2 = f1$ for all values of $z1$ and $f1$. See van Zon and Fuss (2006) for further details.

[26] The interested reader is referred to van Zon and Fuss (2006) for the technical details.

where λ is the relative weight of the variance of the PV of total costs in the objective function. It is further assumed that λ is a non-negative constant.

Producers are supposed to minimize Equation (14.10) by choosing the optimum values of both initial vintage capacity, Y_v^f, per family f for all vintages to be installed during the planning period, and a corresponding 'production plan' (i.e. $X_{v,t}^{f,s}$) for each vintage that one plans to install. Y_v^f and $X_{v,t}^{f,s}$ are chosen conditionally on the expected values and (co)variances of the stochastic variables in this setting, i.e. investment and fuel price growth as well as the proportional rates of change of the capital and fuel coefficients due to embodied technical change.

Given the scenario-specific values of the objective function, the *ultimate* criterion for the electricity investment program is the minimization of the *expected* value of the variance-adjusted costs of buying new and operating total (i.e. both new and 'old') capacity over the entire planning period, i.e. minimization of:

$$\Phi = \sum_s \pi_s \cdot \Phi^s \qquad (14.11)$$

subject to all the constraints mentioned previously. In Equation (14.11), π_s is the subjective probability of scenario s arising.[27] Furthermore, PV^s in (14.11) is evaluated using Equation (14.8) with all S terms set equal to zero to obtain the expected value of the PV of total cost. Equation (14.9) is used to evaluate var(PV^s) in (14.11).

14.4 Simulation results

14.4.1 Technology characterization

This section presents the results obtained using a number of simulations that are meant to illustrate the working of the model. Before describing the outcomes of the various experiments, however, the various production technologies will be broadly categorized in terms of the growth rates and variances of their capital and fuel costs, but also in terms of the growth rates of their capital and fuel productivity and the corresponding variances. The technology characteristics are summarized in Table 14.1.[28]

Such a characterization will be helpful for the interpretation of the results obtained in the simulation runs described below. It follows from Table 14.1 that the combination of low capacity factors and medium to high capital costs is likely to make wind and photovoltaic generation unattractive substitutes for

---

[27]It should be noted that risk aversion could also be introduced at this level of decision making, by amending Equation (14.11) to include the variance in ϕ^s. For reasons of simplicity, this has not been done here. Again for reasons of simplicity, just two demand scenarios have been introduced (1.5% growth versus 2.5% growth), with a priori probabilities 0.75 and 0.25, respectively.

[28]A more precise representation of the dataset can be found in van Zon and Fuss (2006).

Table 14.1 Characterization of technology families

Parameter	Coal	Gas	Nuclear	Hydro	Biomass	Biowaste	Wind	PV
Capital cost	Medium	Low	High	Medium	High	High	Medium	High
Fuel cost	Medium	High	Medium	NR	Medium	Low	Low	Low
Fuel cost growth	Medium	High	Low	NR	NR	NR	NR	NR
Variance fuel cost growth	Medium	High	Low	NR	NR	NR	NR	NR
Fuel-saving technological change	Low	Medium	Medium	NR	High	NR	NR	NR
Variance fuel-saving technological change	Medium	Medium	Low	NR	High	NR	NR	NR
Capital-saving technological change	Low	Medium	Low	Low	High	High	High	Low
Variance capital-saving technological change	Low	Medium	Low	Low	Low	High	High	High
Capacity factor	High	High	High	Medium	High	High	Low	Low

Sources: Anderson and Winne (2004), DTI (2007b) and EIA (2005).
Owing to a lack of data the growth and variance of capital cost are assumed to be equal and low (1%) for all technologies.
PV: photovoltaics; NR: not relevant.

CO_2-intensive technologies, but for the fact that they do not generate any CO_2 emissions. In addition, wind has a high variance in capital-saving technological change because the data also take into account the highly risky offshore wind technologies. Biomass/biowaste has similar cost properties to wind and photovoltaics, but the prospects of cost savings through technological change are more positive than for wind and photovoltaic generation. This also applies for the capacity factor, which is an important determinant of the effective contribution of capital costs to total costs. Hydropower is an established technology in the sense that there is little (capital-saving) technical change, but also little variance in that. Nuclear energy is a relatively secure technology with respect to technical change, which is comparatively slow but not volatile. This counteracts to some extent the high capital costs, next to relatively low fuel costs and low variance in fuel price growth.[29] The latter does not hold for gas, which also suffers from high fuel costs. Therefore, gas can be expected to have a relatively low share a priori, save for the fact that technical change is relatively fast, but also relatively uncertain. However, with a CO_2 emission cap, and given the fact that gas is a peak-load fuel, the share of gas in total electricity production must remain significant. Finally, coal is a 'middle of the road' type of fuel with medium capital and fuel costs and medium or low rates of technical change and variances.

14.4.2 Simulation runs

First, the base run is described, i.e. the outcomes associated with the investment program for a planning period of 30 years that fits all the constraints that were described in more detail before, except that it does not yet contain an emissions cap. The base run is for a value of $\lambda = 0$, thus abstracting from risk aversion. The results are labelled R0 and discussed in Section 14.4.2.1. There are four other runs, called R1–R4, that have the same parameter setting as the base run but then for increasing values of λ.[30]

The results associated with these runs are presented in Section 14.4.2.2. They show how increasing risk aversion influences the optimum composition of the capital stock in terms of technology families. Section 14.4.2.3 contains the run that is based on a value of $\lambda = 1.5E-5$, i.e. R1, and on a cap on CO_2 emissions defined by the emissions path that grows at a constant proportional rate from a level of $150\,MtCO_2$ to a level that is about 25% lower than in R0 at the end of the planning period, i.e. to a level of $(1 - 0.25)\,350 = 245\,MtCO_2$. Section 14.4.3 contains a run,

---

[29] Please note that this analysis refers only to technical and fuel price risks, while the share of nuclear power plants in the national energy mix will always be influenced by considerations of safety risks, decommissioning factors, nuclear waste disposal difficulties and political commitment. Since the primary focus here is technological and fuel price features, we abstract from these risks, but keep in mind that they will probably have an adverse effect on the adoption of nuclear technologies in reality.

[30] This is for $\lambda = 1.5E-5$, $3E-5$, $4.5E-5$, $6E-5$ for R1, R2, R3 and R4, respectively. For $\lambda = 1.5E-5$, the standard deviation of total costs relative to total costs is of the order of 10%, which looks like a reasonable order of magnitude a priori.

i.e. R6, again with $\lambda = 1.5E-5$, but also a doubling of the expected variance in the growth rate of gas prices, from the beginning of the planning period. Finally, Section 14.4.4 contains the results for R7 pertaining to a large shock in the variance of fuel-saving technical change for nuclear energy to see whether renewables could take over from nuclear energy under these circumstances.

14.4.2.1 The base run

For the low demand growth scenario S1 and the high demand growth scenario S2, the production shares are plotted in Figures 14.2 and 14.3, respectively. Both

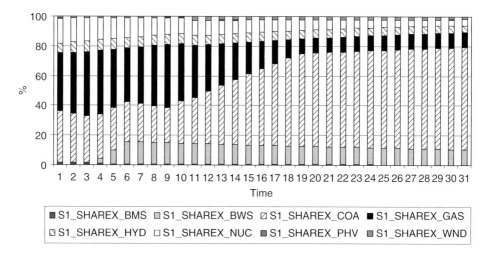

FIGURE 14.2   Production shares for the low-demand growth scenario (S1).

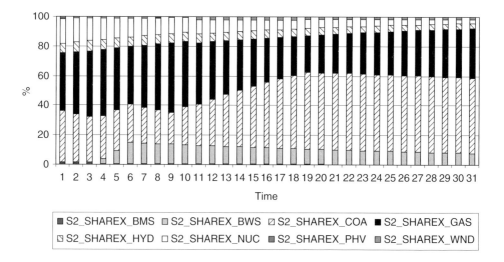

FIGURE 14.3   Production shares for the second (high)-demand scenario (S2).

scenarios start out with the same distribution over technologies. However, it is clear that gas-fired generation takes over to meet the additional demand in the case of the second scenario. This increased production from gas-based electricity generation at the expense of coal also affects hydro and nuclear by a smaller margin. This can be explained by the fact that gas has the lowest investment price and can be installed at a lower additional cost than any of the other technologies. This more than outweighs the disadvantage of being more expensive in terms of fuel, since scenario S2 has a much lower probability of occurring than scenario S1. Indeed, gas is also a peak-load fuel, and as such its higher share in S2 is not surprising.

Figures 14.4 and 14.5 present the underlying capacity composition. Without any risk aversion, the weight of the variance in the objective function is equal to

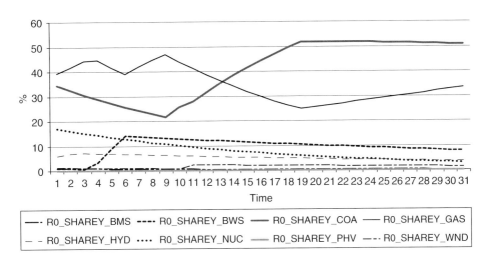

FIGURE 14.4   Capacity shares.

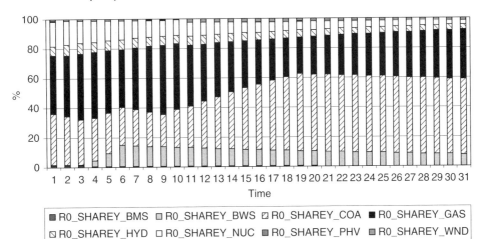

FIGURE 14.5   Capacity shares.

zero. In other words, all technologies become perfect substitutes with gradually evolving cost characteristics per technology driving the intertemporal variations in the composition of the portfolio. Total unit costs evolve according to increases in investment and fuel prices, but may also shrink as a result of either capital- or fuel-saving technical change. Note the importance of the constraints on maximum installable capacity, peak- and base-load characteristics and initial conditions [i.e. production using 'old' vintages may continue, as long as variable (fuel) costs are lower than total unit costs on 'new' vintages, which is essentially Malcolmson's scrapping condition; Malcolmson, 1975].

Figure 14.4 illustrates these points more clearly. Without risk aversion, the technologies are perfect substitutes to the extent that one consistently invests in the cheapest alternative when the PV of total unit costs for one technology falls below that of those of another one. In the beginning gas and hydro are the only technologies that undergo some investment, which is observable from their increasing capacity shares. The other technologies actually see a decrease in their capacity shares, which is due to depreciation, so there is no net investment in those technologies.

This leaves us with the composition of electricity generation equipment shown in Figure 14.5. Overall, gas constitutes the lion's share of the electricity mix, followed by coal, whereas wind and biowastes have gained at the expense of nuclear power, hydropower, biomass and solar photovoltaic generation.

14.4.2.2 Introducing risk aversion

This section presents the results associated with the introduction of increasing risk aversion. Figure 14.6 demonstrates the underlying mechanisms at work: subsequently changing $\lambda = 1.5E{-}5$ in R1, to $\lambda = 3E{-}5$ in R2, to $\lambda = 4.5E{-}5$ in R3 and to $\lambda = 6E{-}5$ in R4, and calculating the corresponding optimal portfolios, effectively traces the convex hull of feasible portfolios, i.e. the set of efficient portfolios. The latter consists of the portfolios generating the lowest variance for a given cost level.

From the shape of the relation between variance and cost, five conclusions can be drawn. (1) The variance falls with increasing cost. (2) The variance is convex in costs. (3) The variance–cost profile defines the set of all cost-efficient portfolios. (4) The point of tangency between the variance–cost profile and the objective function determines the optimum portfolio. (5) Greater risk aversion makes the objective function flatter, i.e. there is a tradeoff between lower variance and higher total cost. These results are quite intuitive from a standard portfolio point of view and show that a portfolio approach combining irreversible investment and embodied technical change still generates the types of result known from financial portfolio theory.

Figures 14.7–14.9 show what happens to the shares of coal, gas and nuclear in total capacity output, as the reduction in variance can essentially only be brought about by a reshuffling of the technology portfolio and the associated production plans. In Figure 14.7 the swings in the share of coal during the base run are stretched out more evenly over time. Shares are lower at the end of the planning period and higher at the beginning. For gas, shares fall structurally below the

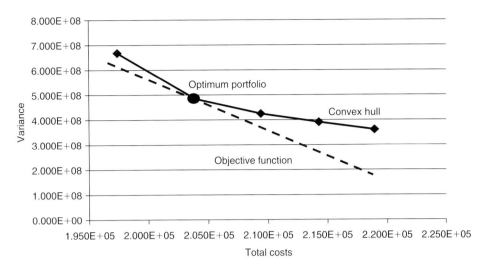

FIGURE 14.6   Variance against present value total costs.

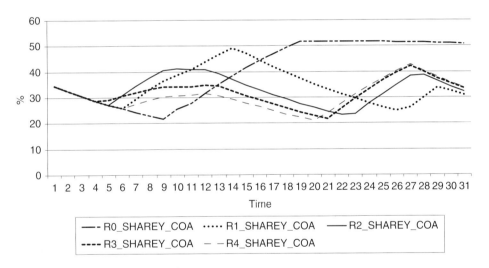

FIGURE 14.7   Capacity shares: coal.

base run in the beginning, picking up at the end again, thus also levelling out fluctuations in shares to some extent. However, one should recall that gas is a fuel with a relatively high variance with respect to both fuel price growth and technical change. Hence, while smoothing out fluctuations is a good strategy for variance reduction, reducing the portfolio shares of high variance technologies is an especially good option here. Figure 14.8 shows both principles at work.

Figure 14.9 shows that risk-averse investors would gladly accept nuclear power as a bridge to a less carbon-intensive future. Of course, concerns about the negative externalities associated with CO_2 emissions should be weighed

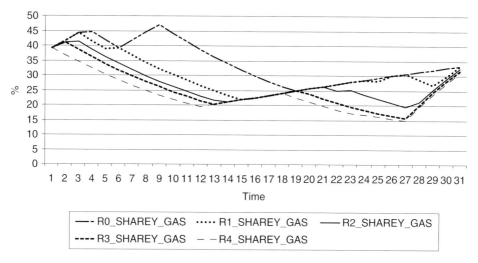

FIGURE 14.8   Capacity shares: gas.

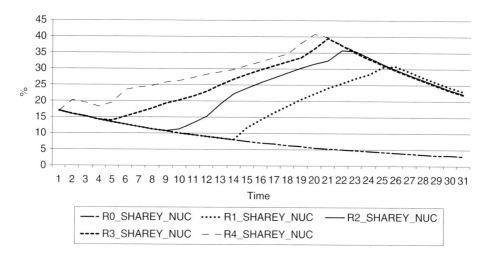

FIGURE 14.9   Capacity shares: nuclear.

against the legitimate concerns about the processing and quasi-permanent storage of nuclear waste material, and the threats of microproliferation among terrorist groups. Nonetheless, nuclear energy is widely regarded as a means to buy time to find the ultimate solution to our energy problems through carbon capture and storage in combination with a more intensive use of renewables. Note from Figure 14.9 that the lower the degree of risk aversion, the later the moment in time within the planning period at which people are starting to 'build the bridge'.

Figure 14.10 shows what happens to emissions owing to the reshuffling of the technology portfolio. The figure contains the probability-weighted averages of

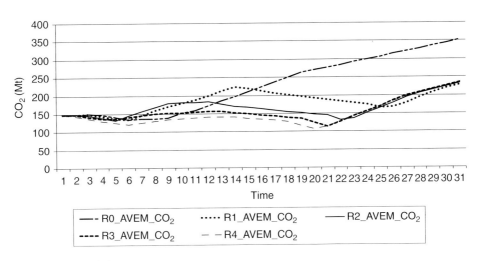

FIGURE 14.10   Emissions.

total emissions in both demand scenarios. The one in the base run generates the highest emissions at the end of the planning period, mainly because coal is still an important portfolio ingredient by then. This is shown quite clearly in Figure 14.10, where fluctuations in emissions follow those in the share of coal in total capacity quite closely (cf. Figure 14.7).

14.4.2.3 Carbon dioxide emission caps
Now, R1 (with $\lambda = 1.5\text{E}-5$) is combined with a cap on CO_2 emissions. Note that even a relatively slight degree of risk aversion generates a reshuffling of the technology portfolio such that in R0, emissions at the end of the planning period are way below emission levels in the base run (by more than 25%). This means that the time path for emissions that begins at $150\,MtCO_2$ and ends at $(1 - 0.25).\ 350 = 245\,MtCO_2$ will not be binding at the end of the planning period (see also Figure 14.10). However, it will be binding in the middle of the planning period, as seen quite clearly from Figure 14.11, in which the flat stretch of emissions for run R5 coincides with the emissions constraint being binding. How the corresponding emission reductions are brought about can be seen from Figure 14.12.

Figure 14.12 shows the absolute differences between the percentage capacity shares in runs R5 and R1. The carbon content of electricity production falls in period 6 owing to a simultaneous reduction in coal and an increase in gas. The more binding the CO_2 emissions cap becomes, the more additional CO_2-free capacity is installed, here nuclear power. Then, as emissions are reduced anyhow from period 15 on (see Figure 14.11), the technology distribution almost reverts to 'normal', except that the share of nuclear energy is slightly above the base run, and the shares of gas and coal are correspondingly lower. This also leads to slightly lower emissions at the end of the planning period, simply because once nuclear capacity has been installed, it will stay around for a relatively long time.

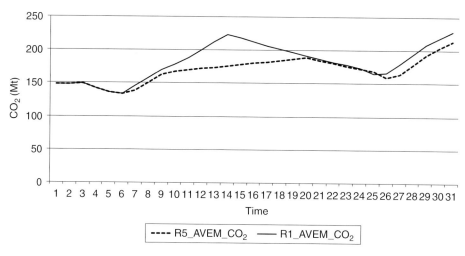

FIGURE 14.11   Emissions with a cap on CO_2.

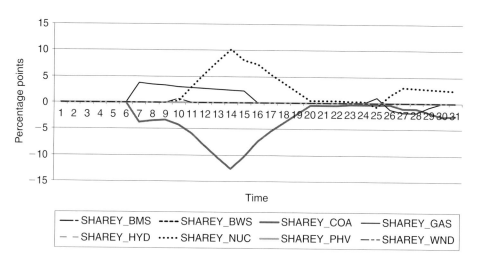

FIGURE 14.12   Absolute differences between capacity shares in R5 and R1.

Therefore, a temporary binding emission constraint can induce semi-structural emission reductions, because the composition of the capital stock changes. In addition, due to the ex post clay character of technologies themselves, and the 'near clay' character of the technology portfolio as a whole, the capital stock only slowly adjusts to a situation where the caps are no longer binding.[31]

---

[31] It should be noted that in this experimental setting this adjustment of the capital stock is taking place as quickly as is technically feasible, since the optimization program 'knows' when the CO_2 emission constraints will become non-binding, and can adapt ex ante (by a suitable adjustment of the portfolio) to this situation by choosing a more CO_2-intensive portfolio.

14.4.3 Fuel price variance

In R6 the variance of fuel price growth has been increased by 100% in the case of gas, for $\lambda = 1.5E-5$. To see what such an increase in fuel price growth variance implies, compare runs R6 and R1, since the latter case has no CO_2 emissions caps either. The absolute differences between capacity shares in this case are presented in Figure 14.13.

The increase in gas price growth variance significantly reduces the portfolio share of gas. Biowaste and somewhat later coal take over, and after a slight dip in the middle of the planning period, nuclear energy is phased in as well. At the end of the planning period, gas has become a very unattractive portfolio component indeed, and coal and nuclear energy have permanently taken over.

14.4.4 Technological variance

For technological uncertainty to have the largest possible impact, it is necessary to implement it as increased variance in the rate of fuel-saving technical change, as fuel consumption associated with a specific vintage is a continuous process, whereas investment in new capacity takes place only at the moment that vintage is installed. Nuclear energy has been chosen for increased technological uncertainty, first of all because controlled fusion has been a technological promise for over 50 years, and it still is. The second reason is that in these simulations, nuclear energy consistently appears to be the 'savior of last resort.' This leads us to wonder whether renewables would stand a chance of taking over this role, if nuclear energy were to become less attractive for some reason. In order to find

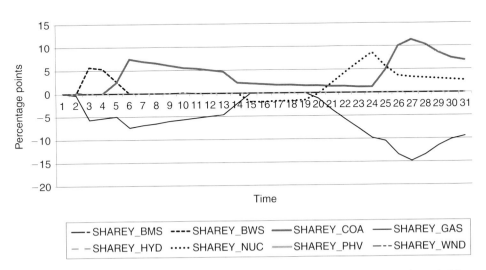

FIGURE 14.13 Effect of fuel price variance: absolute differences between capacity shares in R6 and R1.

this out, an experiment was performed in which the variance of fuel-saving technical change in nuclear energy production was increased by a factor of 100.[32]

Run R7 is the same as R1, except for the shock in the variance of fuel-saving technical change. The results are presented in Figure 14.14. The change in the variance of fuel-saving technical change in nuclear energy production, although implemented from the beginning of the planning period, takes a while before it has an impact. This is due to the fact that nuclear had not been invested in during the first half of the planning period in R1 in the first place. Therefore, only from the period where there was investment in nuclear in R1 (i.e. in year 14) is a negative deviation from the results with respect to R1 observed. Another result is the negative deviation in coal before period 14, which is compensated for by an increase in gas capacity. Since investors are fully aware of the drop in nuclear at the beginning of the planning period already and they know that they will have to compensate for this drop by investing more heavily in coal, they actually have an incentive to decrease their installed capacity of coal earlier on because this will enable them to install a larger amount of more modern and productive vintages when the time has come to replace nuclear capacity. By reducing coal earlier on, investors thus create room for more advanced coal capacity later on. The gap is closed by gas, since gas has relatively low installment costs and can easily make up for the lack of coal in the short run.

This experiment is a good illustration of the working of the model, not only in the technology dimension (i.e. diversification over technology families leads

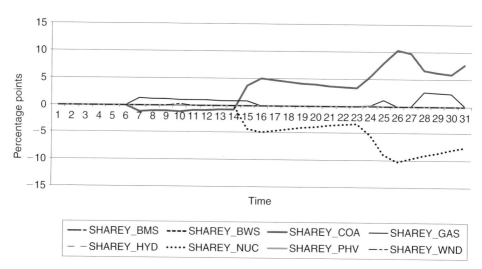

FIGURE 14.14 Effect of technological variance: absolute differences between capacity shares in R7 and R1.

---

[32]Since fuel cost is relatively unimportant in nuclear energy production as compared to coal- and gas-fired power plants, a relatively large shock is needed for its effect to become noticeable.

to substitution of coal for nuclear becoming less attractive through less certain technological prospects), but also in the quality dimension, i.e. it pays off to wait for ongoing technical change to take place and reap the full benefits of being able to install higher quality vintages when investment becomes necessary and thus to substitute investment in coal today for investment in coal at a later point in time.

14.5 Summary and conclusion

This chapter has presented the outcomes of some simulations with a model described in more detail in van Zon and Fuss (2006). The latter model integrates elements from financial portfolio theory with a clay–clay vintage model of production for the electricity sector. The main ideas behind this model are that technical change is embodied in machinery and equipment, and that once installed, the fuel consumption characteristics of power generation equipment cannot be changed. Productivity improvements in electricity production then require investing in the newest equipment that is available on the market: without investment, productivity improvements simply cannot be realized.

However, electricity producers are risk averse. So, investing in a piece of equipment with a given fuel efficiency exposes them to variations in production costs caused by fluctuations in fuel prices. Likewise, capital costs can fluctuate. Since investment is irreversible, electricity producers need to look ahead, and invest sooner rather than later if the future appears uncertain, given the electricity they have to serve. In that case they would want to change their equipment portfolio in favor of vintages with relatively certain consumption characteristics (i.e. the vintage one can invest in now or in the very near future).

In this vintage portfolio model, eight broad technology families are distinguished: coal, gas, nuclear, hydro, biomass, biowastes, wind and photovoltaic generation. Production targets are introduced for renewables, as well as peak- and base-load distinctions between technology families. Moreover, the model allows for uncertainty in demand by specifying different demand scenarios with different probabilities of being realized. Uncertainty surrounding the cost of an investment program over a fixed planning period of 30 years is linked to uncertainty about fuel price growth and uncertainty about the development of fuel- and capital-saving technical change during that planning period. By investing in the newest vintages of each technology and formulating production plans for the entire vintage capital stock, electricity producers can control aggregate uncertainty. Because production technologies are clay–clay, changes in the capital stock only come about through investment and disinvestment at the margin: the room to manoever is limited in a vintage setting, as is the case in reality.

Several simulation experiments have been performed. In the base run without any risk aversion, CO_2 emissions in the UK electricity sector range from about 150 $MtCO_2$ at the beginning of the planning period to 350 $MtCO_2$ at the end of the planning period. When the degree of risk aversion is increased, expected costs will increase, whereas the expected cost-variance of the entire vintage investment program will decrease. The corresponding standard deviation as a fraction of the expected costs also decreases, but less and less so for increasing degrees of risk aversion. As in optimum portfolio theory, the relation between the costs of the entire investment program and the corresponding variance exhibits decreasing returns to variance: a larger variance generates a less than proportionally higher rate of return in ordinary portfolio theory, and in this case a less than proportionally lower expected cost of the investment program.

Changes in fuel price growth variances or technological variances have the expected effect. Increased variance with respect to some technology family reduces investment in that family, while increased risk aversion reduces fluctuations in investment over time. Furthermore, with increased risk aversion, electricity production becomes more diversified over technologies, to such an extent even that CO_2 emissions would be significantly reduced, mainly by switching toward nuclear energy production rather than toward renewables. In an experiment where a cap on CO_2 emissions is introduced, nuclear energy turns out to be the 'savior of last resort', but gas also increases in importance. When nuclear energy production is 'punished' by increasing its technological variance, gas and coal take over, rather than renewables. In addition, the anticipation of a switch toward another technology in the future makes producers want to invest less in that technology now and more in a substituting technology. In this way, they can benefit more from the cumulative nature of (ongoing) embodied technical change until the moment they will actually execute the switch. Nonetheless, the fact that gas and coal will take over from nuclear energy in this case suggests that, given the data used here, none of the renewables is strong or promising enough[33] to take over from nuclear energy or coal on a large scale. This will only occur when initial costs are lower or when technological uncertainties surrounding renewables are reduced, or both.

Further research could explore the nature of technical change by investigating not only the effects of ongoing (exogenous) technical change, but also the connection of costs and cumulative production through the inclusion of learning curves. Related to this, a distinction could be made between internal learning by doing and industry-wide spillovers, which could lead renewables (which are currently presumably on the initial parts of their respective learning curves) to advance more quickly to levels where they become more competitive, especially as the more mature, carbon-intensive technologies enter the flatter stretches of their own learning curves.

References

Aghion, P. and Howitt, P. (1992). A model of growth through creative destruction. *Econometrica*, 60(2), 323–351.

Anderson, D., Winne, S. (2004). *Modeling Innovation and Threshold Effects in Climate Change Mitigation.* Working Paper 59. Tyndall Centre for Climate Change Research.

Awerbuch, S. (2006). Portfolio-based electricity generation planning: policy implications for renewables and energy security. *Mitigation and Adaptation Strategies for Global Change*, 11(3), 693–710.

---

[33] Note that some of the renewables such as hydropower would in principle be advanced enough to take up a larger share of electricity production; however, large-scale hydropower is severely constrained through geographical feasibility.

Awerbuch, S. and Berger, M. (2003). *Applying Portfolio Theory to EU Electricity Planning and Policy-Making*. Paris: Working Paper EET/2003/03 International Energy Agency.

Awerbuch, S. and Sauter, R. (2006). Exploiting the oil–GDP effect to support renewables deployment. *Energy Policy*, 34, 2805–2819.

Bar-Lev, D. and Katz, S. (1976). A portfolio approach to fossil fuel procurement in the electric utility industry. *Journal of Finance*, 31(3), 933–942.

Dixit, A. and Pindyck, R. (1994). Investment under Uncertainty. Princeton, NJ: Princeton University Press.

DTI (2007a). *Energy White Paper: Meeting the Energy Challenge*. Available at http://www.dti.gov.uk/energy/whitepaper/page39534.html

DTI (2007b). *UK Energy Sector Indicators 2007*. Available at http://www.dti.gov.uk/energy/statistics/publications/indicators/page39558.html

Energy Information Administration (2005). *Other International Electricity Data*. Available at http://www.eia.doe.gov/emeu/international/electricityother.html

Fortin, I., Fuss, S., Khabarov, N., Obersteiner, M. and Szolgayova, J. (2008). *An Integrated CVaR and Real Options Approach to Investments in the Energy Sector. Journal of Energy Markets*, 1(2): 61–85.

Fuss, M. A. (1978). Factor substitution in electricity generation: a test of the putty–clay hypothesis. In Fuss, M. A. and McFadden, D. (Eds), *Production Economics: A Dual Approach to Theory and Applications*. Amsterdam: North-Holland, pp. 187–213.

Humphreys, H. and McClain, K. (1998). Reducing the impacts of energy price volatility through dynamic portfolio selection. *Energy Journal*, 19(3), 107–131.

Johansen, L. (1959). Substitution versus fixed production coefficients in the theory of economic growth: a synthesis. *Econometrica*, 27(2), 157–176.

Kaldor, N. and Mirrlees, J. A. (1962). A new model of economic growth. *Review of Economic Studies*, 29, 174–192.

Madlener, R., Kumbaroglu, G. and Ediger, V. (2005). Modelling technology adoption as an irreversible investment under uncertainty: the case of the Turkish electricity supply industry. *Energy Economics*, 27(1), 139–163.

Malcolmson, J. (1975). Replacement and the rental value of capital equipment subject to obsolescence. *Journal of Economic Theory*, 10, 24–41.

Markowitz, H. (1952). Portfolio selection. *Journal of Finance*, 7(1), 77–91.

Phelps, E. S. (1962). The new view of investment: a neoclassical analysis. *Quarterly Journal of Economics*, 76(4), 548–567.

Phelps, E. S. (1963). Substitution, fixed proportions, growth, and distribution. *International Economic Review*, 4(3), 265–288.

Pindyck, R. (1991). Irreversibility, uncertainty and investment. *Journal of Economic Literature*, 29(3), 1110–1148.

Pindyck, R. (1993). Investments of uncertain cost. *Journal of Financial Economics*, 34, 53–76.

Rockafellar, R. and Uryasev, S. (2002). Conditional value at risk for general loss distributions. *Journal of Banking and Finance*, 26, 1443–1471.

Roques, F., Newbery, D. and Nuttall, W. (2006). *Fuel Mix Diversification Incentives in Liberalized Electricity Markets: A Mean-Variance Portfolio Approach*. Working Paper 2006/33. RCAS of the European Institute, Florence School of Regulation.

Salter, W. E. G. (1960). *Productivity and Technical Change*. Cambridge: Cambridge University Press.

Schumpeter, J. (1942). *Capitalism, Socialism and Democracy*. Harper and Row, New York.

Solow, R. M. (1960). Investment and technical progress. In Arrow, K. J., Karlin, S. and Suppes, P. (Eds), *Mathematical Methods in the Social Sciences, 1959; Proceedings*. Stanford, CA: Stanford University Press, pp. 89–104.

Solow, R. M., Tobin, J., von Weizsäcker, C. C. and Yaari, M. (1966). Neoclassical growth with fixed factor proportions. *Review of Economic Studies*, 33(2), 79–115.

van Zon, A. and Fuss, S. (2005). Irreversible Investment and Uncertainty in Energy Conversion: a Clay–Clay Vintage Portfolio Selection Approach. Research Memorandum RM2005-013. UNU-MERIT.

van Zon, A. and Fuss, S. (2006). Irreversible Investment under Uncertainty in Electricity Generation: a Clay–Clay-Vintage Portfolio Approach with an Application to Climate Change Policy in the UK. Working Paper 2006-035. UNU-MERIT.

Shimon Awerbuch was a lovely man. This book is a tribute to his contributions to financial economics in the energy sector.

Dr. Awerbuch was a financial economist specializing in utility regulation, energy and the economics of innovation and new technology. Before joining SPRU (at the University of Sussex), he served as Senior Advisor for Energy Economics, Finance and Technology with the International Energy Agency in Paris. For the decade preceding this assignment, Awerbuch maintained an independent financial economics practice specializing in energy, utility regulation and the economics of innovation and new technology. Awerbuch had taught graduate and undergraduate courses in Corporate and Regulatory Finance, Capital Budgeting and Technology Adoption. He served as Chief, Economic and Policy Studies, Utility Intervention Office, New York State Executive Department, where he developed expert testimony on various aspects of regulatory economics and finance. He served with the Management Consulting Service of Ernst & Young and held various economic and policy analysis positions with the New York State Executive Department and Legislature.

Shimon had 30 years of experience in finance, and regulatory and energy economics, involving the private sector and all levels of government. He had an international presence based on an extensive record of publication and research. He was co-author of *Unlocking the Benefits of Restructuring: A Blueprint for Transmission* (PUR, November 1999, to be republished by Elsevier) and co-editor of *The Virtual Utility: Accounting, Technology and Competitive Aspects of the Emerging Industry* (Kluwer, 1997). He was Series Editor, Elsevier Topics in Global Energy Economics, Regulation and Policy, and a member of the Editorial Board of Energy Policy. He contributed over 50 papers to such journals as *American Economic Review* and *Journal of Regulatory Economics and Energy Policy*. He presented his research findings to delegates of the United Nations Commission on Sustainable Development in New York and testified in numerous regulatory proceedings.

Awerbuch successfully guided investment, strategy and technology decisions to yield added value for multinational corporations and some of the world's largest utilities. He also advised the UN, World Bank, and energy ministries in Europe. Asia and Latin America. His most recent research and consulting (and the subject of this tribute book) focused on portfolio approaches for enhancing energy security and for valuing renewables. He will be deeply missed.

All royalties from the sale of this Tribute Book will be donated charity

The donations will be made in memory of Dr. Awerbuch, and given to charities undertaking innovative work in the least developed countries in the finance and energy sectors. The Elsevier Foundation will match these donations. The initial organisations and programmes to be funded are:

Grameen Shakti

Grameen Shakti (GS) was initiated in 1996 by the founders of Grameen Bank. Inspired by the vision of Professor Yunus, GS aims to create a synergy between renewable energy technology and micro-credit in order to give the rural people a chance to improve their quality of life and take part in income generating activities. The book proceeds will be focused on helping to transform the lives of rural women and create 100,000 green jobs through its innovative Grameen Technology Center (GTC) program. These Centres train women on renewable energy technologies and help them to set up their own businesses. To find out more, go to www.gshakti.org.

Concern

Concern Worldwide is an international, non governmental, humanitarian organisation dedicated to the reduction of suffering and working towards the ultimate elimination of extreme poverty in the world's poorest countries. Founded in 1968, Concern currently works in 28 countries in the Developing World. Proceeds from this book will go towards AMK micro-finance bank in Cambodia, a subsidiary of Concern. AMK aims to increase the opportunities available to poor people to earn a living through the provision of financial services. To find out more, go to www.concern.net.

Solar-Aid

SolarAid is a new and innovative charity that helps poor communities in developing countries use solar power to fight poverty and climate change. The money raised by this book will go towards their work to convert kerosene lamps in Africa into solar lamps, leading to reductions of CO_2 and mitigating the serious health hazards caused by burning kerosene. To find out more, go to www.solar-aid.org.

Dr. Morgan Bazilian holds a political appointment as the Special Advisor on Energy Security and Climate Change to the Minister of Energy in Ireland. Prior to this post, he was the Head of Energy and Climate Policy at Ireland's national energy agency. He holds two Masters degrees and a Ph.D. in areas related to the techno-economic aspects of the energy sector, has been a Fulbright Fellow, and maintains an adjunct post at the Electricity Research Centre at University College Dublin. He is a member of the UNFCCC's Expert Group on Technology Transfer (EGTT), a member of the EU's Energy Economists Expert Group, the Chair of the Renewable Energy and Energy Efficiency Partnership (REEEP) Programme Board, and acts as a contributor to the International Energy Agency's World Energy Outlook.

Fabien Roques is a Senior Energy Economist in the Economic Analysis Division of the International Energy Agency (IEA), and an Associate Researcher of the Cambridge University Electricity Policy Research Group. Prior to joining the IEA, Fabien held a number of research and consulting positions, including the UK Parliamentary Office of Science and Technology and the Cambridge-MIT Electricity Project. Fabien's research spans a large range of issues in energy and environmental economics, including electricity market design and regulation, optimization of utilities' generation mix under uncertainty, and integrated modeling of economic, energy use, and emissions scenarios. Fabien holds a PhD in Energy Economics from the University of Cambridge and an MSc. in Engineering from Ecole Centrale Lyon.